Construction Insurance, Bonding, and Risk Management

Construction Insurance, Bonding, and Risk Management

William J. Palmer
Ernst & Young LLP

James M. Maloney
Willis Corroon Construction

John L. Heffron, III
Ernst & Young LLP

McGraw-Hill

New York   San Francisco   Washington, D.C.   Auckland   Bogotá
Caracas   Lisbon   London   Madrid   Mexico City   Milan
Montreal   New Delhi   San Juan   Singapore
Sydney   Tokyo   Toronto

Library of Congress Cataloging-in-Publication Data

Palmer, William J.
    Construction insurance, bonding, and risk management / William J.
Palmer, James M. Maloney, John L. Heffron, III.
       p.   cm.—(Construction series)
    Includes index.
    ISBN 0-07-048594-1 (hardcover)
    1. Construction industry—Insurance. I. Maloney, James M.
II. Heffron, John L. III. Title. IV. Series: Construction series
(McGraw-Hill, inc.)
HG8053.7.P35 1996
368—dc20
                                           96-14592
                                               CIP

McGraw-Hill

A Division of The McGraw-Hill Companies

  2 3 4 5 6 7 8 9 0 DOC/DOC 9 0 1 0 9 8 7

ISBN 0-07-048594-1

*The sponsoring editor for this book was Larry Hager, the editing
supervisor was Virginia Carroll, and the production supervisor was
Pamela Pelton. It was set in Century Schoolbook by North Market
Street Graphics.*

Printed and bound by R. R. Donnelley & Sons Company.

McGraw-Hill books are available at special quantity discounts to use as
premiums and sales promotions, or for use in corporate training programs. For more information, please write to the Director of Special
Sales, McGraw-Hill, 11 West 19th Street, New York, NY 10011. Or contact your local bookstore.

 This book is printed on recycled, acid-free paper containing a
minimum of 50% recycled de-inked fiber.

Contents

Contributors

Jeff Albada (CHAP. 4) is a senior vice president and managing director of Willis Corroon Construction in San Francisco. He has over 23 years' experience in the insurance and surety bonding industry and has been with Willis Corroon for 18 years, specializing in arranging and maintaining surety bond credit facilities for international and domestic contractors. He has a B.S. degree from the University of Oregon and is a member of the Environmental Committee, AGC of California; past director, Construction Financial Management Association, California Chapter; past director, Contract Bond Committee, Surety Association of America; and past president, Surety Forum of Northern California.

Richard S. Beck (CHAP. 4) has a B.S. degree from Ashland College and over 25 years' experience in the insurance and surety industries. He directs the surety operations for Willis Corroon's Pittsburgh office, where he also serves as national co-chairman of the Minority Enterprise Team (MET). He is a past president of both the Insurance Club and the Rotary Club of Pittsburgh, past director and current membership chairman of NASBP, and serves on two national committees of AGC of America.

Paul R. Becker (CHAP. 6) is the managing director of the Construction Risk Management Group for Willis Corroon in Nashville, Tenn., with responsibilities in marketing, safety management consulting, claims management, statistical support and coverage, and contractual issues. He has over 17 years' experience in the insurance industry, including positions with Aetna and CIGNA.

Richard Carris, CPCU, CLU (CHAP. 9) is a senior manager in the New York City Risk Management and Claims Consulting practice of Ernst & Young LLP. He has conducted numerous workers' compensation risk financing and claim studies for contractors, construction managers, and project owners. He has designed workers' compensation insurance programs for multibillion-dollar construction programs as well as for individual subcontractors. He is a graduate of the University of Florida and holds an M.B.A. degree from Pace University, New York City.

James H. Costner, CPCU, ARM (CHAP. 7) is senior resource consultant and senior vice president of the Advanced Risk Management Services Division,

Willis Corroon Corporation and serves as chair of Willis Corroon's Property Interests Group, consulting on matters related to property insurance coverage design, marketing, and loss adjusting. He is a graduate of Arkansas State University and the Graduate School of Sales Management and Marketing at Syracuse University, and has taught at the American University in Washington, D.C., and Nashville Tech in Tennessee.

Steven D. Davis, CPCU, ARM (CHAPS. 1, 5, 16) is executive vice president and managing director for Willis Corroon Construction. He has specialized in the construction industry for the past 17 years, focusing his efforts on the development of alternative risk programs, such as captives, self-insurance, and other loss-sensitive programs designed for contractors. He is a nationally recognized speaker on cost of risk engineering for construction insurance and risk management and has authored numerous articles and books. He holds a Bachelor of Business Administration degree in Insurance and Risk Management from the University of Texas at Austin.

B. Calvin Deaner (CHAP. 15) is a vice president of claim administration for Willis Corroon Construction in Charlotte, N.C. Prior to joining Willis Corroon, he was claims manager for the Maryland Casualty Company. He attended the College of Insurance Management program, has served as president of the District of Columbia Metropolitan Claim Manager's Council, and has more than 35 years of claims administration and adjusting experience.

David J. Dybdal, CPCU, ARM (CHAP. 10) is a nationally recognized expert and corporate resource available to Willis Corroon account executives and clients who are designing and implementing risk management programs that have environmental liability loss exposures. He has served on the U.S. Environmental Protection Agency, Contractor Indemnification Technical Review Panel, and a focus group, Identifying Opportunities for Integrating Pollution Prevention into the Insurance and Risk Management Professions. He has also provided technical information on environmental insurance issues to the Departments of Defense and Energy. He has BBA and MBA degrees from the University of Wisconsin-Madison.

Carl H. Groth III, ARM (CHAP. 8) is a vice president and managing director of advanced risk management services of Willis Corroon Americas. He has 10 years' progressive experience in risk financing and captives for the large account market, serving the manufacturing, retail, transportation, construction, textile, paper, food, security, industrial cleaning, and publishing industries. He previously worked with Liberty Mutual Insurance Group. He has a bachelor's degree from the University of Washington and an MBA from George Washington University.

Douglas H. Hartman (CHAP. 14) is a principal of Ernst & Young LLP in New York, where he directs the risk management consulting practice, a field in which he has over 20 years' experience. Mr. Hartman graduated from Calvin College with a BA in History, and Michigan State University, where he graduated with an MBA degree in Finance.

John Heffron III (CHAPS. 3, 5) is a partner and the national co-director of construction consulting for Ernst & Young LLP in Philadelphia, Pa. He has more than 20 years' experience in construction assisting contractors, owners, and bonding companies in project controls, dispute resolution, workouts, financial management, and strategic planning. He has BA and MBA degrees from Rutgers University and attended Rutgers University–Camden School of Law. He is the coauthor of *A Construction Course for Owners* and a contributing author of *The Handbook of Construction Law and Claims*. He has been a featured speaker on construction-related topics.

Robert D. Heuer (CHAP. 4) is a senior vice president and managing director of the Great Lakes Region for Willis Corroon Construction, with over 22 years' experience in insurance beginning with his career as a casualty underwriter for Fireman's Fund Insurance after graduation from the University of Michigan. He has written and spoken extensively on insurance and surety issues for contracting organizations and has been a guest lecturer at Stanford University's Graduate School of Engineering.

Steven J. Kothe (CHAP. 4) is vice president of Willis Corroon Construction's Pittsburgh, Pa., office. He is a graduate of Washington and Jefferson College and a 28-year veteran of the surety industry, including nine years as an underwriter with the Travelers Indemnity Company in Florida and Pennsylvania. He is a regional resource for Willis Corroon in noncontract surety, environmental surety, and environmental impairment liability.

James M. Maloney (CHAP. 3) is deputy chairman of Willis Corroon Construction with overall responsibility for surety operations. His career started with Aetna Life and Casualty Bond Department and he has over 30 years of surety and insurance experience. A graduate of Providence College, he is active in many civic and professional organizations. In 1992, he was elected president of the National Association of Surety Bond Producers and now serves on the executive committee.

Douglas R. McPherson (CHAP. 17) is president of McPherson Enterprises Limited, a Towson, Md.–based corporate registered investment advisor, providing financial counsel exclusively to construction company owners, and Douglas R. McPherson & Associates, Inc., a marketing firm for insurance products. He has over 25 years' experience in financial counseling and is a recognized estate and tax planning authority. He is a contributing author of *The McGraw-Hill Construction Business Handbook,* 2d ed., and numerous construction industry publications and newsletters.

Ronald D. O'Nan, CPCU, ARM (CHAP. 8) is an independent risk management and insurance consultant specializing in risk financing methods, including captive insurance companies. His 40 years of experience in the insurance business includes underwriting for two major insurers, sales and marketing for agents and brokers, corporate risk management for a large international service company, and consulting for Willis Corroon in its Chicago and Nashville offices. He has been active with many insurance-related organizations and was Director of Faculty at the Insurance School of Chicago, where he also taught CPCU and ARM courses. He has been a speaker at numerous AMA, CPCU, and RIMS meetings throughout the United States. He earned a Bachelor of Science in Economics at Illinois Institute of Technology, and has authored a textbook, *A Study Guide to CPCU Part I*, published by the Insurance Society of Chicago.

William J. Palmer (CHAP. 13) is a retired partner and past national director of construction industry services for Ernst & Young LLP. He has headed the audit management team for one of the largest engineering/construction companies in the world, and was vice chairman of the American Institute of Certified Public Accountants Committee on Construction Accounting. He is a coauthor of *Construction Accounting and Financial Management* and *The Construction Management Form Book* (McGraw-Hill), *Businessman's Guide to Construction* (Dow Jones Books), and *Construction Litigation—Representing the Contractor* (Wiley Law). He is a graduate of the University of California at Berkeley and is a faculty member of the Haas Graduate School of Business there.

Salvatore J. Perrucci (CHAP. 11) is executive president and managing director of Willis Corroon Construction, leading the firm's global initiative in the architect/engineer, design/build, and construction management industry sector. A 14-year insurance industry veteran, he is a recognized specialist in professional liability, environmental liability, alternative funding mechanisms, and complex single-project insurance programs. He is a graduate of the State University of New York at Stonybrook.

Jeffrey A. Segall, CPCU, ARM (CHAP. 8) is a senior vice president in the Construction Risk Management Group of Willis Corroon Construction. As such, he functions as a national coordinator for claims, safety management, statistical analysis, and service support between the firm's local, regional, and national offices. He is also part of Willis Corroon's national research team, with which he develops programs, writes articles, and delivers speeches on insurance topics relative to the construction industry. He graduated from the University of Texas in El Paso with a BBA degree, and worked for the St. Paul Insurance Company as a casualty underwriter before joining Willis Corroon.

Mark A. Smith, Ph.D., PE (CHAP. 2) is a partner and national co-director of construction consulting for Ernst & Young LLP in San Francisco. He has more than 15 years' experience in assisting clients in optimizing their construction programs and reducing risks of overruns and costly disputes, and developed the construction management graduate program for the University of Maryland College of Engineering, where he was an assistant professor. He received his BS and MS in civil engineering from Oklahoma State University and his Ph.D. from the University of Texas at Austin. He is a licensed professional engineer as well as a certified public accountant.

William F. Ward III, ARM (CHAP. 12) is vice president of Willis Corroon Construction's National Resource Team. He is responsible for new product development, alternative risk financing programs, cost-of-risk studies and allocation systems, and risk management consulting for the construction industry. Formerly with Fluor Corporation, his experience includes risk management responsibility for engineering and construction projects from Saudi Arabia to Thailand. He is a graduate of Walla Walla College, Wash., and holds an Associate in Risk Management certification.

Series Introduction

Construction is America's largest manufacturing industry. Ahead of automotive and chemicals, it represents 14 percent of this country's gross national product. Yet construction is unique in that it is the only manufacturing industry in which the factory goes out to the point of sale. Every end product has a life of its own and is different from all others, although component parts may be mass-produced or modular.

Because of this uniqueness, the construction industry needs and deserves a literature of its own, beyond reworked civil engineering texts and trade publication articles.

Whether the topic is management methods, business briefings, or field technology, it is covered professionally and progressively in this series. The working contractor aspires to deliver to the owner a superior product that is ahead of schedule, under budget, and out of court. This series, written by constructors for constructors, is dedicated to that goal.

M. D. Morris, P.E.
Series Editor

Preface

For a number of years, each of us has felt the need for a book on the subject of insurance and bonding and risk management, which could be used as a primer for those starting a construction company, ascending to management in a medium- or larger-size construction company, and who serve the construction industry—the certified public accountant, a banker, and, yes, the insurance agent and surety. It took a telephone call from M. D. "Dan" Morris to bring the three of us together. Dan had had a similar thought that there was a noticeable void in an area of information on insurance and bonding written in simple terms for your average "construction stiff" and those who serve him.

After agreeing that there was a need and that the three of us would try to fill it, we then had to convince our senior editor, Larry Hager, of the need for such a book. "Aren't there any books already out on this subject?" he asked. While we agreed to do a search, we immediately answered a resounding no! In writing, we carefully convinced him that the only books on the subject were by lawyers writing for other lawyers on insurance law or insurance agents writing for other insurance agents on how to sell the insurance and maximize their premiums. Nowhere could we find a book written for the contractor, who is the one who must buy the insurance, determine the needs for insurance, maintain adequate records, account for the cost of the premiums, and document and file claims in a timely manner.

It is unusual to find the contractor who is also trained in the fields of insurance and bonding. Yet if his or her business is to grow and succeed, the contractor must learn to understand insurance and bonding and use the help that an agent can provide. Also, contractors should find this book useful in helping to understand and evaluate the recommendation of the agent and other professional advisors and be able to put those recommendations into practical operation. Finally, contractors ought to have the basic knowledge to determine what insurance is necessary and whether it is worth what it costs.

The passage of time has changed the image of the construction industry as a purely masculine domain. Nowadays, women are occupying an increasingly important place in the construction industry. We owe a debt of gratitude to these women, as it was the National Association of Women in Construction who published *A Dictionary of Construction Industry Terms*. All of us made liberal use of this dictionary in the early days of our construction careers so we didn't make fools of ourselves when dealing with contractors.

It is our earnest hope that this book not only will prove to be as useful as it was planned for anyone whose activities touch on the construction industry but also may help to give the individual contractor a better understanding of an increasingly important part of the operations that too often in the past has been looked upon as a necessary evil.

In a sense, this book is more a compilation of knowledge obtained over many years of working with contractors and with those in the insurance industry. For years, we have been shamelessly picking the brains of everyone we've met who would talk about construction insurance and bonding. Many of our former associates may recognize some of their own thoughts and ideas in these pages. Examples, too, have been collected from many sources. To all of these friends, companies, trade associations, and others, we owe a large debt of gratitude. A special thank you to the National Association of Surety Bond Producers, American Institute of Certified Public Accountants, Associated General Contractors of America, and Associated Builders and Contractors for the knowledge and training they have provided over the years.

Last, our special thanks to Willis Corroon and Steve Davis for coordinating the authors provided by Willis Corroon and to Ernst & Young LLP's support in providing the skills of Jeanette McCarver, Kathleen Mongan, Suzie O'Mura, Peggy Wall, and Ana Alegria-Malendez. Also, our special thanks to M.D. "Dan" Morris and Larry Hager of McGraw-Hill who patiently put up with our tardiness in delivering the final manuscript for this book.

William J. Palmer
James M. Maloney
John L. Heffron, III

Construction Insurance, Bonding, and Risk Management

Steven D. Davis, CPCU, ARM
Willis Corroon Construction

1.1 Introduction

In evaluating the construction risk management/insurance market, it is important to take a step backward to truly understand the magnitude of the changes that have occurred in every facet of this business.

Historically, contractors, agents/brokers, and insurance companies alike have chosen to focus on the premiums associated with risk transfer rather than losses, the basis by which all premiums are derived. Contractors who purchased insurance in the seventies and eighties generally opted for the more traditional insurance products such as guaranteed cost, with no associated deductibles or self-insurance. As a result of this buying philosophy, market cycles tended to give the definite impression that things were not in the contractors' control, thereby creating uncertainty, frustration, and a lack of trust among contractors and the general insurance market.

Of course, when one evaluates the products and services that have been provided, as well as the agency and brokerage marketing mentality and the legal environment of the day, it can be derived that very little was actually done in the form of education to change the buying habits within the construction industry. Why? Most agents/brokers who marketed construction insurance services really (1) did not understand the needs of contractors, (2) were not qualified, and (3) were not comfortable treading into new and unknown areas. This aversion to change existed on both the contractor level and the insurance industry level because it was simply easier to remain in the comfort zone of "this is how we've always done it."

In 1983, however, with insurance capacity problems escalating and rate hikes and insolvency issues making the headlines, a complete paradigm shift began. This shift was created out of need rather than opportunity. Contractors began to demand changes, demand information, make decisions, and insist on rewriting the agents/brokers' role to include education. Today, this new philosophy has created opportunities like risk management profit centers, contractor-controlled insurance programs, group captives, and other sophisticated funding techniques whereby additional margins may be recognized. It certainly appears that, due to this shift in the buying philosophy, contractors have tremendous opportunities to economically capitalize on new methods of managing their risks.

These three areas (capacity, rate hikes, and insolvency) have brought sophistication to the buying philosophy within the industry, which in itself demands leading-edge creativity by service providers, and they will continue to revolutionize every facet of this business.

The new order of things may be described as the Cost of Risk Engineering (CORE). This philosophy has allowed the insurance industry, and particularly construction-oriented insurers, to realign its focus on the issues that truly impact a contractor's long-term cost.

Cost of Risk Engineering is (1) the identification of the total risk picture, (2) the resulting costs in both dollars and intangibles, (3) a guide to handling risks, and (4) assurance that the risk management function is a true profit center.

The Cost of Risk Engineering may be divided into the following elements, each having a varying role depending on the sophistication, risk appetite, and management philosophy of the contractor:

- Program design
- Information systems
- Brokerage
- Risk financing
- Claims management
- Safety engineering and management
- Risk management

1.2 Program Design

The design of any insurance and risk management program should begin with defining the overall objectives of the contractor, taking into consideration his or her appetite or aversion to risk.

Just as construction operations change from one year to the next, the design of the insurance/risk management program should be reevaluated annually by considering the following:

- Coverages
- Limits of liability
- Location issues
- Severity of risk/frequency issues
- Historical losses and projected losses
- New operations (i.e., design-build/CM)
- Financials—balance sheet

Overall, a comprehensive design of the insurance/risk management services enables contractors to better align the dispersal of funds (losses) with the internal supply of funds (premiums).

1.3 Information Systems

In order to be in a proactive mode when managing the cost of risk, it is imperative that contractors utilize state-of-the-art information systems that are (1) reliable, (2) accessible, and (3) user friendly.

These highly technological systems play a large role in developing existing costs as well as future costs—both in the form of loss dollars and their relation to funds set aside in the form of premiums. Without the proper information systems, the contractor continues to react to market trends, loss costs, and experience modifiers, and is unable to fully embrace a proactive role.

Remember one axiom: Today represents the first indication of what your cost of risk will be for the next five years.

1.4 Brokerage

Traditionally, agents/brokers have focused on the brokering element of CORE as a measurement of the success of any risk management program. However, it is now apparent that this area represents only 5 percent of the cost of risk to contractor, with the other 95 percent representing loss and loss derivative areas such as adjustments and legal costs, as well as insurance company expenses which average 30 percent.

Contractors have renewed their focus on the loss cost arena, whereby proper management and risk assumption can be redefined as additional profits through the Profit Center Model for CORE.

1.5 Risk Financing

This area is subject to varying degrees of creativity ranging from depressed pay-in to captive insurance programs and self-insurance. Again, the risk financing tools employed must relate to the program design as well as the objectives of the contractor.

Risk financing mechanisms, such as premiums and operations, need to be continually evaluated based upon the financial position of the contractor. As the alternatives available are expanding because demand remains high, consider the following:

- Retrospective rating—incurred loss
- Paid loss retrospective rating
- High deductible/S.I.R. plans
- Self-insurance—workers' compensation
- Captives—group and single parent
- Rent-a-Captives
- Chronological stabilization plans
- Variable retention/dividends
- Investment retrospective
- Expected loss retrospective

1.6 Safety Management

As a direct element of CORE, safety management continues to be of high interest when discussing the reduction of the contractor's overall cost of risk.

Preventing losses through education is primarily the key, both at the executive level as well as the staff level. Key areas under review are:

- Management commitment
- Supervisory direction and control
- Pre-job self-assessments
- Employee orientation
- New employee verification systems
- Performance measurement and accountability systems
- Employee safety and health training
- Subcontractor administration
- Substance abuse programs

- Return to work
- Disaster planning
- Ergonomics

Note: Over two-thirds of accidents occurring in the construction industry involve ergonomics. Specific action plans should be developed to address these issues before they occur.

1.7 Claims Management

This area, combined with safety management, represents the largest opportunity for contractors to improve their margins (i.e., Cost of Risk Engineering).

With approximately 65 percent of each dollar of premium being spent to fund losses, a mere 10 percent improvement in this area will favorably impact the cost of risk by as much as 25 percent. This is why many contractors have rededicated internal and external resources to claims management and are holding a much higher standard to insurance companies that adjust claims. Areas of focus are:

- Reserves and reserve analysis
- Claim reviews
- Preferred provider organizations
- Medical/bill review and utilization
- Experience modifiers
- Claim reporting systems
- Claimant contact
- Alternative dispute resolutions

1.8 Risk Management

Risk management dates as far back as the Old Testament with the Egyptian pharaoh and Joseph predicting seven years of plenty followed by seven years of famine. Today, risk management in the construction insurance arena has expanded to new dimensions which include but are not limited to:

- Contract document analysis
- Subcontract document reviews
- Certificate auditing and maintenance systems
- Minority mentor programs

- Contractor coordinated insurance programs
- Loss analysis, forecasting, and trending
- Exposure analysis
- Allocation of costs and accruals
- Training and education
- Plan evaluations

Through implementing CORE and balancing the various elements to reflect a profit center approach to risk management, contractors are taking control of their destiny by managing one of the largest cost items in their financials. In doing so, the earned benefits include improved margins through the effective management of both direct and indirect costs of accidents. Direct costs, being much easier to identify, include medical and indemnity expenses. Indirect costs, however, include the following:

- Time lost from work by injury
- Loss of full earning power
- Economic loss to injured's family
- Lost time by fellow workers filing for injured
- Loss of efficiency due to disruption in process
- Cost of temporary worker
- Cost of educating new worker
- Damage to tools, equipment, property, etc.
- Spoiled work
- Loss of actual production
- Failure to complete work

These items of indirect costs can increase direct costs by a multiple of at least 200 percent.

The pioneers in the construction industry who have either implemented CORE or are in the process of doing so have searched the market and are beginning to develop opportunities by creating partnerships with insurance companies and agents/brokers. Three major trends for the 1990s that will prove to be methods of improving through CORE are:

- *Contractor-controlled insurance programs (CCIP).* This program resembles a quantification to the true *partnering* concept between owners and contractors and will provide major opportunities for many contractors and their project(s). The reason: approximate loss ratios with standard/traditional programs average 60 to 65 percent, while properly structured and engineered CCIP programs between general contractors and subcontractors rarely exceed 30 percent.

- *Minority mentor programs.*   When contractors express their biggest issues for the 1990s, more times than not minority business enterprises (MBEs) are at the top of the list. These requirements, coupled with a future shock shortage of talented labor by the year 2000, create a tremendous opportunity for contractors who wish to be at the cutting edge of competition. Brokers, such as Willis Corroon, through their Minority Enterprise team, have taken this initiative to create a network of contacts within the minority construction industry and its existing clients for future projects and services.

- *Niche marketing and consolidation.*   It is evident that insurance companies and agents/brokers are continuing to focus on areas where they have major market presence. This is especially true in the construction industry where insurance companies and agents/brokers are realigning their expense structure along the lines of practice groups. This trend will continue into the year 2000 with expense reductions being managed through consolidations. In essence, partnering with insurance companies and agents/brokers will become the norm rather than the exception.

It is increasingly apparent that the walls of tradition are beginning to crumble in the construction insurance and risk management industry. To remain both competitive and profitable, insurance companies must reengineer their operations. This restructuring must focus on quality client-focused services as well as education, for together they will make a difference in the overall costs.

By shifting the paradigms that have conveniently allowed traditional methods to address new issues, we are opening an entirely new spectrum in the construction insurance and risk management field. Brokers and agents must change and assume a leadership role with their customers and confront new challenges with new solutions. It is time to either lead, follow, or step out of the way. Most important, it is time to grasp the opportunities that redefine the circumstances by which competition exists, both in the construction industry and the insurance industry.

As Niccolò Machiavelli once said in reference to the genesis of change: "There is nothing more difficult to take in hand, more perilous to conduct, or more uncertain in its success than to take the lead in the introduction of a new order of things."

2

Understanding Construction Risks

Mark A. Smith
Ernst & Young LLP

Construction is a diverse industry requiring enormous amounts of capital and resources to function efficiently. Because of the size and diversity of the construction industry, the major industry players—owners, designers, and contractors—frequently appear to be in conflict. These conflicts can be brought on by any number of situations, but they usually result from a lack of consideration about the real risks involved with a construction project.

Construction problems are exacerbated because the construction industry is so fragmented. The Business Roundtable determined through its landmark study, *Construction Effectiveness*, that construction is a $445 billion industry comprising nearly 1 million contractors, more than 70 national contractor associations, more than 6.5 million workers, and an indeterminable number of owners. Without a single group to unite the various fragments, it is no wonder that construction disputes arise on nearly every project.

Although major construction problems occur on many projects and result in significant cost and schedule overruns, many of them can be avoided. Knowledge of the construction management process is essential to control the many factors that influence project results.

Both the owner and contractor should realize that a disputed claim on a major project is likely. The principal objective of both parties should be to avoid the problems and related litigation altogether. Failing to prevent and mitigate such damage, both parties should carefully track and document the impact of specific problems. In all cases, both parties should work toward an equitable resolution of all problems.

2.1   What Is Risk?

Inevitably, there will always be construction disputes primarily because there is always uncertainty involved in construction. When problems arise on a project, it is usually a combination of the owner's, designer's, and contractor's faults. In construction, as in most businesses, there is risk. In fact, most contractors are in the business because they are risk takers.

Risk is defined in Webster's dictionary as the chance of injury or damage or loss. While this obviously is a general definition of risk, construction risk can also be related to the chance of loss associated with three primary constraints: time, cost, and quality. Only two of these constraints are within the contractor's power to control—time and cost.

The constraint of time is defined not only by how fast a project can be completed, but also by the likelihood of finishing a project according to the agreed-on schedule. The risk constraint of cost is defined as the variance in the actual costs to complete a project compared to the budgeted costs. If contractors are to manage risk—and make a profit—they must concentrate on managing these two risk constraints.

The third constraint, quality, is determined by the designer through conformity with the acceptable standards and specifications. This constraint is normally outside the control of the construction contractor.

2.2   Pitfalls in Construction

The reason that construction, when compared with other business industry segments, has an increased propensity for risk situations is because of the numerous pitfalls involved with this industry. These pitfalls represent recurring problems that significantly affect both cost and schedules for almost any type of construction project. Table 2.1 presents an overview of the types of problems that can be related to the major parties of a construction project—owner, designer, and contractor. How these pitfalls affect the success or failure of a project is discussed in more detail as follows.

2.2.1   Failure to fund

Adequate funding for construction projects, in the current economic period, is becoming much more difficult for the majority of the construction industry. There is significant aversion by financial institutions to adequately funding small- to medium-sized construction projects. Lenders are now enforcing stringent guidelines for construction owners. The banking industry is becoming much more selective in the kinds of projects in which it is willing to invest. The construction

TABLE 2.1 Pitfalls in Construction—Owner, Designer, and Contractor Perspectives

Owner	Designer	Contractor
Failure to fund	Defective plans and specs	Slow to mobilize
Owner-furnished materials not available	Shop drawing review and material approval	Failure to staff project
Major changes in requirements	Improper or delayed change orders	Failure to provide sufficient equipment
Failure to make progress payments	Failure to coordinate between primes	Failure to coordinate
Interference	Inadequate information	Inadequate project management controls

contractor must be concerned about whether the owner will be able to reimburse or pay progress payments in a timely manner.

By far, the most common cause of contractor bankruptcy is not failure to earn a profit over the long term, but failure to receive payment in a timely manner. Most contractors go bankrupt because they cannot meet short-term obligations, resulting from being unable to manage their current obligations.

2.2.2 Owner-furnished materials

Owners typically want to procure and furnish their own equipment for unique projects, such as educational facilities, hospitals, and research facilities, because of the special needs associated with such projects. Most owners fail to understand that contractors have specific materials management plans for the project—plans for when each material will be installed and where it will be stored once it arrives at the site. Most owners have the mistaken notion that whenever owner-supplied material arrives on the site, the contractor can efficiently store or install the material. Contractors usually have strict installation schedules that they must follow. This type of misunderstanding leads to significant material management problems.

2.2.3 Major scope changes

The owner is ultimately responsible for defining the scope of a project. During the conceptual phase of the project, the scope will probably change quite frequently without causing undue hardship. However, changes that occur during the construction phase, or even the detailed design phase, inevitably make life difficult for the construction contractor. Scope changes result from a number of situations, such as project owner desires, design review, constructibility issues, safety requirements, and operations and maintenance concerns.

2.2.4 Defective plans and specs

There have been few instances where an engineering design was so complete that a project could be built to the exact specifications contained in the original design documents. Errors or omissions in the design documents will result in some modifications to the originally planned construction process.

2.2.5 Slow to mobilize

Many contractors do not perform sufficient planning in advance to effectively mobilize for the project. Often the owners or their representatives inform the contractor that the site is available and ready for mobilization to begin. Most contractors do not move onto a site in an efficient or timely manner because of insufficient planning regarding the equipment and labor needs that are required.

2.2.6 Failure to allocate resources

Earning a profit is obviously essential to any contractor's long-term strategy. How does a contractor make a profit? A profit is normally earned by managing a project with the least number of people and the least amount of equipment on site. The more workers and equipment situated on the site, the more cash outflow is required. Managing limited amounts of personnel and equipment efficiently and cost-effectively is more easily said than done, especially if a contractor is dealing with several projects at once.

What frequently happens is that a contractor either does not have the right mix of people or lacks the right equipment on a particular project because he or she uses these limited resources on other projects. Labor and equipment availability is an important consideration to take into account when preparing a bid on various types of construction projects.

2.2.7 Inadequate project management controls

Project management areas that cause the greatest concern include:

- Lack of meaningful cost control information
- Project decisions that are not made in a timely manner or not promptly implemented
- Decisions affecting the total project that are made by one group unilaterally
- Confusion regarding the apparent authority and responsibility of project personnel
- An apparent lack of integrated planning

2.2.8 Insufficient work definition

This has a far-reaching effect because a project's scope must be defined in enough detail for comprehensive scheduling, estimating, and resource allocation to be reasonably accomplished. Whenever the owner is unwilling to make a concerted effort to define the construction project in the early stages, problems will undoubtedly occur throughout the entire project life cycle.

2.2.9 Unrealistic schedules and underestimated costs

Whenever time schedules and/or project costs are based on optimistic predictions, the project will be burdened with missed milestone dates, cost overruns, and even low morale among project personnel. Sufficient time must be allocated in the planning stage of a project to realistically estimate time and cost requirements.

2.2.10 Inadequate cost control and accounting practices

These problem areas are associated with imperfect management planning and control. Possible explanations of why estimated costs for project activities are so inaccurate when compared to actual costs include:

- Lack of planned information systems
- Poor correlation of work definitions
- Improper control of schedules and costs segregation
- Inaccurate cost allocation methods

2.3 Identifying Construction Risk Areas

To properly plan and organize an execution of any construction project, a concerted effort must be undertaken to identify risk areas that can affect the project's success or failure. No two construction projects are alike. Thus, an assessment of the risk areas must be performed for each new undertaking to fully understand which situations may cause undue hardships in completing a project on time and within budget.

Table 2.2 summarizes the types of risk factors and their variable levels of risk that are significant when assessing the likelihood of critical problems occurring on any given construction project. The major components of construction risk are categorized into the areas of:

- Timing
- Project nature
- Approvals

- Owner structure
- Project delivery
- Project controls

TABLE 2.2   Identifying Risk Factors and Level of Risk

Risk factor	Level of Risk		
	High	Moderate	Low
1. Timing			
Preconstruction time	Weeks	Months	Years
Construction pace	Fast-track	Selected overtime	Normal
2. Project nature			
Type of construction	State-of-the-art	Government work	Standard office
Relative size	Large	Normal	Small
Complexity	Rehab; process	Moderate	Spec. building
Location	New geography	Regional	Local
Status of plans and specs	Insufficient for bid	Essential components show	Complete
3. Approvals			
Regulations and codes	Many agencies and rules	Standard safety codes	Basic building codes
Politics	Adverse community reaction	Conformity standards	Community support
4. Owner Structure			
Decision process	Building boom	Planned growth	Undeveloped
Scope changes	Many committees and users	One small committee	One person
Design/construction experience	Many preliminary designs	General program known	Direct approach
Monitoring talent	Nobody	Use of consultant	Construction staff
Fund resources	No money for overruns	Need to borrow for overruns	In-house funding
5. Project delivery			
Contract type	Fast-track	GMP	Fixed-price
Project management	Owner acts as contractor	Construction manager	Design-build
6. Project controls			
Owner's representative	None	Architect/ engineer	Project management oversight
Schedule	No formal schedule	Bar charts	Critical path method
Cost control	Schedule of values	Job cost system	Cost loaded schedule
Contracts	None	Standard forms	Custom
Change order management	Resolve at end of job	Resolve postjob	Resolve prejob

The risk area of timing applies not only to the amount of up-front time that is spent in planning a construction project, but also to the project's progression during physical completion of the work. If virtually no preplanning has been undertaken for an upcoming project and the work is to be done in a fast-track manner (construction is begun before a complete design is finished), the level of risk is sufficiently high that problems will undoubtedly occur on this project.

Project nature relates to the overall definition of the anticipated project, which includes such variables as type, size, complexity, location, and completeness of design plans. Typically, commercial buildings that are designed to normal specifications are less prone to risk of problems and disputes than complex industrial process plants that are built to unique design criteria. Each risk factor obviously can have a varying degree of risk depending on the circumstances. A standard office building can be quite risky if it is built in a new geographical location and the plans and design specifications are incomplete prior to beginning construction.

All construction projects have various levels of approvals that must be secured before starting construction fieldwork. Typically, projects that must meet stringent environmental regulations (and be reviewed and approved by numerous levels of government agencies) are prone to time delays and cost overruns. Likewise, projects that are adverse to the general public perception, such as nuclear power plants, will encounter significantly more approval problems than publicly supported projects, such as the building of a public park or recreational facility.

The type and experience of the owner greatly impact the variability of risk associated with a given construction project. The owner is the single most important entity in determining the likelihood of future problems on a planned project. An owner organization that has little construction experience or does not engage qualified construction management consultants creates significantly more risk for a project than an organization that has a defined set of policies and procedures for monitoring the construction work. Most sophisticated owners fully understand that changes to the scope of work create havoc among construction contractors who try to complete the job on time and within budget. In addition, owners who have not adequately budgeted—with an appropriate level of planned contingency funds—create significant risk hazards.

The category of project delivery is related to the type of contractual relationship between the parties and the organizational makeup for overseeing the work. Fixed-price or lump-sum types of contracts are inherently less risky than fast-track construction that requires a cost-plus type of contract arrangement. In addition, owners who attempt to fulfill the responsibility of acting as their own general contractors, by subcontracting all or a significant portion of the work, are more likely to have costly problems occurring during physical completion of the project.

Project management controls are a risk area that has critical impact on ensuring that problems are discovered on a project before they become significant cash outflows. The control mechanism with the lowest risk levels is one in which the schedule and cost control components are linked together by a cost-loaded schedule. This type of control ensures that ongoing work activities are monitored accurately by determining the timing of their physical completion, as well as the direct costs associated with the individual work items.

In addition, when change orders are dealt with up front during the ongoing construction of the project (instead of after the project is finished), risk associated with the planned undertaking is reduced.

Finally, all risk factors should be critically evaluated to determine the overall uncertainty of the construction project. Some factors, such as owner structure and project controls, may be more prone to risk than others. Once the risk factors have been identified and evaluated, an appropriate course of action can be developed to ensure that the risk for the project is reduced to an appropriate level.

2.4   Minimizing Risks to Avoid Disputes

The period of initial planning and the periods before and immediately following the award of the contract are extremely important in preventing disputed claim situations altogether. These periods also represent the ideal time to provide for systematic collection and retention of information needed to manage the project or to support a future claim, if necessary.

Effective management begins with the planning of the project. It requires a coordinated effort among owners, designers, engineers, and experienced construction personnel, with heavy emphasis on the involvement of construction expertise in the design stages. Three areas warrant special attention:

1. Selecting and installing management systems

2. Carefully reviewing the contract and subcontracts

3. Establishing proper records and filing systems

2.4.1   Selection and installation of management systems

The first step toward preventing a major overrun and related claims disputes on a project is to assure from the start that it will be properly managed and controlled. In past years, with fewer government regulations and other external influences, and with projects being generally smaller, better defined, and less technically complex, establishing effective controls generally was not the problem it is today.

Now, with increased emphasis on larger projects, "fast-track" approaches, and increased use of negotiated or cost-plus types of contracts, establishing control has become more difficult and complex—and an even more essential requirement.

The construction industry generally has been slow to recognize the importance of effective integrated management systems, although such systems have been shown to be the first step in preventing major overruns and related claims disputes. By nature, most contractors seem to avoid heavily structured management approaches, priding themselves on being entrepreneurial personalities with the ability to scramble through a myriad of adversities. While this attitude can be highly effective, such shirt-pocket, seat-of-the-pants, and other rudimentary management systems based on the abilities of one or a few individuals are totally inadequate for major construction projects.

Key components of an effective management system must include:

1. *Comprehensive schedules.*   A study by the Business Roundtable concluded that on average, the duration of most construction projects could be reduced by 10 percent through effective scheduling practices, with the resulting cost savings of approximately 3 percent of the total project cost. At the same time, the study found a strong reluctance by contractors to use and maintain effective scheduling systems, particularly network systems.

Contracts for all major construction projects should require some form of network scheduling system that is actually used to direct and control the project. Moreover, the requirement for such a system should be a provision of the contract. Owners should assure themselves that contractors selected for major portions of the work have the capability to provide and operate under this and other key management systems. Day-to-day management is the responsibility of both the owners and the various contractors hired to perform the work. The owner is generally held responsible for ensuring the coordination of contractors so as not to impede the progress of any individual contractor. Failure to provide for appropriate controls, or to ensure the competence of any contractor hired, could raise serious questions regarding fulfillment of this responsibility by the owner or a general contractor.

2. *Effective estimating and cost control systems.*   Proper scoping of the job and its risks, effective budgeting, and cost accounting systems can have a dramatic effect on performance of the job and on subsequent claims. Yet these areas remain substantially neglected in many, if not

most, construction projects. In fact, the combined allocation to all of these functions typically ranges from 1 to 2 percent of project costs. Poor scoping of a project alone is estimated to increase costs by 2 to 4 percent.

3. *Bid estimates.*   In preparing to bid or estimate a package and initial project schedules, contractors should recognize that each element may be critical later in proving that claimed damages did not result from unreasonable estimates or schedules, or from the contractor's own negligence. Where anticipated project costs are sufficient to warrant concern, the contractor always should ensure that estimates and schedules are based on detailed workpapers supporting their reasonableness.

Standard preparation and management review procedures should be used to ensure that all estimates have been developed from a thorough investigation of available facts, that any productivity and cost standards are reasonable, and that calculations are accurate.

The contractor must ensure that all specifications used in the estimate either match those provided by the owner or are otherwise specifically justified. Management should also ensure that jobsite conditions represented by the owner are accurate.

4. *Accounting systems.*   Accounting systems appropriate to the particular job's complexities are essential. Most important to potential claim situations is a system that ensures that every extra work item that is not the fault of the contractor be subject to a separate account identifying all labor, equipment, material, and overhead costs. Important tools in this regard are labor and equipment daily timecards, purchase order and receiving reports, and the cost accounting system.

Every contractor performing a major construction project should have cost accounting systems and procedures that permit separate tracking of costs for any work outside the scope of the contract, or for work which results from problems imposed by the owner. The chart of accounts should be sufficiently flexible to provide this capability.

During the initial planning phase, and later during mobilization, each area of the job should focus on high-cost items and on items that are otherwise considered to be high risk. Attention should be given at this point to methods for measuring the impact of problems occurring in any area and to the documentation needed to support a future claim for each cost category. Recognizing these potential areas beforehand will help to establish a suitable accounting and reporting system to deal with these problems when they occur.

The cost accounting system should be tailored specifically to accommodate these potential problems and accounting requirements. At a minimum, the system should provide for a separate cost account or subaccount for labor, material, and equipment on each disputed or extra work activity that occurs.

5. *Materials management systems.* Neglect of materials management systems is one of the most common weaknesses found in the planning and execution of most construction projects. Manufacturing and construction both require the same basic resources: materials, physical space, equipment and tools, trained labor, and usually some form of energy. The absence of any one of these elements can stop production completely. An inadequate supply of appropriate resources when needed can delay production and increase costs substantially.

The construction industry is generally far behind manufacturing operations in recognizing the importance of materials management as part of production planning and control. Considering that materials account for approximately 60 percent of all construction costs, it is surprising that more attention is not given to this function. According to industry estimates, improved materials handling systems could reduce average labor costs (currently 25 percent of total cost) by approximately 6 percent. Yet this function consistently remains a stepchild of the organization and management of most jobs.

Every construction contract that involves substantial handling of material should require a detailed plan and schedule for materials management from the contractor (who should be required to demonstrate the ability to establish and operate an effective materials management system before being awarded the contract). The materials management plan should be integrated with the overall project and should provide for the following:

- Assignment of qualified personnel, reporting relationships within the project organizational structure, and any necessary on-site training
- Procedures for vendor selection, purchasing, receiving and inspection, materials storage, accounting, security, and scheduling for timely delivery to the job site
- Space and facility requirements for storage
- Lead items, order points, and expediting

2.4.2 Review of the contract and subcontracts

Lack of understanding of the contract and failure to perform a careful review of its contents are major shortcomings that often prevent the contractor from making a reasonable assessment of risk involved and are substantial contributors to future claims situations.

For example, does the contract contain a Change clause? Does it establish the right of the contractor to collect for additional costs under changed conditions (no clause, no recovery)? Is there a Termination for Convenience clause that may preclude the contractor from collecting lost or anticipated profits? Are there exculpatory clauses, that is, no damages for delay (most are considered basically valid)? What are the notification requirements (such as EPA general conditions) included? And perhaps most important, how are the courts where the work will be performed likely to interpret the contract language and clauses? All of these factors should be matters of concern and careful review by the owner, the contractor, and their respective legal counsels during the bidding or negotiation stage after the contract is received for signing.

After receiving the awarded contract, contractors and their legal counsel should conduct a thorough review and analysis of its contents to ensure that the scope for the work described accurately reflects the work that was bid. Particular attention should also be given to all areas incorporated by reference, so that all rights and obligations are thoroughly understood. Several articles typically are of special significance with respect to future claims:

- Changes
- Changed conditions
- Variations in quantities (unit cost)
- Terminations for default
- Damages for delay
- Time extensions
- Escalation of labor or material

The prime contractor should also perform a careful review of the forms and contracts provided to subcontractors. Work in these documents must encompass all work required by the owner's contract; all that is expected from each subcontractor must be described completely and clearly. For example, what is the meaning of "installation"? Keep in mind that the contract will be interpreted against the party that drafted it.

Does each subcontract incorporate all the terms and conditions of the contract? Is the incorporation by a general inclusion statement or by specific reference? Are the courts in the area where the work will be performed likely to accept incorporation by reference, or will they require specification of the terms and conditions? All of these are legitimate concerns with respect to future claims and are matters that should be discussed with the contractor's legal counsel before the subcontract forms are prepared and distributed.

2.4.3 Records and filing systems

Standard record and filing systems, along with standard reporting procedures, should be established and maintained for each contract to provide required information. Good project management, preparation of claims, and defense against claims require systematic collection, preservation, and retrieval of at least the following:

- Request for quotation or bid, specifications, test reports, and other documents provided by the owner
- The contract and all amendments
- Meeting notes and correspondence with the owner and engineer
- Original estimate and estimate workpapers plus any revisions
- Project schedule and budgets, including all changes
- Subcontractor files containing contract, bids, bonding information, and correspondence
- Progress reports and time-phased photographs of construction progress and job conditions

In establishing policies and procedures governing the items that will be retained, contractors and owners alike should bear in mind that project file contents most likely will be subject to discovery by the opposition in future litigation. Potential embarrassment or damage to a future case should be a guiding consideration in preparing written materials and decisions regarding their retention.

2.5 Conclusion

Minimizing construction risk requires a multifaceted approach that must be shared by all parties associated with a construction project. All projects contain some degree of uncertainty; thus risk is inherent in every decision made on the project. With proper identification and evaluation of significant risk factors, many troublesome areas can be discovered and action taken prior to costly mishaps.

Contract disputes can be reduced when sufficient efforts are made by the principal parties to the contract to understand the magnitude of risk that is associated with a given construction project.

Surety Bonds for Construction Contracts

John L. Heffron, III
Ernst & Young LLP

James M. Maloney
Willis Corroon Construction

3.1 Introduction

A surety bond is a guarantee. It is a three party agreement among the owner, contractor, and surety company. Under the terms of the bond, the contractor and surety guarantee to the owner that the construction project will be completed as provided for in the plans and specifications and construction contract. In every surety transaction, in addition to these three parties there is an underlying agreement, typically the contract that the bond supports.

A surety bond is not an insurance policy. Although bonds and insurance may appear to be similar to the untrained eye, there are basic inherent differences in their function. The surety is a guarantor, not an insurer. The surety company guarantees to the owner that the contractor will perform required obligations and that if not performed, the surety company will do so. In essence, the surety bond is a credit device similar to a co-signed note for a bank. The basic theory of suretyship, as in any case involving the extension of credit, presumes that there will be no loss. Surety bond premiums are service fees for the extension of financial data and are not based on actuarial or statistical probabilities of loss.

Some of the key differences between surety bonding and insurance are:

- The surety agreement is a three-part agreement where a third party (the surety) guarantees the performance of a contractor to an owner.

- Insurance is a two-part agreement where the premiums are calculated based on the probability of loss.

3.2 Why Are Bonds Needed?

Construction is a risky business. Bankruptcy and failure are commonplace. The owner, when contracting for a project, faces a number of risks including:

1. The apparent low bidder will not accept the award at the bid price or furnish the required performance and payment bonds.
2. The contractor will not be able to complete the project in accordance with the plans and specifications for the contract price.
3. Incurred labor and material bills will not be paid by the contractor, resulting in loss to the owner.

These risks can be mitigated and transferred through use of bonds executed by a competent and financially sound surety company.

3.3 Types of Surety Bonds

There are a variety of bonds. The most common ones are bid bonds, performance bonds, labor and materials payment bonds, combination performance and payment bonds, and supply bonds.

3.3.1 Bid bonds

The bid bond is the basic instrument of prequalification for many contract bids. Unless otherwise specified by the owner, the bond may be secured from any qualified bonding company, and its purpose is to validate the bid price submitted by the contractor to the owner. This bond is submitted with the contractor's bid. If the bid is accepted by the owner, the contractor must:

- Enter into a contract.
- Provide a sufficient bond for the performance of the terms.

If the contractor fails to meet either one or both of these requirements, the bid bond is forfeited. Depending upon the terms, the penalty assessed will be either the difference between the defaulting contractor's bid price and the next lowest price or the penal sum of the bond, whichever is less. Typically, the penal sum of the bond is either 5 or 10 percent of the contract price.

As an example, Contractor A submits a bid for the construction of a project in the amount of $3,000,000 accompanied by a bid bond in the amount of 10 percent. Contractor B submits a bid for the same project in the amount of $3,250,000. The owner awards the contract to Contractor A. Contractor A refuses to enter into the contract or does not enter because of some unrelated reason. The owner then awards the contract to Contractor B at a price of $250,000 more than A's price of $3,000,000. Under the bid bond, the surety is responsible for paying the owner up to 10 percent of the bid (.10 × 3,000,000) or $300,000. In this instance, it would pay the $250,000 for the difference in price and any related rebid costs up to the $300,000 ceiling.

The issuance of bid bond is not merely a formality. As far as the surety is concerned, the issuance and execution of a bid bond are of utmost importance. The underwriting analysis of the contractor has generally been completed prior to the surety issuing bid bonds.

3.3.2 Performance bonds

The performance bond is issued after a proposal has been accepted. It provides security in the amount of the face value, which is usually the contract price. Its purpose is to guarantee the completion of the work in accordance with the plans and specifications and at the contract price.

The wording of the usual form is very simple. Essentially, it states that if the contractor faithfully performs all the conditions required, the bond will be null and void. If, however, the contractor does not comply with the conditions, the bond will come into effect.

The protection provided by a performance bond is occasionally misunderstood. It is designed to assure to the owner only that the project will be completed as specified. This normally includes protection from financial loss such as would be caused by property liens resulting from unpaid bills incurred by the contractor. The performance bond itself does not, however, guarantee to creditors of the contractor that unpaid obligations are directly covered. That protection is provided by the labor and materials payment bond discussed in the following section. A performance bond is generally executed for either 50 percent or more commonly 100 percent of the contract price. The premium cost of the bond is based on the total contract price.

3.3.3 Labor and materials payment bonds

Labor and materials payment bonds are generally issued in conjunction with performance bonds. Typically, each one has the same penalty as the performance bond issued for a specific project. A payment bond guarantees that the contractor will pay all accounts arising from the job, thereby allowing the owner to take possession of a lien-free project at completion.

To the owner, a labor and materials payment bond makes it possible for the suppliers and subcontractors to provide their products and services at the lowest cost by reducing the credit risk. Similar to the performance bond, the labor and materials payment bond is generally executed in the same amount, 100 percent of the contract price.

3.3.4 Combination performance and payment bonds

Sometimes a contractor will be required to furnish a bond which includes both the performance and payment obligations in one form. The face amount of the bond must adequately cover both the loss the owner might suffer and any and all unpaid bills which an insolvent contractor might be unable to pay.

3.3.5 Supply bonds

Supply bonds are similar in intent to performance bonds. They are issued for contracts to supply materials, goods, or machinery at a specified time and place.

3.4 Underwriting Requirements for a Contract Surety Bond

The contract surety bond underwriting process is a source of confusion and mystery to many contractors. A contractor who applies for his or her first bond approaches the insurance agent to buy a bond for a construction project expecting to fill out an application form, pay a premium, and walk away with the bond. Instead, the agent asks for information on every aspect of the contractor's business, including copies of financial statements and tax returns for the three prior years, details on his or her personal financial position, the amount of life insurance, the past job experience, banking relationships, resumes on key employees, and a breakdown of all work currently underway. The shellshocked contractor wonders who could possibly want all this information about the company, and why?

3.5 The Underwriting Process

The objective of all surety companies is to extend bond credit selectively to contractors on construction projects who present little risk of loss to the surety. Those contractors and projects are identified by the surety through the underwriting process, which is a process of investigation and analysis resulting in a judgment being made as to the acceptability of a particular risk. The contract underwriter works with the contractor's independent bond agent to gather and evaluate information in the underwriting process. There is no uniform approach to underwriting that fits every account or every bond. Underwriting cannot be reduced to a simple checklist; it is a judgment process. Each situation will be unique, and the facts of each situation must be assembled and then judged accordingly. However, there are certain essential areas which should be explored for all contract bond risks. In the simplest terms, contract bond underwriting can be boiled down to three key areas: character, capacity, and capital.

Character deals with the most qualitative aspect of the contractor's business. It is unquestionably the most essential element of the underwriting decision process. The character of the individuals who own and operate the construction firm must reflect qualities of honesty and integrity. These qualities are a large measure of their willingness or determination to meet the obligations of the construction company even in the face of adversity. These qualities will be reflected in their business and personal backgrounds, relationships, reputations, and habits.

Capacity represents the contractor's ability to do the actual work as described in the contract. Determining a contractor's capacity is often the most difficult evaluation that the underwriter must make. Factors to be considered include the contractor's prior experience, the depth of his or her organization, the quality of management and controls, the type of work, the size and location of the job, the subcontractors involved, the number and size of other jobs concurrently underway, the adequacy of the contract price, etc. Even when all the facts have been assembled and the questions have been answered, it comes down to a judgment call by the underwriter as to the contractor's ability to execute the work under the contract without default.

Capital refers to the contractor's ability to finance the work undertaken and to absorb any losses or financial setbacks which might occur during the course of the work. Contractors often enter into fixed-price or lump-sum contracts, which means that even before work starts, the contractor guarantees the price for the work. The price is based on the contractor's estimate of what the work will cost and expected markups. The estimate is made prior to the start of work. A lot can change between the time that the contractor originally estimates the cost of the work and the date of the project's final completion. Therefore, a contractor must have sufficient capital not only to pay for labor and material between progress payments from the owner on the job, but also to absorb any additional costs which may not be reimbursed by the owner, or losses on the work, or delays in payment from the owner. This requires both liquidity in the form of working capital and equity in the form of net worth.

There are two distinct kinds of contract underwriting decisions. The first kind of decision is a determination as to the acceptability of a contract account in general. The second kind is an appraisal concerning the acceptability of a specific bond for the account. Before any underwriting decision can be reached, however, all of the details must be assembled, reviewed, and evaluated. The underwriter obtains information from many sources including the contractor, the certified public accountant preparing the statements, the bank, various credit services, and the business community as a whole. The following discussion focuses on the items of information needed, the importance of each item in the underwriting process, and the sources of this information.

3.6   Underwriting the Contractor

The contractor underwriter wants to know everything possible about a contractor account. This includes the company's background, the current status and situation, and future plans. The underwriter's goal is to understand the strengths and the weaknesses of the contractor account. This understanding is needed to properly prequalify a contractor, evaluating not only what a contractor can do if everything goes right, but also what resources a contractor has to absorb losses if something or everything goes wrong. Both the short-term and long-term loss absorption power of the contractor are important, and extensive financial analysis is performed to evaluate the strength of the contractor's financial position.

As part of the underwriting process, the underwriter will often put conditions on a contractor account to address any weaknesses which are identified. These conditions are designed to provide a buffer or contingency for any problems encountered. They are intended to protect the contractor as well as the surety. Because the surety's liability under the bond will arise only after the contractor somehow fails— either to perform according to the contract terms or to pay a subcontractor—the surety tries to foresee potential problems that the contractor may encounter on a job and build in safeguards to avoid or address these problems.

3.6.1   Ownership

There are various legal forms of doing business available to the contractor, including a proprietorship, a general or limited partnership, a corporation, a closely held Subchapter S tax corporation, and a limited liability company. Each form of business has obvious legal ramifications, not the least of which is the owner's liability for the debts of the business. Most small and midsize contractors are closely held businesses with the owners actively participating in the day-to-day running of the business. In these cases, the surety needs to know about the owners and their financial interests outside of the construction company. Specifically, the underwriter looks for items which may be used to strengthen the overall case, such as strong personal cash position, or which might detract, such as another investment which needs cash to supplement its operations; the concern in that case is that the construction company may be used as a financing vehicle for the other company or investment. This concern is a very real one given the demise of numerous contractors and developers in the late eighties and early nineties. The contractor needs to have liquidity to handle short-term problems and a strong equity base to act as a loss buffer for the long term. Any funds taken out of the contracting entity to finance other personal interests deplete these buffers. Conversely, the strong and liquid personal statement may be a source of funds for the contracting entity if its financial position needs bolstering for a particular project or program of work.

Therefore, the prudent underwriter will request and review personal financial statements of the owners in most situations. Typically, the owners will be asked to execute an indemnity agreement in which they agree to indemnify the surety for any losses it may incur on work bonded by the surety. Spouses of individual owners are also required to sign an indemnity since personal assets are often in joint title or easily transferable from one spouse to the other. The spousal indemnity is a safeguard to insure that the assets shown in the personal statement as supporting the contracting entity will remain accessible to the surety in the event of a loss.

Fundamentally, in requiring the indemnity of the owners, the surety is only asking them to do what the surety itself is doing: standing behind the business and guaranteeing that it will meet its contractual obligations and indemnifying the project owner against any loss arising out of the contract.

3.6.2 Organization

A surety typically will require the resumes and background information of all key personnel in the contracting whether or not they have any ownership in the business. Underwriters look for a well-rounded background in both business and construction, but more specifically, they look for prior experience in managing the type of work and the size of jobs and programs which the contractor is currently entertaining.

In addition to assessing the individual members of the organization, the underwriter evaluates how the organization functions as a team. If key members of the organization have been with the firm for only a short period of time, the fact that the organization is not seasoned may weigh heavily in the evaluation.

The strength of a contractor's organization is critical to a contractor's success. Therefore, the surety underwriter evaluates not only the current strengths and weaknesses but also the future potential, confirming that the organization includes people who can act as backups to those in key positions. If a key employee should leave the business, become disabled, or die, the company should have another person who can take the responsibilities of that position.

Most of these organizational details are developed through a surety questionnaire completed by the contractor and bond agent at the beginning of the surety relationship. However, the underwriter needs to be kept up to date on any significant changes which may occur in the organization as they could impact the overall underwriting of the account.

3.6.3 Background and construction history

The underwriter will track a contractor's evolution with specific emphasis on the last five years to gain a complete understanding of an account. Any changes in the management or the scope of the company's operations occurring during the last five years are evaluated to determine the impact of that change on the company. Consideration is given to areas where the contractor has been most successful as well as those areas where problems may have developed. Careful scrutiny is given to the causes of any problems and the contractor's success in dealing with them.

In reviewing a contractor's construction history, the underwriter looks at

- Type of work
- Size of the contracts
- Geographical location of work performed
- Owners and architects on completed work
- Profitability of the individual jobs
- Size of work programs handled in the past

Specific information is developed on any unusual or complex projects as well as any problem jobs, and the underwriter tries to establish which individuals in the organization were involved on these projects. It is important for an underwriter to know that the management talent involved on large jobs completed in the past is still available to work on the large job being considered currently. Experience that employees may have had with prior companies can also be factored into the equation. Again, most of this information is provided directly by the contractor and agent.

3.6.4 Continuity

The underwriter is interested in the potential financial and managerial impact on the contracting company if one of the owners should die or leave the business. The underwriter looks for a well-thought-out and consummated plan of continuation which addresses both these aspects.

Continuity planning may involve:

- Formal buy-sell agreements
- Employment contracts
- Life insurance payable to the company
- Testamentary provisions
- Trust agreements
- Simple handshake deals

The details of each of these are evaluated carefully. The surety underwriter wants assurance that if a key individual is removed from the operations at a critical time, the company will be able to continue its work without significant disruption. If a company does not have an adequate continuity plan, bond credit may be limited or denied due to the level of risk this creates. Usually the surety underwriter works with the contractor and his advisors to put together a viable plan for the continuation of the business, but the ultimate decision is the contractor's.

Since many construction firms are owned by a single individual or small group of people, a continuity plan should provide that adequate funds will be available in the form of life insurance or personal finances to cover any estate taxes due in the event of an owner's death, and furthermore to provide for an orderly transition of ownership, without the need to invade the company's financial base. There should also be adequate backup in the business from a managerial standpoint, and if there is not, then other provisions should be made to have additional funds available—either to hire someone to fill this position permanently or to obtain the services of another consulting contractor to finish the work.

3.6.5 Current operations

The underwriter's evaluation of the account looks at the current operations and future plans of the company, including the type of work under way and under consideration, plans for expansion into new geographical areas, the targeted owner base of either public or private entities, and the contractor's status as a union or nonunion company. Many contractors now include this information in annual operating plans which are made available to the surety for review. The underwriting process ties the current and future plans back into the prior experience to make sure that the company's direction reconciles with its abilities and resources.

The surety is also interested in the operating policies and procedures of the contractor. For example, if a contractor works frequently in the private sector, the underwriter wants to determine what steps the contractor takes to verify the availability of financing for the work on those projects. Or if a significant amount of the contractor's work is subcontracted, the contractor's policy regarding the selection and bonding of subcontractors becomes an important issue. In both of these situations, the underwriter is looking for assurances that the contractor's business is protecting itself at the outset of the work to avoid problems later.

On private work, the contractor should verify the owner's financing prior to beginning work under the contract. The surety wants assurances of the amount and form of the financing as well as any terms or conditions which might be involved if the financing is from a lender. Since contractors invest their own resources in the form of labor and materials in the job prior to billing the owner, contractors should always confirm that the owner will have the funds readily available to reimburse them in a timely manner. Such verification should be done again during the course of the work as any change orders or extras arise since the original financing package may not have contemplated the extra work. The surety's requirement that financing be verified before the bond is issued can provide the contractor with leverage to obtain this information from an otherwise uncooperative owner.

Many contractors subcontract portions of the work under their contract with the owner. A general contractor building a new school may subcontract the mechanical, electrical, and excavation portions of the job, for example. The general contractor remains liable for the performance of this work under his contract with the owner. Therefore, the surety underwriter often requires that major subcontractors be bonded, and many contractors have established their own policies regarding the review and bonding of subs. Even when a subcontractor is bonded, it is important to prequalify the surety company providing the bond. Many surety companies enter and leave the marketplace each year, some involuntarily. If a subcontractor defaults and the surety is not financially able to respond under the bond, then the contractor has lost this protection and will probably suffer a loss on that work.

Factors to consider in prequalifying a subcontractor's surety would include the A.M. Best rating of the surety's financial strength, profitability, management, and operations, as well as its reputation within the community. Sureties licensed to do business with the federal government appear on the Treasury Department's list, commonly referred to as the T list. It is also wise to look into the claims handling policies of a surety company to determine how it will handle any problems which would involve the bond. The contractor's own surety company can be a valuable source of information when the contractor is reviewing the subcontractor's bonding company, and in many situations the surety will ask to review the bonding companies being used by subs. Again, the sub bond protects not only the contractor but his surety as well.

Certain areas of construction inherently pose a much greater risk to the contractor and, therefore, to the surety. These include jobs involving hazardous materials, design liability, exposure in foreign countries, and financing of the work by the contractor. The potential legal and practical ramifications of a contractor's engagement in these types of contracts may prevent a surety from bonding an account even if all other underwriting factors are satisfactory. The negative aspects of each of these areas will be touched upon in the discussion on individual bond underwriting, but it is important for the underwriter to know up front, if a contractor is involved in high risk activities or plans to pursue such work in the future, whether the work is bonded or not, as it will impact the decision whether or not to provide the account's bonds.

3.6.6 Financial condition

The contract bond underwriter generally requires the following financial information:

- CPA-prepared fiscal year-end statements for the last three to five fiscal years, including schedules of completed and uncompleted work. These statements should preferably be audited statements with unqualified opinions from the CPA.

- An interim financial statement and job schedules as of a current date.

- Current and prior two years' business tax returns.

- Current financial statements on the owners and any related companies.

The purpose of the underwriter's financial analysis is twofold. The financial information provides a basis for determining the overall soundness of the contractor's financial position as well as the adequacy of the assets to absorb any loss that may arise in the course of construction activities. In the analysis, the contract underwriter is looking for both a strong working capital and net worth position.

Working capital, or *net quick,* is defined as the difference between the total current assets, which are assets that can be readily liquidated, and the total current liabilities, which are liabilities due within a one year period, as shown on the contractor's balance sheet. This is an indicator of the contractor's overall liquidity in relation to his or her short-term debt. These liquid assets are essential for meeting unanticipated cash needs as well as the day-to-day requirements of the business.

Net worth represents the owner's equity in the business, determined by subtracting the total liabilities from the total assets shown in the balance sheet. There may be hidden equity in the balance sheet if the true market value of certain assets, such as equipment or real estate, exceeds the book value shown in the balance sheet. Such hidden equity will be recognized in the underwriting process but may require substantiation in the form of asset appraisals. The net worth is an indicator of the business's overall staying power and ability to absorb losses.

An underwriter's analysis of the contractor's financial statement is very detailed and will include close scrutiny of all the balance sheet items as well as the profit and loss statement and the job schedules included in the report. The balance sheet is the basis for the net quick and net worth analysis.

Typically, the balance sheet accounts are scrutinized in the process. The first area is usually accounts receivable. Accounts receivable usually comprise a significant amount of the contractor's current assets. The underwriter wants assurances that those receivables are collectible within a reasonable time limit and do not contain any unapproved receivables or claims in connection with work under way or completed work.

The underbilling, or the costs and earnings in excess of billings, are investigated to confirm that they have subsequently been billed and collected. A significant buildup of underbillings on the balance sheet raises questions such as whether this item might include unapproved costs on a job or why the work hasn't been billed. If any receivables or underbillings are not readily collectible, they will not be treated as current assets by the underwriter in calculating working capital.

The underwriter looks for strong cash balances as they are the ultimate source of liquidity for funding operations. The underwriter often talks to the contractor's banker to verify the cash balances and determine what the average cash balances are during the course of the year.

Other significant assets in the working capital evaluation include investments, inventory, and fixed assets. Investments will be classified as either recurrent or long-term based on the ease of liquidating those assets. Many contractors own marketable securities or bonds. The underwriter wants to determine the current market value of these items versus the value shown at the financial statement date as part of the underwriting process. Inventory will generally receive treatment as a current asset if it is composed of materials readily available for work under way and the inventory turns several times during the course of the year. Any stockpiled inventory will be considered a long-term asset.

Fixed assets include equipment and real estate, including the contractor's plant. These are often assets essential to doing business, and while they provide long-term loss absorption power, they cannot be readily liquidated. Therefore, they are counted in the equity analysis but are deferred in the working capital analysis.

Just as accounts receivable are a significant current asset item, the offsetting accounts payable are a significant part of the current liabilities. The accounts payable are the amounts due to subcontractors and suppliers on work they have completed for the contractor, and generally the receivables should exceed the payables to indicate that the contractor will be able to retire these debts as contract balances are collected.

Bank debt is another very significant liability item. Generally, any bank debt which is not secured by fixed assets or which is due within a 12-month period will be considered a current liability. Since contractors often have short-term cash needs, the contractor's ability to borrow from the bank to generate additional cash is very important. Most banks issue letters outlining the borrowing facilities available to a contractor. The surety wants copies of the current bank letters. Most contractors need a short-term loan facility for working capital purposes and an additional facility for purchasing fixed assets on a long-term secured basis. If a contractor's financial position relies heavily on bank borrowing, the underwriter usually obtains copies of the loan agreements in order to review the terms and conditions before making a decision on the account.

The contractor's overbillings, or billings submitted for costs and profits not yet earned on a job, can be a vehicle for the contractor to generate some cash flow. In essence, the contractor is borrowing the owner's funds, but this money must later be put back into the jobs or earned as profit. This can be a smart way to improve a contractor's cash flow, but the underwriter wants to be certain that the overbillings are offset by sufficient cash balances and accounts receivable to generate the funds to be put back into the job later.

Income tax provisions can have a major impact on the working capital of a company. In addition to current taxes which may be due, there should be a reserve for any taxes which have been deferred due to the use of different timing methods for tax reporting and financial statement preparation. Therefore, it is common practice for an underwriter to request a copy of the contractor's federal tax return.

This becomes more significant when the contractor is a Subchapter S corporation for tax purposes or a partnership. In these situations, the individual stockholders or partners are personally liable for the income tax and no tax provision is included in the business balance sheet. However, the most likely source of funds for retiring the personal taxes is the business, and, therefore, a reserve should be set up for anticipated withdrawals to be made to satisfy personal tax obligations. Such a reserve will impact the business's working capital and, therefore, affect the amount of bonding which will be available to the contractor.

The most common long-term liability on the contractor's balance sheet will be bank debt against fixed assets. The underwriter wants to know the terms of the debt, including the security and repayment provisions. The underwriter will evaluate the ratio of total debt to equity on the business's balance sheet.

After all the working capital and net worth analyses have been done, the underwriter looks at these figures in relation to the contractor's total work program. Typically, the surety underwriter will look at the net quick position in relation to the total work program. From an equity standpoint, the underwriter also will evaluate the tangible net worth in relation to the total program as well. These ratios will vary widely, however, depending upon the other underwriting factors involved in a case and the risk appetite of the particular surety company.

The *profit and loss* (P&L) statement of the financial statement reflects the amount of revenues that the contractor has performed within the one-year period along with the general and administrative expenses (G&A), tax consequences, and bottom-line profit or loss of the company for the period. The P&L is used to track the contractor's profitability currently and over a period of time. It is one measure of the contractor's success and ability to handle the type of work he or she is performing. The surety will track the contractor's profitability based on different revenue scenarios and will also use historical G&A numbers for use in forecasting future profitability. In order for contractors to make a profit, they must control their expenses, particularly those that are fixed, such as the G&A. The surety underwriter also uses these numbers to predict future success.

Most contractors prepare schedules reflecting the current status of the jobs they have under way. The jobs are evaluated as to the contract price, estimated total costs on the job, the costs to date versus the billing to date, and the profit anticipated on the project. These schedules, together with similar schedules on completed contracts, allow the underwriter to track the progress of individual jobs as well as to focus on overall trends, such as overbilling on work, profit fades on jobs as they near completion, or overall low profit margins. These trends and the job specifics can be used by the underwriter to forecast future profits based on the contractor's estimates in the schedule. Often, the surety underwriter is the first one to spot a problem area and bring it to the contractor's attention in time for the contractor to make adjustments to either the expenses or the job costs. It is because of the value of this tool that the underwriter wants not only fiscal schedules but also schedules prepared at interim periods during the course of the year.

3.6.7 References

References and credit reports are invaluable in the underwriting process. At the time a surety underwriter first underwrites an account, he or she will check the references provided in connection with work that the contractor has previously performed, which would include the owners, architects, engineers, and other contractors. Contractors may also provide the name of their bankers as well as members of the community as references. Any negative feedback which is received may warrant additional investigation prior to the surety's committing on an account.

Credit reports are also good sources of information regarding the contractor's background, payment habits, and legal proceedings. Any negative items reflected in the credit report will generate additional dialogue between the underwriter and the contractor as there are often explanations for these items.

Upon development and review of all this information, the underwriter will make an assessment of the account. If the references have been satisfactory and the analysis reflects adequate experience and capital in relation to the desired jobs and programs with all other factors being satisfactory, a bonding relationship will be established with agreed-upon parameters. The continued underwriting process allows the underwriter to be flexible in the overall account guidelines as factors change. During the course of the year after the initial underwriting has been done, the underwriting in connection with individual bond requests gives the underwriter the opportunity to update the account underwriting to confirm that there has been no deterioration or other significant changes. If there are, then the underwriter will evaluate the ramifications of the changes on the overall account.

3.7 Specific Job Underwriting

The surety and the contractor establish an overall relationship based on the underwriting of the account in general, but the surety's actual liability derives from the bonds executed on specific contracts undertaken by the contractor. The amount of investigation required in connection with a specific job depends upon its size, the type of work and its location in relation to work previously done by the contractor, the terms of the contract itself, the bond form, and the contractor's other work on hand: in other words, the risk involved in the job. The more risk potential there is in a project, the more underwriting information that will be required to evaluate the risk. To the extent that each job is unique, it can generate its own set of underwriting questions. However, the typical areas investigated are as follows.

3.7.1 Job description

The underwriter wants a complete description of the work to be performed under the contract as well as the location of the work. The underwriter looks for the contractor to be well experienced in all aspects of the work undertaken. If a road contractor wants to perform a highway job similar to one previously performed by his or her organization, then it is a relatively easy call for the underwriter. If the same highway contractor wants to erect a building next to the highway, then more detail needs to be developed to determine the contractor's ability to handle that work.

The geographical location of a project is important since there are additional risks to contractors who venture into a region with which they are not familiar. In a new area, there are different players and the contractor is not operating on familiar turf. Subcontractors, suppliers, and even owners tend to favor local contractors and can often make life miserable for the outsider. Contractors also need to be aware of local laws and regulations which could affect their work or obligations under the contract. Since surety companies have seen many losses as a result of a contractor moving too far from the former area of operation, this one factor can be very critical.

3.7.2 Job size

If a contractor has successfully completed jobs similar in work and in the same price range, the risk is less than if the job size represents a significant increase over projects previously completed. The surety underwriter looks for controlled growth in the size of jobs undertaken by a contractor. Simply, the surety looks for the contractor to walk before running. Surety companies have seen many contractors run into problems when they have bitten off more than they can chew in the way of larger jobs.

The increase in job size over a contractor's experience increases the risk, but there may be other factors involved in the underwriting which will compensate for that risk and convince the underwriter to bond the project. The evaluation process also helps the contractor think out the situation before moving too quickly into a project larger than any performed before.

3.7.3 Work program

The same principles involved in job size apply to the size of the work program which a contractor is seeking. When there is a material jump in the amount of work a contractor is handling at one time, there can be extraordinary strain on the organization. Again, the surety is looking for controlled growth.

Another facet of the contractor's work on hand is the risk to the account that might be present in unbonded jobs. For example, if the contractor does a lot of private work, the underwriter will be interested in the financing on those projects even though they were not bonded. The contractor who has collection problems on a private job may run into financial difficulties, and the resulting cash flow deficiency can, of course, affect the ability to finance the bonded jobs. This is just one example of how the other work under way can affect the decision on the current proposition from an underwriting perspective.

3.7.4 Contract terms

Many contracts, such as the AIA documents, follow industry standards, but owners are increasingly deviating from the standard contract terms and introducing provisions onerous to the contractor and the surety. The underwriter wants to know the basic contract terms, including time for completion, payment provisions, liquidated damages, and hold harmless provisions. The underwriter can be very helpful in reviewing contract terms and recommending changes where there are terms unfavorable to the contractor. Whereas a performance bond generally guarantees the performance of the contract terms, the underwriter is not inclined to simply overlook any provisions that are significantly negative unless there are strong countervailing factors involved.

3.7.5 Bond forms

The performance and payment bond forms required by owners are falling victim to the same onerous changes by owners as contract terms. Since the surety executes the bond form, the underwriter will review these documents very carefully and negotiate terms where necessary. Standard industry forms continue to be the preferred bond forms, but even these require careful review as the owner will sometimes start with an industry form and insert changes.

3.7.6 Miscellaneous job provisions

The underwriter will want a list of the subcontracted amounts on the project and will often require that the subs be bonded or at least reviewed with the underwriter for possible bonding. As previously noted, the subcontractor's surety itself should be reviewed as well to evaluate the extent of the protection being given by the sub bond.

3.7.7 Underwriting conditions

To the extent that the underwriting may fall short on a given proposition, the underwriter will require that certain conditions be met, such as bonding subcontractors or generating additional working capital through fixed asset refinancing. Job conditions are designed to reduce the risk on a proposition not only for the surety but also for the contractor. While the surety may handle a specific account without the indemnity of the owners or an affiliate company, for example, there may be a requirement of that indemnity for a specific job which is outside the usual operating parameters established for the account. Also, there may be changes in the overall underwriting, such as a major operating loss which depletes the working capital and net worth. This might require indemnity or additional capitalization to handle a specific job or perhaps the account in general. The conditions imposed on an account or a specific bond approval should be well thought out and justified by the underwriter, and the underwriter will generally be open to negotiation if the contractor has another means of diminishing the risk involved.

There are special job considerations which require intense underwriting detail and which often pose risks viewed by the surety as undesirable. These include contracts which involve hazardous waste, design liability, financing of the work by the contractor, and foreign work.

Hazardous waste remediation may subject the contractor and the surety to potential liability under CERCLA which is strict and joint and several. The surety can be exposed to liability far in excess of its bond limits under the current structure of the law, making it almost impossible to underwrite and handle this type of risk.

When a contractor designs as well as builds a project, the surety may pick up the design liability which is normally an insurance risk. Even if there is insurance in effect, it may be difficult to distinguish between the design liability covered by the insurance and the construction liability covered under the bond. However, this has not been prohibitive if the other underwriting factors are satisfactory and the underwriter can obtain assurances regarding the experience of the firm doing the design and the insurance coverages involved.

In the situation where the contractor is providing the financing on a bonded job, the surety runs the risk of having a completion obligation if the contractor defaults. Normally, when a surety is required to take over a job, the owner has contract balances available to fund much of the work. However, when the contractor is providing the financing, this is not the case and the surety could incur a much greater loss as a result. Again, this is not absolutely prohibitive, but the underwriting scrutiny will be much greater and the standards higher to compensate for the additional risk.

A contractor engaging in foreign work must be versed with all of the local laws, work force, customs, and currency. There are surety companies which specialize in foreign work, but with the exception of the contractors who operate internationally, the risk is difficult for any surety to handle when it is a U.S. contractor venturing outside the protection of the United States.

3.8 Reinsurance

The purpose of reinsurance is the same as that of insurance: to spread risk. Reinsurance is basically a transaction whereby one insurance company, the reinsurer, agrees to participate in part of a risk written by another insurance or bonding company, the reinsured or primary company, with the result that the reinsurer indemnifies the primary company against all or part of the loss that the company sustains under a bond which it has issued. Reinsurance is a mechanism that allows the bonding industry to function more efficiently.

There are various types of reinsurance; the kind of reinsurance used is determined by the needs of the primary company. There are two basic types of reinsurance contracts: facultative and treaty. Facultative reinsurance is given on an individual risk or bond basis with the reinsurer actually underwriting the risk along with the primary company. Under a reinsurance treaty, the reinsurer must accept all of the primary company's business within the agreed-upon guidelines. Therefore, facultative reinsurance spreads the risk on an individual bond or account while treaty reinsurance provides protection on an entire book of business.

There are a number of reasons for a bonding company to purchase reinsurance. In the surety industry, capacity is a major issue as the size of the bonds and construction projects grow. Reinsuring part of the risk allows the bonding company to entertain these larger obligations. Additionally, the reinsuring company is a source of underwriting information and expertise which can supplement the surety company's own underwriting efforts.

Reinsurance plays an important financial role in managing the surety's business as it protects the reinsured against a single catastrophic loss and helps smooth the primary company's overall operating results from year to year, thereby stabilizing year-end results. The bonding company's financial position and operating results will affect its A.M. Best rating and treasury limit which defines the size bond a surety may write on a federal project. Many other government bodies and private owners require a certain level of Best rating and status on the treasury list before they will accept a surety's bond.

The availability of reinsurance is a significant factor in the overall surety bonding industry and may dictate whether or not a primary company can handle a larger risk. Therefore, it is important to have a basic understanding of the items outlined here.

3.9 Summary

In summary, the surety is looking for full disclosure regarding the contractor's business. The financial information is particularly vital and should be of good quality and high reliability. Once the surety has a full understanding of the contractor and his or her operations, the underwriter is in a better position to address any bond needs and to advise the contractor. A weakness in the underwriting is not necessarily fatal to the bonding relationship as the underwriter can recommend actions to soften or even overcome the weakness. The surety underwriter and the contractor should work together with the agent as a team to ensure the best results for everybody. One important point that bears repeating is the need and requirement for honest communication between the contractor and surety. Of the three Cs, character is the most crucial in establishing and maintaining a good business relationship.

Preparing to Be Bonded—
Contractors and Subs

Jeff Albada, Bob Heuer, Richard Beck, and Steven Kothe
Willis Corroon Construction

When a construction company is formed, it is merely a matter of time before that company will be asked to furnish a bid bond, a performance bond, and/or a labor and materials payment bond, guaranteeing the completion of a contract it plans to take and the payment of all labor and material bills pertinent to that contract. In preparing to be bonded, a contractor will assemble information about the company for submission to the surety bond market. The quality of that information and the manner in which it is presented will dramatically influence whether or not surety credit is granted and to what quantitative extent. With proper preparation and guidance, emerging contractors can do a great deal to maximize their surety bond credit.

Talk to almost any established contractor and you will likely find that he or she has relatively vivid memories of the process of getting the first bond. Seldom will the contractor tell you that the process was easy. More likely you will hear a story of frustration, aggravation, and time-consuming effort, very likely trying to meet a contract signing or bid date under the pressure of stringent time limitations. Worse yet, there may have been some feelings that the surety was the adversary. On reflection, the storyteller will probably recall that it was the beginning of a new partnership that, with continuing effort, became a decades-long, cherished relationship.

In preparing to begin that relationship, the contractor should assemble information concerning:

1. Reputation and experience

2. Finances

3. Game plan and vision

4.1   Reputation/Experience

There are many fine construction organizations in business today that started out with one hardworking, entrepreneurial individual with a hammer and a saw and a pickup truck. However, in today's highly competitive and sophisticated world, typical surety underwriters look for much more to hang their hats on when asked to extend surety bond credit on behalf of a new company. In other words, what the contractor has accomplished in the months and years before he or she even considers obtaining a bond will be closely examined by the surety industry. A person who has been in the contracting business for some time will furnish specific references for past projects from sources including owners, engineers, subcontractors, and suppliers. Representatives from each of these companies will be contacted and asked to comment on the bond candidate's past performance. A firm list of solid references, including names, addresses, and phone numbers, will assist the underwriter in gaining some comfort and insight into the character and capabilities of the applicant. The longer and stronger this list of respected members of the construction community, the more likely underwriters will be to continue their investigation in a positive light.

If the contractor is in the early formation of the business and has previously worked for others, the references may come from past employers or from owners, general contractors, engineers, or subcontractors who have had experience with the contractor in his or her role as a superintendent or project manager. In either case, reputation and experience are a critical foundation for establishing the contractor's bondability.

Any other documentation to establish reputation and/or experience will further enhance the presentation. As always, a picture is worth a thousand words, and photographs or brochures documenting where contractors have been and what they have done are a significant enhancement. Video film has also been effectively used on many occasions.

4.2   Finances

In the area of finances, the contractor's most important means of preparation is to accumulate working capital to support the work program needs of the company. For companies that have been in business for a period of time, this is achieved by operating profitably and leaving significant percentages of profit in the corporation. This is often contrary to the principal's initial thoughts of preferring to leave relatively little cash in the corporation in order to minimize the tax burden. And who wouldn't rather have spendable cash than pay more taxes? Here is where good financial consulting from the contractor's bonding agent and bonding company will lead to the conclusion that the accumulation of working capital and net worth on a consistent basis, year after year, will maximize surety bond credit.

A contractor who is newly in business will have to set aside personal monies for the capitalization of the new venture. Alternatively, he or she may have a financing source for help in this area. However, it should be noted that borrowed startup capital also produces debt, leaving little, if any, positive net worth.

The net worth and working capital requirements of the sureties will vary, depending on the work program sought. For an emerging contractor, the surety will be conservative and will tend to seek to underwrite to a 10 percent case—this means working capital equal to 10 percent of the total uncompleted work on hand, both bonded and unbonded. For example, if you are looking for a $5,000,000 work program, the surety will be looking for at least $500,000 in working capital. (The surety will calculate working capital by taking cash and accounts receivable—any other current assets will be scrutinized and often discounted—less all current liabilities.)

Established accounts of the surety company will qualify for work programs significantly greater than 10 times working capital. As greater amounts of working capital are accumulated, the leverage factor tends to increase significantly. This, combined with building a track record with the surety through a growing list of successfully completed projects, will all add to building greater surety support based on more than the pure financial analysis.

Not enough can be said about the importance of the financial reporting provided to the surety. First, it must be prepared in proper form and format by a CPA who also carries a solid reputation and recognition within the surety industry. There needs to be a continuing flow of financial reports during the course of the year, ideally quarterly.

This information allows the surety to track the progress of the company and the progress of the work. Over time, this will further demonstrate the contractor's capabilities in forecasting and in identifying problems early, avoiding disappointing surprises to the surety underwriter which can shake the underwriter's confidence and level of surety support.

In addition to providing financial statements on the business entity (ideally, the past three year-end financial statements), the surety will ask for personal financial statements of the individual owner(s). Additional assets outside the business entity (whether a corporation, partnership, or sole proprietorship) will be given consideration, since the personal guarantees of all owners and spouses will be asked for, at least initially, to guarantee the obligations of the principal. Personal net worth notwithstanding, the surety will seek to be sure its business applicant has a financial statement considered adequate to support the cash flow and funding needs of the business in pursuing its work program.

4.3 Game Plan/Vision

In addition to reputation, experience, and adequate capitalization, the contractor seeking surety must have a game plan. In today's specialized construction world, the surety industry will look skeptically at the emerging contractor who does not have some level of specialization or at least an angle for differentiating his or her business in a positive way from the competition. This angle may take the form of a special relationship with certain owners, proven abilities in value engineering, or expertise in certain construction methods. This is not to be confused with having a one-of-a-kind specialty that will be difficult for the surety to evaluate.

The surety, in its subjective evaluation, will be trying to assess your point of view as a contractor, i.e., your attitude toward bidding and obtaining work. It will be looking for a conservative point of view toward taking risks. Despite their reputation as riverboat gamblers, most successful contractors are remarkably conservative when it comes to risking their assets on projects which are inherently risky. The good contractor looks for a sensible risk/reward ratio. The riverboat gamblers have a high mortality rate.

The creation and submission of a formal business plan will most favorably assist the surety's evaluation of the contractor's point of view or vision. This plan should help the surety follow where the contractor wants to go and how he or she proposes to get there. The surety will generally seek to evaluate the quality of this business plan. It will want to see reasonable parameters the contractor has established as to types of work, geographical location, method of financing, and how it will be accomplished.

In evaluating the stability and capabilities of the company, the surety underwriter will want to see an organization chart. The small emerging business will not be expected to show a highly layered level of management (and overhead). However, if the chart begins and ends with one individual, it will lead to concern about who will finish the work if that individual is suddenly not able. In this regard, life insurance should be purchased with estate/probate tax liability of the business owner, and the business should be infused with enough extra cash to complete all existing projects.

Getting bonds need not be a frustrating, adversarial experience, and it certainly can be a long, mutually rewarding partnership with a dependable surety company.

4.4 Subcontractors and the Bonding Process

Surety bonds are the most commonly invoked device for the transferring of the risk in the contracting process. For a fee (premium) the surety accepts the risk of nonperformance and nonpayment by its contractor (or subcontractor) customer. The fee is ultimately paid by the beneficiary (obligee) of the performance and payment bonds, whether that beneficiary is the owner of the project or the general contractor, since the bonded party will include the bond premium in its bid.

In the standard lump-sum general contracting arrangement, the owner can elect (or under public works contracts is required) to transfer 100 percent of the risk to the contractor and his or her surety by requiring, and paying for, performance and payment bonds, each typically in the amount of 100 percent of the contract price.

Under construction management (CM) contracts, the owner is still able to transfer the risk of contracting, but the process is more complicated. While there are many variations on the CM concept, they share certain common characteristics. The construction manager acts as a representative of the owner, providing supervision for the project and other management services but not actually contracting to perform the construction work. Under this scenario, the CM doesn't even guarantee a maximum contract price. All of the actual construction work is performed by prime contractors (and their subcontractors) each of whom is responsible directly to the owner under individual and separate contracts. It therefore becomes the responsibility of the owner to require appropriate surety bonds of the prime contractors. Unbonded prime contractors thus become the sole risk of the owner.

In a variation on the CM theme, the contract may compel the construction manager to agree to a guaranteed maximum price or to assume responsibility for the performance of the prime contractors. In these instances, the CM is no longer acting as a fee-for-services agent of the owner but has become a risk-taking general contractor.

The vast majority of work performed under contract involves the general contractor taking 100 percent of the responsibility and risk for the work to be performed. However, most general contractors (particularly in the building trades) subcontract substantially all of the work to trade and specialty contractors. Few modern general contractors have any significant direct labor other than their supervision cost. They are very much dependent on the performance of their subcontractors for the successful completion of the project. In addition, the general contractor accepts full responsibility for their performance and payment to their subs under his bonds. By requiring those bonds, the owner has transferred the risk to the general contractor and his or her surety.

Most good general contractors have a formal written subcontracting policy to which they make few exceptions. Sureties would see their risk greatly diminish if every general contractor bonded all subcontractors, or at least those whose contracts exceeded a specified dollar value. If this were easily accomplished, general contractors would gladly comply. Unfortunately, contractors obtain their work via competitive bids and do not have large profit margins to work with. Gross profits exceeding 5 percent are rare, and 2 to 3 percent is closer to the rule. So a general contractor who bonds subcontractors risks losing the contract to a competitor who is not bonding subcontractors or eroding the gross profit by 50 percent or more.

What should a contractor do to limit the risk in the subcontracting of work? There are a number of established practices other than bonding to control subcontractors during the project. Retainage can be increased so as to hold back contract funds for the potential rainy day. Waivers can be required from suppliers before payment. Joint checks to subcontractors and their suppliers, escrow accounts, and the possible use of letters of credit can be used as alternatives to bonds. The general contractor can also work closely with trade unions to insure that payments are being made in a timely manner for wages and benefits. Most of these efforts are directed to the issue of making certain that subcontractors meet their obligations for labor, material, and other direct costs of the contract. Certainly this represents the most frequent problem in the failure of a subcontractor, but it is not the only risk.

Although not as frequent a problem as nonpayment, there is always the possibility that the subcontractor is unable to perform or fails financially before completing the work and is therefore actually in default of the contract. Holding back retainage or holding a letter of credit can help provide dollars to remedy a default by a subcontractor. However, a performance bond is the only available method of securing the performance of the subcontractor which extends beyond the completion of the contract even after full payment to all parties. Unlike the letter of credit, which ceases to exist upon its return or cancellation, the bond remains in full force and in effect usually for a period of up to two years after the date on which final payment fell due. Also, specific warranties extended under the contract are automatically included under the bond (unless specifically excluded in the contract or bond). It is also possible to obtain a separate warranty against defective workmanship or material called a maintenance bond which is usually written for a specific term of one or more years.

Clearly, it would be wonderful if general contractors had the ability to bond all of their subcontractors, but it is not a reality in today's construction market. General contractors have to have a plan for each of their projects which will enable them to limit the risk accepted and to determine what risks they will transfer to bonds provided by their subcontractors. Those elements of the project which are critical to its success and those which involve subcontractors who may be difficult or impossible to replace ought to be most seriously considered for subcontractor bonding. It may also be necessary to avoid using subcontractors who traditionally refuse or are unable to provide bonds. All of the other methods available can also be used but the central point is to have a plan and to put it in place before the bid, not after the award.

Whether it is the owner who is transferring risk to the contractor or the general contractor who is trying to limit risk during and after the construction process, surety bonds continue to represent the best available means. No other vehicle offers both performance and payment protection and includes extended liability beyond completion guaranteed by a third party.

The prudent and strategic use of subcontractor bonds in combination with other means of managing risk represent an important part of the risk evaluation process. It is essential that this process result in the orderly selection of risks to be transferred and risks to be accepted in the contracting process.

4.5   Elements of a Contract Surety Bond Submission

The elements of a contract surety bond submission include the following:

- Contractor's questionnaire, including credit and performance references (see Fig. 4.1)
- Résumé on owners and key personnel (see Fig. 4.2)
- Bank reference letter (see Fig. 4.3)
- Work-in-progress spreadsheet information (see Fig. 4.4)
- Personal financial statements on owners and spouses (see Fig. 4.5)
- Dun & Bradstreet Report (not included)
- Bid bond order form (see Fig. 4.6)
- Performance bond order form (see Fig. 4.7)

CONTRACTOR'S SURETY SURVEY

NAME_____   [ ] Corporation
                                               [ ] Partnership
ADDRESS_____   [ ] Limited Partnership
                                               [ ] Proprietorship

I. ORGANIZATION AND BACKGROUND

A. Date Business Formed_____   B. Date Incorporated_____

C. If SUCCESSOR to Prior Business, Name of Predecessor_____

D. List of Officers/Owners and Key Personnel

NAME	POSITION AND RESPONSIBILITY	% OF OWNERSHIP	AGE	YEARS & EXPERIENCE IN CONSTRUCTION

E. List of Affiliated, Subsidiary or Related Companies In Which This Firm or Its Stockholders Have An Interest

NAME & ADDRESS	STOCK OWNERSHIP	SCOPE OF OPERATIONS	ENDORSEMENT BY PRINCIPAL OR STOCKHOLDERS

F. Name surety company presently providing contract bond and through which agency.

_____

G. If change desired by contractor, why? _____

Figure 4.1 Contractor's questionnaire.

H. What company (or companies) was Surety prior to present one _____

_____

I. State limits and Carrier of Liability and Compensation Insurance _____

_____

J. In regard to present work-on-hand:

    1. Were the bids in line with other bidders? _____

    2. Are projects all on schedule? _____

    3. Are there any delays or disputes? _____

K. In regard to the contractor's equipment:

    1. Is equipment adequate for work program desired? _____

    2. If not, what expenditures are anticipated? _____

    3. Is equipment owned? _____ Or leased? _____

III. CREDIT INFORMATION

A. CREDITORS: List of Suppliers From Whom Contractor Buys Most Materials

NAME	STREET ADDRESS	CITY & STATE

B. Are bills paid in a discount/prompt manner? _____

    If not, why? _____

_____

Figure 4.1 *(Continued)*

C. Bank

NAME AND ADDRESS	BANK OFFICER	LINE OF CREDIT ESTABLISHED	NATURE OF SECURITY AND/OR NAME OF ENDORSER

D. Life Insurance

AMOUNT	INSURED	BENEFICIARY	INSURER	CASH SURRENDER VALUE

II. SCOPE OF OPERATION

Type of Construction_____Territory_____

A. What percentage of work is as: 1.prime?_____

2. sub?_____

B. How much of an average job is: 1.subbed?_____

2. Made up of materials? _____

C. Are bonds required from subcontractors?_____

When?_____

D. What was the largest work-on-hand handled in the past?_____

When?_____

E. What size jobs and total work program does contractor feel best able to handle? _____

_____

Figure 4.1   *(Continued)*

F. Approximately what percent of work requires contract bonds?_____

G. Is this a union or non-union contractor?_____

H. Has contractor or any of the owners ever:

            1. Defaulted on a contract?_____If yes, give details_____

_____

_____

            2. Caused a Surety to pay a loss?_____If yes, give details_____

_____

_____

l. List of Largest Jobs Completed

Contract Price	YEAR	TYPE WORK	NAME AND ADDRESS OF ARCHITECT/ENGINEER	Owner or General Contractor	DESCRIPTION OF JOB

•••••ATTACH "STATUS OF CONTRACTS" (UNCOMPLETED WORK SCHEDULE)

CONCURRENT WITH FINANCIAL STATEMENTS FURNISHED (ESPECIALLY LATEST FISCAL

YEAR-END) AND CURRENT REPORT, IF FINANCIAL STATEMENT MORE THAN THREE (3)

MONTHS OLD•••••

Figure 4.1   *(Continued)*

IV. FINANCIAL DATA

•••••ATTACH COMPLETE, LAST THREE (3) FISCAL YEAR-END FINANCIAL STATEMENTS (IF
NOT FULL CPA AUDITS, ATTACH SCHEDULES OF ALL BALANCE SHEET ITEMS)•••••

A. What is fiscal year-end?_____

B. If statements are not audits, will one be considered?_____

C. What method of accounting is used in preparing statements?_____

     [ ] Completed Contract   [ ] % of Completion   [ ] Simple Accrual

D. On what basis of accounting are taxes paid?_____

     [ ] Completed Contract   [ ] % of Completion   [ ] Accrual   [ ] Cash

E. Have stockholders elected to be considered a "Subchapter 'S' Corporation?"_____

F. In what year was contractor last checked by I.R.S.?_____

G. What portion of inventory shown on financial statement is material for jobs in

progress?_____

H. Is personal indemnity of the owners/stockholders available? _____

     Whose?_____

     •••••ATTACH PERSONAL STATEMENTS OF INDEMNITORS

     CONCURRENT WITH FISCAL YEAR-END OF CONTRACTOR•••••

I. Is a buy-sell agreement in effect?_____

J. Have operations been profitable since statement data?_____

Figure 4.1 *(Continued)*

K. Have any changes occurred since statement data such as acquisition of additional equipment, purchase of fixed assets, loans to officers, investments, withdrawals, or dividends that would significantly affect financial condition of contractor?

_____

_____

_____

_____

_____

L. Are any new ventures or investments contemplated? _____

_____

V. NEEDS

A. Desired annual volume_____

B. Desired maximum uncompleted work-on-hand at any one time_____

C. Desired maximum size of single job_____

The above answers are true to the best of my knowledge and belief

_____ Date_____
          SIGNATURE

Figure 4.1 _(Continued)_

RESUME

Principal_____Home Address_____

City, State, Zip_____Telephone__(___)_____

PERSONAL DATA

Date of Birth_____ Social Security #_____

Driver's License #_____ Marital Status_____

Spouse's Name_____ Spouse's Employer_____

_____
(Name, Address, Position, and Length of Employment)

EDUCATION

Did you graduate high school?   Yes   No

College - 19_____ to 19_____

SPECIAL EDUCATION RELATING TO CONSTRUCTION AND/OR TO YOUR TYPE OF

PROFESSION:_____

_____

_____

BUSINESS AND PROFESSIONAL EXPERIENCE RELATING TO CONSTRUCTION AND/OR

YOUR TYPE OF PROFESSION: Indicate firm name, length of time employed, occupation, largest

project you were involved in and reason for leaving: _____

_____

_____

_____

PERSONAL REFERENCES (name, address, phone number, length of time acquainted)_____

_____

_____

_____

USE REVERSE SIDE FOR ADDITIONAL INFORMATION

Figure 4.2   Résumé on owners and key personnel.

INSTRUCTIONS - BANK REFERENCE LETTER

Please have your bank(s) provide the following information in a letter

format <u>on their bank letterhead.</u> Must be original signature.

Letter should contain the following:

- Account number (s)

- Length of time doing business

- Average account balances

- Loan information:

    a) Amount of loan,

    b) Status,

    c) Type of security (i.e. A/R, signature, none, etc.)

- Line of Credit information:

    a) Amount of line

    b) Amount in use

    c) Type of security (i.e. A/R, signature, none, etc.)

Figure 4.3   Bank reference letter.

XYZ Bank
200 Main Street
New York, NY 10036

212-555-1212

November 5, 19—

Mr. John A. Doe
John Doe Construction Company
100 Broad Street
Anytown, NY 12345

Dear Mr. Doe:

I am pleased to inform you that *XYZ Bank* has approved the following loan accommodation:

Borrower: John Doe Construction Company

Amount $100,000.00   (One Hundred Thousand Dollars)

Type: Master Demand Note Line of Credit

Expiration: October 31, 19—

Collateral: Unsecured

This loan accommodation is subject to the following:

- Continued receipt of annual CPA reviewed financial statements
- The Bank's continued satisfaction of the operating and financial performance of the company
- Any other information the Bank may request from time to time or on a continuous basis

If you have any questions about this loan accommodation, please feel free to call me at above number.

Sincerely,

James R. Smith
Assistant Vice President
Small Business - Northeast

Figure 4.3   *(Continued)*

Work In Progress Spreadsheet

CONTRACTOR: _____

DATE: _____

UNCOMPLETED CONTRACTS

Job No.	CONTRACT DESCRIPTION	Date Started Mo./Yr.	A Contract Price Including Approved Change Orders	B Original Estimate of Gross Profit	C Total Amount Billed to Date Inc. Retainage	D Amount Retained to Date	E Total Costs Incurred to Date	F Estimated Cost to Complete Remaining Work	G Revised Estimate of Profit	H Est. Comp. Date Mo./Yr.
1										
2										
3										
4										
5										
6										
7										
8										
9										
10										
TOTALS										

CONTRACTS COMPLETED SINCE LAST REPORT DATED _____

CONTRACT DESCRIPTION	Contract Price Including Approved Change Orders	Original Estimate of Gross Profit	Final Gross Profit (Loss)	Gross Profit Earned in Prior Years	Gross Profit Earned This Fiscal Year
TOTALS					

(Use reverse side for listing additional contracts completed)

Do any billings include unapproved claims or disputed items?   [ ] Yes [ ] No   If yes, attach complete explanation.

Are any contracts behind schedule and subject to penalty?   [ ] Yes [ ] No

Completed by _____   Date _____

Figure 4.4   Work-in-progress spreadsheet.

59

CONFIDENTIAL PERSONAL STATEMENT

TO_____

Name (s)_____Phone_____

Address_____

        (Number)   (City)   state)   (Zip Code)

For the purpose of procuring and maintaining credit from time to time in any form whatsoever with the above named   for claims and

demands against the undersigned, the undersigned submits the following as being a true and accurate statement of its financial condition on

the following date, and agree that if any change occurs that materially reduces the means or ability of the undersigned to pay all claims or

demands against it, the undersigned will immediately and without delay notify the said   and unless the   is so notified it may continue to

rely upon the statement herein given as a true and accurate statement of the financial condition of the undersigned as of the close of business.

                                 Date_____, 19_____.

Please do not leave any questions unanswered.

ASSETS				LIABILITIES			
Cash on hand and in Banks				Notes payable to Banks			
U.S. Government securities				Secured			
Cash surrender value of Life Insurance--see Schedule IV				Unsecured			
Listed Securities--see Schedule IV				Notes payable to relatives			
Unlisted Securities--see Schedule IV				Notes payable to others			
Accounts and Notes Receivable Due from relatives and friends				Accounts and bills due			
Accounts and Notes Receivable Due from others--good				Taxes on Real Estate			
Accounts and Notes Receivable Doubtful				Other unpaid taxes			
Real Estate owned-- see Schedule IV				Mortgages payable on Real Estate--see Schedule II			
Automobiles				Chattel Mortgages and other Liens Payable			
Personal Property				Other debits--itemize			
Other assets--itemize							
				TOTAL LIABILITIES			
				NET WORTH			
TOTAL ASSETS				TOTAL LIABILITIES AND NET WORTH			

Figure 4.5   Personal financial statement on owners and spouses.

SOURCE OF INCOME		PERSONAL INFORMATION
Salary		Business or occupation Age
Bonus and commissions	$	Partner of officer in any other venture
Dividends	$	Married Spouse's Name
Real Estate income	$	Children
Other income--itemize	$	Single Dependents
TOTAL	$	
CONTINGENT LIABILITIES		GENERAL INFORMATION
As endorser or co-maker	$	Are any assets pledged?
On leases or contracts	$	Are you defendant in any suits or legal actions?
Legal claims	$	Personal bank account carried at
Provision for Federal Income Taxes	$	Have you ever taken bankruptcy? Explain:
Other special debt	$	

Figure 4.5   *(Continued)*

I—SCHEDULE OF STOCKS AND BONDS OWNED

Description	In name of	Market Value

II. SCHEDULE OF REAL ESTATE OWNED

Description of Property and Improvements	Date of Acquisition	Title in name of	Cost	Market Value	Mortgage Amount	Mortgage Maturity

III—SCHEDULE OF MORTGAGES OWNED OR REAL ESTATE SOLD ON CONTRACT

Description of property covered	Date of Acquisition	In name of	Maturity	Amount

IV—SCHEDULE OF LIFE INSURANCE CARRIED

Amount	Name of Company	Beneficiary	Cash surrender value	Loans

GIVE NAMES OF BANKS OR FINANCE COMPANIES WHERE CREDIT HAS BEEN OBTAINED

Name	High Credit	Basic

Do you have a will? _____ If so, who is the executor? _____

_____ 19 _____

_____
(Signature)

Figure 4.5   (Continued)

BID BOND ORDER FORM

Legal Name and Address of Bidder: _____

_____

_____

_____

Legal Name and Address of Obligee: _____

_____

_____

_____

Title or Description of Job and Portion to be Bonded: _____

Date of Bid: _____ Time: _____

Bid bond Percentage: _____ Estimated Contract Price: _____

Completion Time Stipulated in Specs (If none, Contractor's Completion Time): _____

Starting Date: _____ New Construction_____ Renovation_____

Penalty or Liquidated Damages for Non-Completion on Time: _____

Maintenance Period: _____ Percentage of Montly Retainage: _____

Special Bid Bond Form (if so, please attach a copy): _____

If final bonds wil be required, percentage amount of performance, labor and material, and if required,

maintenance bonds: _____

If copies of proposed final bonds are available, please attach.

Current backlog (total uncompleted work on hand): _____

Date bond requested: _____ Requested by: _____

Figure 4.6 Bid bond order form.

Chapter Four

PROJECT DETAIL

NARRATIVE DESCRIPTION: _____

_____

_____

_____

LOCATION: _____

_____

_____

MAJOR SUB-CONTRACTED WORK, INCLUDING DOLLAR OR % ESTIMATE. CAN SUBS
BE BONDED?: _____

_____

_____

IF APPLICABLE, QUANTITIES, MATERIALS, DESCRIPTION, NATURE OF SOIL, DEPTH
OF CUT, ETC.: _____

_____

_____

_____

_____

Figure 4.6   *(Continued)*

PERFORMANCE BOND ORDER BLANK

NAME AND ADDRESS OF CONTRACTOR: _____

_____

NAME AND ADDRESS OF OBLIGEE: _____

_____

JOB DESCRIPTION: _____

_____

_____

_____

DATE OF CONTRACT: _____ CONTRACT PRICE: $_____

AMOUNT OF BOND: _%Performance:_____ %Labor & Materials_____ %Maintenance**

**NAME & ADDRESS OF PARTY WITH WHOM CONTRACT WAS ORIGINALLY WITH:

_____

**REQUIRED TERM FOR MAINTENANCE BOND: _____

COMPLETION TIME: _____ /Starting Date:_____

PENALTIES OR LIQUIDATED DAMAGES: _____

OTHER BIDDERS: _____

_____

SPECIAL BOND FORM: _____

HOW MANY EXECUTED: _____

WAS BID BOND REQUIRED: _____ WHEN BID:_____

DATE BOND REQUESTED:_____

REQUESTED BY:_____

ORDER TAKEN BY:_____

Figure 4.7 Performance bond order form.

Insurance Coverages Checklists for the Construction Industry

Steven D. Davis
Willis Corroon Construction

John L. Heffron III
Ernst & Young LLP

There are a variety of standard and nonstandard coverages that are available and required by contractors in every type of construction business. These include:

- General liability
- Builder's risk
- Auto
- Excess liability
- Umbrella/excess coverage
- Workers' compensation
- Contractor's pollution liability
- Professional liability
- Miscellaneous coverages

All of these major coverages will be covered in detail in other chapters of this book. Understanding the relationship between common risks or hazards in contracting with methods of control is useful in assessing the potential coverages required by a specific contractor. We have listed these hazards and coverages in Table 5.1.

TABLE 5.1   Risks and Insurance Controls

Common risks or hazards in construction contracts	Who is exposed?	Method of control
Loss of income due to death or injury to worker	Worker and dependent	Workers' compensation coverage—state and/or federal statutes (Longshoremen's Harbor Worker's Act and Jones Act)
	Employer	Voluntary compensation and employer's liability for employees not required to be covered by workers' compensation law
Lawsuit brought by injured employee or dependent	Any employer	Most workers' compensation acts eliminate the employee's right to sue employer within scope of act Employer's liability (Coverage B or part two of standard workers' compensation policy would provide employer's liability coverage) Stopgap insurance (employer's liability in monopolistic state fund states)
Lawsuit brought by a member of the public or a party to a construction contract for damages arising out of bodily injury, death, or damage to property of others and personal injury		Commercial general liability insurance Include coverage for liability assumed under contract Professional liability insurance
Lawsuit brought by a member of the public or a party to a construction contract for damages arising out of a professional error or omission	Architects, engineers Construction managers Consultants, attorneys Certified public accountants Contractors on design-build projects	Architect's and engineer's professional liability errors and omissions insurance
Lawsuit to recover damages for bodily injury or damage to property of others arising out of the use of an automobile owned by the defendant	Any auto owner or driver	Automobile liability insurance

TABLE 5.1   Risks and Insurance Controls *(Continued)*

Common risks or hazards in construction contracts	Who is exposed?	Method of control
Lawsuit to recover damages for bodily injury or damage to property of others arising out of the use of a vehicle not owned by the defendant	Contractors Subcontractors Owners Architects Engineers	Nonowned automobile liability insurance
Physical loss or damage to a project under construction	Owner Contractors Subcontractors Material suppliers Mortgage companies	Builder's risk or course of construction insurance Named peril form All risk builder's risk
Low bidder unwilling to enter into a contract at the bid price and supply final bond(s)	Owner General contractor	Bid bond
Contractor (or subcontractor) unable to turn over to owner (or general contractor) a project completed in accordance with the plans and specifications at the agreed price	Directly: Owner (or general contractor) Indirectly: Commercial bank mortgage company	Performance bond
Contractor fails to pay accounts of those having a direct contract with him/her to supply materials or perform services	Directly: Subcontractors, suppliers, and employees Indirectly: Owners, from public relations standpoint	Labor and materials payment bonds
Owner fails to pay contractor	Contractor	Lien acts Lawsuits
Construction delayed due to any unforeseen circumstance	Owner Contractor	Penalty-bonus clause or liquidated damages
Construction delayed due to a physical loss such as fire, flood, loss of major item in transit, loss of special contractor's equipment not readily replaceable, causing increased construction costs and extended financing charges	Owner Contractor (loss of profit or increased costs)	Consequential loss insurance or builder's risk soft costs coverage

TABLE 5.1   Risks and Insurance Controls *(Continued)*

Common risks or hazards in construction contracts	Who is exposed?	Method of control

Accident frequency—any type	Contractor	Contractor's equipment insurance
		  Specified peril form
		  All risk form
		  Auto physical damage insurance
		Force majeure, efficacy, or systems performance insurance coverage

Construction delayed due to strike, lockout, non-delivery or nonperformance of key equipment	Owner and contractor	

Special risks or hazards in construction contracts	Who is exposed?	Method of control

Use of aircraft	Contractor
	Subcontractor
	Owner
		1 Owned and nonowned coverage
		  a Hull physical damage coverage
		  b Liability—including passenger liability
		2 Waiver clause under owner's insurance or asked to be named as additional insured on the aircraft owner's policy as respects both Hull and liability insurance
		3 Accident insurance—death and disability
		4 Cargo insurance (if not covered by all risk builder's risk insurance)

Use of watercraft	Contractor Subcontractor Owner	1 Delete watercraft exclusion from general liability policy 2 Marine insurance to cover hull, P & I, cargo
Marine construction projects	Contractor Subcontractor Owner	U.S.L. & H and Jones Act coverage as required

Dangerous operations:
  Blasting
  Pile driving
  Excavation (damage to underground utilities, etc.)
  Underpinning, shoring or removal of support
  Demolition
		Confirm no XCU exclusions exist within policy or endorsements

TABLE 5.1   Risks and Insurance Controls *(Continued)*

Special risks or hazards in construction contracts	Who is exposed?	Method of control
Foreign operations	Contractor Subcontractor Owner	Varies with laws of country involved. A close check should be made with your insurance broker to determine type of workers' compensation coverage necessary and that the liability and automobile insurance carrier is approved or admitted in the country involved. This is extremely important. Also, workers' compensation laws vary from country to country. In many instances, it is necessary to have two separate programs on workers' compensation insurance—one to cover national hires and one to cover stateside hires.

The checklists in Fig. 5.1 were developed by Willis Corroon to quickly identify special or miscellaneous coverages which may be applicable to a given contractor's circumstances. These checklists include:

- Business interruption
- Crime coverage
- Boiler and machinery
- Computer insurance
- Director's and officer's liability
- Fiduciary liability
- Aircraft liability
- Property coverage
- Contractor's equipment
- Office contents

Business Interruption
Insurance Check List

Insured _____

_____

Addt'l Named Insureds _____

_____

_____

_____

Insurance Co._____

Policy No. _____

Cancellation Notice _____ Days

Does Policy Cover the following:

Business Interruption	Yes ( )	No ( )
a. Gross Earnings with Full Payroll	Yes ( )	No ( )
b. Gross Earnings with Ordinary payroll Excluded	Yes ( )	No ( )
c. Earnings Insurance	Yes ( )	No ( )
Contingent Business Interruption	Yes ( )	No ( )
a. Contributing	Yes ( )	No ( )
b. Recipient	Yes ( )	No ( )
c. Leader Property	Yes ( )	No ( )
Insuring Agreement	Yes ( )	No ( )
a. Actual Loss Sustained	Yes ( )	No ( )
b. Valued _____	Yes ( )	No ( )
Loss of Rents	Yes ( )	No ( )

Figure 5.1   Insurance checklists.

Time of Element Insurance

Check List

Leasehold Interest	Yes ( )	No ( )
Extra Expense	Yes ( )	No ( )
Blanket	Yes ( )	No ( )
Agreed Amount (No coinsurance)	Yes ( )	No ( )
On Premises power interruption	Yes ( )	No ( )
Off Premises power interruption	Yes ( )	No ( )
a. Underground	Yes ( )	No ( )
b. Above ground	Yes ( )	No ( )
Perils	Yes ( )	No ( )
a. All Risk	Yes ( )	No ( )
b. Including EQ and Flood	Yes ( )	No ( )
c. Other _____	Yes ( )	No ( )
Selling price clause	Yes ( )	No ( )
Profits or commissions exposure	Yes ( )	No ( )
Deductible(s):		
_____	Yes ( )	No ( )
_____	Yes ( )	No ( )

HAVE YOU CONSIDERED:

1. Maximum probable loss	Yes ( )	No ( )
2. Coverage adequate for extended period of restoration	Yes ( )	No ( )
3. Business recovery extension (extended period of indemnity)	Yes ( )	No ( )

Figure 5.1 *(Continued)*

Time of Element Insurance

Check List

4. Business interruption due to
   contingent liability because of
   building laws, zoning or
   environment laws   Yes ( )   No ( )

5. Tuition fees (schools, clubs, etc.)   Yes ( )   No ( )

6. Use of deductibles   Yes ( )   No ( )

7. Boiler & Machinery business
   interruption exposure   Yes ( )   No ( )

8. Contractors equipment business
   interruption exposure   Yes ( )   No ( )

9. Builders risk business interruptions   Yes ( )   No ( )

10. Does client understand form, time
    limitations, coinsurance, etc.   Yes ( )   No ( )

11. Time element reporting clause -
    grace period of _____ days   Yes ( )   No ( )

Figure 5.1   *(Continued)*

CRIME COVERAGE CHECK LIST

Insured_____

_____

Addt'l Named Insureds _____

_____

Insurance Co. _____

Policy No.

_____

Policy Period _____

Cancellation Notice _____ Days

Policy Form	_____	3-D Policy
	_____	Blanket Crime Policy
		or
	_____	Combination Crime Policy

Limits $ Employee Dishonesty
                            Forgery or Alteration
                            Theft, Disappearance & Destruction
                            Robbery & Safe Burglary

Options $ Premises Burglary
                            Computer Fraud
                            Premises Theft & Robbery - Outside
                            Lessees of Safe Deposit Boxes
                            Securities Deposited with Others

Deductible $

Endorsements
      Co- Indemnity   Yes ( )   No ( )
      State requirements   Yes ( )   No ( )
      Cancellation   Yes ( )   No ( )
      Directors & Trustees   Yes ( )   No ( )
      Non-Compensated Officers   Yes ( )   No ( )
      Partner Infidelity   Yes ( )   No ( )
      ERISA provisions   Yes ( )   No ( )
      Volunteer workers   Yes ( )   No ( )
      Definite Term   Yes ( )   No ( )
      Agents coverage   Yes ( )   No ( )

Figure 5.1   *(Continued)*

Crime Coverage Check List

Credit Card Forgery	Yes ( )	No ( )
Money Orders and	Yes ( )	No ( )
Counterfeit Currency	Yes ( )	No ( )
Territorial extension	Yes ( )	No ( )
Extended Merger Notification	Yes ( )	No ( )

Have you considered:
1. Adequacy of limits
2. Adding coverages not presently afforded
3. Choice of deductibles
4. Suggesting loss control measures when appropriate
5. Superseded suretyship intact
6. Omnibus clauses for company insureds and for employee benefit plans
7. Extended time periods for 1) cancellation of policy, insuring agreement or employee; 2) merger notification; and 3) coverage on ex-employees
8. Availability of 3-year premium terms
9. Premium financing
10. Co-surety or layered programs
11. Potentially unfavorable exclusions, such as for losses involving precious metals, warehouse receipts, bills of lading, etc.
12. Policy warranties would be avoided whenever possible
13. Client must be aware of "manifest intent" requirements in fidelity coverage, and of possible cover for losses caused by unidentifiable employees
14. Are joint ventures (50/50) covered or specifically excluded
15. Adequate marketing of crime insurance program
16. Suggesting related coverages, such as K & R and mail insurance

Figure 5.1 *(Continued)*

BOILER & MACHINERY POLICY CHECK LIST

Insured _____

_____

Addt'l Named Insureds _____

_____

Insurance Co. _____

Policy No. _____

Cancellation Notice _____ Days

Policy Limits:

    a.   Combined _____

    b.   Separate _____

Coverage Form:

Comprehensive Standard	Yes ( )	No ( )
Extended	Yes ( )	No ( )
All Risk	Yes ( )	No ( )
Blanket Group Plan	Yes ( )	No ( )
Groups Covered _____		

Insurable Objectives: (Client has exposure)

    Boilers and Fired Pressure Vessels

Steam	Yes ( )	No ( )
Hot Water	Yes ( )	No ( )
Unfired Pressure Vessels (Air Tanks, (LP Tanks, Sterilizers, Vulcanizers, Diegesters, etc.)	Yes ( )	No ( )

Figure 5.1 *(Continued)*

Boiler & Machinery Policy
Check List

Electrical Objects (Generators, Transformers,
Motors, Miscellaneous
Electrical Apparatus)   Yes ( )   No ( )

Refrigeration & Air Conditioning

Objects   Yes ( )   No ( )

Mechanical Power Objects

(Turbines, Pumps, Motors, Engines,
Compressors, Fans, etc.)   Yes ( )   No ( )

COVERAGE EXTENSIONS

Definition of Accident:

Limited   Yes ( )   No ( )

Broad   Yes ( )   No ( )

Repair & Replacement (New for Old)   Yes ( )   No ( )

Expediting Expense

In the Limit   Yes ( )   No ( )

Sub-Limit   Yes ( )   No ( )

Extra Expense   Yes ( )   No ( )

Business Interruption Needed   Yes ( )   No ( )

Actual Loss Sustained   Yes ( )   No ( )

Valued   Yes ( )   No ( )

Contingent   Yes ( )   No ( )

Consequential Loss (spoilage due to
temperature change)   Yes ( )   No ( )

Contamination Loss-Ammonia   Yes ( )   No ( )

Off Premises Power   Yes ( )   No ( )
(Service Interruption)

Figure 5.1   *(Continued)*

Boiler & Machinery Policy
Check List

Production Machinery (only necessary on Standard Comprehensive Policy)	Yes ( )	No ( )
Testing Coverage	Yes ( )	No ( )
Broad Named Insured	Yes ( )	No ( )
Joint Loss Agreement with Property Carrier	Yes ( )	No ( )
Extra Expediting Expense	Yes ( )	No ( )
Delete in Use/Connected Ready for use	Yes ( )	No ( )
Automatic Coverage	Yes ( )	No ( )
Demolition and Increased Cost of Construction	Yes ( )	No ( )
EDP Control Equipment used in conjunction with Insured Objects	Yes ( )	No ( )
Hazardous Waste Inclusion	Yes ( )	No ( )
Brands and Labels	Yes ( )	No ( )
Telephone Service (On and Off Premises)	Yes ( )	No ( )
Additional Extra Expense	Yes ( )	No ( )

Figure 5.1 *(Continued)*

COMPUTER INSURANCE CHECK LIST

Insured _____

_____

Insurance Co. _____

Policy No. _____

Policy Period _____

Cancellation Notice _____ Days

Does Policy Cover the Following:

1.	Hardware	Yes ( )	No ( )
2.	Media	Yes ( )	No ( )
3.	Stored data on discs	Yes ( )	No ( )
4.	Breakdown coverage	Yes ( )	No ( )
5.	Reproduction coverage	Yes ( )	No ( )
6.	Extra Expense	Yes ( )	No ( )

HAVE YOU CONSIDERED:

1.	Any Exposure at other locations	Yes ( )	No ( )
2.	Any transit exposure (i.e. moving of terminals/equipment between locations)	Yes ( )	No ( )
3.	Business Interruption	Yes ( )	No ( )
4.	Any warranties regarding Fire Proof Safes	Yes ( )	No ( )
5.	Power Interruption	Yes ( )	No ( )
6.	Media Back-up Procedures	Yes ( )	No ( )
7.	Emergency Planning	Yes ( )	No ( )
8.	Vandalism (employee tampering)	Yes ( )	No ( )
9.	Cost of Research & Development for programs _____		

Figure 5.1   *(Continued)*

DIRECTOR & OFFICER LIABILITY CHECK LIST

Insured _____

_____

Addt'l Named Insured _____

_____

_____

Insurance Co. _____

Policy No. _____

Policy Period _____

Cancellation Notice _____ Days

Limits _____

     Defense included in Limits   Yes ( )   No ( )

Retentions:

Each D&O _____

In the Aggregate_____

Corporate Reimbursement _____

Automatic coverage for all Directors & Officers, including subsidiary corporations (named insured)	Yes ( )	No ( )
Reporting of new acquisitions and Directors & Officers _____ days	Yes ( )	No ( )
Continuity of coverage (short form renewal application)	Yes ( )	No ( )
95% participation clause eliminated:	Yes ( )	No ( )
a. Excess of $1,000,000	Yes ( )	No ( )
b. Excess of retention	Yes ( )	No ( )

Figure 5.1 *(Continued)*

Discovery Clause - Election Period

    Extended to _____   Yes ( )   No ( )

FAILURE TO MAINTAIN:

Insurance Exclusion eliminated   Yes ( )   No ( )

Does coverage apply to innocent directors
    or officer even though some other
    director(s) or officer(s) made mistake   Yes ( )   No ( )

Outside directorship exclusion eliminated   Yes ( )   No ( )

Prior Acts exclusion eliminated and/or
    Retroactive date   Yes ( )   No ( )

Pending & prior Litigation Exclusion eliminated   Yes ( )   No ( )

Coverages:

    To indemnity   Yes ( )   No ( )

    Pay on behalf of   Yes ( )   No ( )

Hostile Take Over exclusion eliminated   Yes ( )   No ( )

Green Mail Exclusion eliminated   Yes ( )   No ( )

Advance Defense Costs   Yes ( )   No ( )

Figure 5.1   *(Continued)*

FIDUCIARY LIABILITY CHECK LIST

Insured _____

_____

Addt'l Named Insured _____

_____

_____

Insurance Co. _____

Policy No. _____

Policy Period _____

Cancellation Notice _____ Days

Broad Named Insured	Yes ( )	No ( )
Sponsors _____ Plans _____	Yes ( )	No ( )
Notice: Canc. _____ Non-Ren. _____		
Reduc. _____	Yes ( )	No ( )
Notice of Occ. Modification	Yes ( )	No ( )
Cov. for Unintentional E&O in App/Dec.	Yes ( )	No ( )
Satisfactory Other Insurance Clause	Yes ( )	No ( )
Automatic Coverage:		
Fiduciaries _____ Administrators _____	Yes ( )	No ( )
Party in Interest _____	Yes ( )	No ( )
Exceptions: _____	Yes ( )	No ( )
Changes Must be Reported _____	Yes ( )	No ( )
Defense Provided		
Not Subject to Deductible	Yes ( )	No ( )
In Addition to Limit	Yes ( )	No ( )
For Non-Pecuniary Relief	Yes ( )	No ( )
Covers:		
Wrongful Acts - Not limited to ERISA	Yes ( )	No ( )
Administrative Errors & Omissions	Yes ( )	No ( )
Including: W.C. _____ Soc. Sec. _____	Yes ( )	No ( )
Dis. Ben. _____ Unemploy. Com. _____	Yes ( )	No ( )
5% Civil Penalties - Sec 502 (1)	Yes ( )	No ( )

Figure 5.1   *(Continued)*

Fiduciary Liability Checklist

Covers Claims Made Subsequent to Policy   Yes ( )   No ( )
    Period IF Incident Reported Prior to Expiration

Discovery Extension Available:
      _____ Months for _____ of Annual Premium   Yes ( )   No ( )

Right of Recourse Waived

_____   Yes ( )   No ( )

_____   Yes ( )   No ( )

_____   Yes ( )   No ( )

_____   Yes ( )   No ( )

Excludes:
     Fines _____ Penalties _____
     Punitive Damages _____
     Willful _____ Reckless _____ Knowing _____
     Violation of ERISA _____
     Liability Assumed Under Contract _____
     B.I. _____ P.D. _____ P.I. _____
     Libel _____ Slander _____ Discrimination _____
     Failure to Maintain Insurance

Figure 5.1 *(Continued)*

AIRCRAFT CHECK LIST

Insured _____

_____

Addt'l Named Insured _____

_____

_____

Insurance Co. _____

Policy No. _____

Policy Period _____

Cancellation Notice _____ Days

Owned Aircraft liability	Yes ( )	No ( )
Non-Owned Aircraft Liability	Yes ( )	No ( )
Passenger Liability	Yes ( )	No ( )
Injury to Property	Yes ( )	No ( )
Medical Payments	Yes ( )	No ( )

Hull

Valued Form	Yes ( )	No ( )
All Risk While	Yes ( )	No ( )
a. In Flight	Yes ( )	No ( )
b. Not in Flight	Yes ( )	No ( )

Limits: BI & PD

        With Passenger _____

        Without Passenger _____

    Medical Payments _____

    Hull _____

Figure 5.1 *(Continued)*

Aircraft Check List

Policy Form _____

Deductible _____

Coverage Extensions:

Rotor Wing Coverage	Yes ( )	No ( )
Waiver of Subrogation - Pilot	Yes ( )	No ( )
Pilot Warranty - all will have some sort of pilot warranty and it should be noted	Yes ( )	No ( )

Figure 5.1 *(Continued)*

PROPERTY COVERAGES CHECK LIST

Insured _____

_____

Addt'l Named Insured _____

_____

_____

_____

_____

Insurance Co. _____

Policy No. _____

Policy Period _____

Cancellation Notice _____ Days

LIMITS: (Specify if Blanket,　　　Values　　　　　Coins%
　　　　　　Specific or Loss Limit)

　　　　Buildings _____　　　_____　　　_____

　　　　Contents _____　　　_____　　　_____

　　　　Stock _____　　　_____　　　_____

　　　　　　　Specific _____

　　　　　　　Reporting_____

　　　　Rental Income _____　　　_____　　　_____

　　　　Business
　　　　　　Interruption _____　　　_____　　　_____

　　　　Earthquake _____　　　_____　　　_____

　　　　Flood _____　　　_____　　　_____

Figure 5.1　*(Continued)*

Property Coverages Check List

Perils Covered:

Fire	Yes ( )	No ( )
Extended Coverage	Yes ( )	No ( )
Vandalism & Malicious Mischief	Yes ( )	No ( )
Special Form	Yes ( )	No ( )
Broad Form Perils	Yes ( )	No ( )
All Risk	Yes ( )	No ( )
Difference in Conditions	Yes ( )	No ( )
Sprinkler Leakage	Yes ( )	No ( )
Earthquake	Yes ( )	No ( )
Earthquake Sprinkler Leakage	Yes ( )	No ( )

Deductibles:

Per Location	Per Occurence
Building _____	_____
Contents _____	_____
Earthquake _____	_____
Flood _____	_____

Valuation - Buildings:

Actual Cash Value	Yes ( )	No ( )
Replacement Cost	Yes ( )	No ( )
Permission to Rebuild at other than same location	Yes ( )	No ( )

Valuation - Contents & Equipment:

Actual Cash Value	Yes ( )	No ( )
Replacement Cost	Yes ( )	No ( )

Figure 5.1   *(Continued)*

Property Coverages Check List

Valuation - Stock:

Actual Cash Value	Yes ( )	No ( )
Selling Price Clause	Yes ( )	No ( )
Personal Property of Others	Yes ( )	No ( )
Personal Effects of Employees	Yes ( )	No ( )
Contingent Liability from operation of building laws	Yes ( )	No ( )
Increased Cost of Construction	Yes ( )	No ( )
Business Interruption Form	Yes ( )	No ( )
Demolition Cost	Yes ( )	No ( )

Limit _____

Food Spoilage

Refrigeration Failure	Yes ( )	No ( )
Contamination	Yes ( )	No ( )
Signs	Yes ( )	No ( )
Attached	Yes ( )	No ( )
Detached	Yes ( )	No ( )
Dams & Dikes	Yes ( )	No ( )
Piers & Docks	Yes ( )	No ( )
Fine Arts	Yes ( )	No ( )
Foundations	Yes ( )	No ( )
Accounts Receivable	Yes ( )	No ( )
Valuable Papers	Yes ( )	No ( )

Property Coverages Check List

Agreed Amount Clause	Yes ( )	No ( )

Expiration Date _____

Figure 5.1 *(Continued)*

Property Coverages Check List

Automatic Mutual Waiver of
    Subrogation Clause   Yes ( )   No ( )

Salespersons Samples   Yes ( )   No ( )

Automatic Coverage for Newly
    Acquired Property
    Limit _____   Yes ( )   No ( )
    No. of Days _____

Automatic Coverage for Unnamed Locations   Yes ( )   No ( )

Leasehold Interest   Yes ( )   No ( )

Improvements & Betterments Cov   Yes ( )   No ( )

Debris Removal Limit _____   Yes ( )   No ( )

Vacancy Clause Days _____   Yes ( )   No ( )

All "Insurable Interests" covered   Yes ( )   No ( )

Joint Loss Agreement with Boiler and
    Machinery carrier   Yes ( )   No ( )

Computers: (see computer check list for
         more detail)

    Software   Yes ( )   No ( )

    Hardware   Yes ( )   No ( )

    EDP Coverage   Yes ( )   No ( )

    Update of Values   Yes ( )   No ( )
        Date _____

Transit (See Inland Marine Check List)

Brand & Labels   Yes ( )   No ( )

Control of Damaged Goods   Yes ( )   No ( )

Figure 5.1 *(Continued)*

Property Coverages Check List

Glass Coverage   Yes ( )   No ( )

       Blanket   Yes ( )   No ( )

       Specific   Yes ( )   No ( )

       Lettering   Yes ( )   No ( )

Valuation Update

       Building -   Date _____

       Contents -   Date _____

       Rents -   Date _____

       Bus Interruption -   Date _____

Perils Excluded

Mortgagee Requirements:

HAVE YOU CONSIDERED:

Adequacy of Limits   Yes ( )   No ( )

Proper Valuation - Appraisal   Yes ( )   No ( )

Maximum Probable Loss   Yes ( )   No ( )

Blanketing   Yes ( )   No ( )

Use of Larger Deductibles   Yes ( )   No ( )

Is Occupancy Description correct   Yes ( )   No ( )

Peak Season Endorsement   Yes ( )   No ( )

HPR   Yes ( )   No ( )

Consequential Loss Assumption Clause
Needed (Clothing or Refrigeration Risk)   Yes ( )   No ( )

Figure 5.1 *(Continued)*

Property Coverages Check List

Property of Others on Premises	Yes ( )	No ( )
Property Delivered Under Conditional Sale	Yes ( )	No ( )
Step Down Clause - XS DIC Policies	Yes ( )	No ( )
Pollutant Clean-Up Restrictions	Yes ( )	No ( )
Will Improvements & Betterments Form that exhausts pro rata with term of lease fully compensate for loss	Yes ( )	No ( )
Radioactive Contamination (Hospitals, Labs, Manufacturers, etc.)	Yes ( )	No ( )
Property of Insured on Another Premises	Yes ( )	No ( )

On 3 year policies does Quake & Flood annual
aggregate coincide with policy date or is it "any
12 consecutive months"?

Policy Date	Yes ( )	No ( )
12 months	Yes ( )	No ( )

Figure 5.1 *(Continued)*

CONTRACTOR'S EQUIPMENT CHECK LIST

Insured _____

_____

Addt'l Named Insureds (include Sponsored Joint Ventures)

_____

_____

_____

Insurance Co. _____

Policy No. _____

Policy Period _____

Limits _____

_____

Policy Form _____

Blanket Coverage   Yes ( )   No ( )

All Risk   Yes ( )   No ( )

Named Peril   (list) _____

_____

Cancellation Notice _____ Days

Deductible _____

Actual Cash Value   Yes ( )   No ( )

Replacement Cost   Yes ( )   No ( )

Equip Rented or Leased from Others Covered   Yes ( )   No ( )
Contractor's Equipment Check List

Equip Rented or Leased to Others Covered   Yes ( )   No ( )

Radios or Cellular Phones Covered   Yes ( )   No ( )

Figure 5.1 *(Continued)*

Contractor's Equipment Check List

Quake & Flood covered	Yes ( )	No ( )
Deductible _____		
Boom or Crane Exclusion	Yes ( )	No ( )
Overweight or Lifting Exclusion	Yes ( )	No ( )
Coverage over Water	Yes ( )	No ( )
Damage to Property while Underground	Yes ( )	No ( )
Newly Acquired Equip Covered	Yes ( )	No ( )
Rental Reimbursement/Expense Covered	Yes ( )	No ( )
Loss of Earnings Covered	Yes ( )	No ( )
Co-Insurance Requirement	Yes ( )	No ( )
Amount _____		
Recent Equip Valuation	Yes ( )	No ( )
Date _____		
Swing Clause Included	Yes ( )	No ( )

Other Coverage Exclusions (if any) _____

_____

Territorial Limits _____

_____

Figure 5.1 *(Continued)*

6

General Liability Insurance

Paul Becker
Willis Corroon Construction

6.1 Introduction

The area of insurance collectively known as *liability coverage* is among the most important insurance a contractor purchases yet at the same time is probably the least understood. When huge jury awards are publicized, major construction site accidents occur, pollution incidents are discovered, or employees sue for some alleged wrong which occurred on the job, the first coverage thought of is liability insurance. Each of these examples may in fact be covered by liability insurance of some type, however, usually not on one policy.

The purpose of this chapter is to introduce liability insurance briefly, address in some detail the more commonly purchased coverages, and review the intent, coverage, exclusions, and conditions which make up the commercial general liability form used today. In addition, frequently purchased policies for umbrellas, excess, owner's and contractor's protective, and railroad protective liability coverages will be reviewed.

6.2 Commercial General Liability

This policy form was introduced in 1986 under the name *commercial general liability* and has been amended several times since. It is the successor to and represents a revision of the policy form which was entitled *comprehensive general liability*. The coverage under the new commercial general liability policy is essentially the same as that of the previous comprehensive general liability policy by an endorsement known as the *broad form comprehensive general liability endorsement*. While the newer commercial general liability was created to be the standard insurance industry form for a wide variety of businesses, it is neither specific nor broad enough to address the needs of most contractors.

Under a commercial general liability policy, the insurance company agrees to pay those sums that the contractor becomes legally obligated to pay because of bodily injury or property damage claims brought against the contractor by third parties. In addition, the policy provides for defense of legal action by others against the insured contractor. The insurance company pays the costs of investigation and defense in addition to the policy limits under most policies.

There are two versions of the commercial general liability policy, the *occurrence* form and the *claims-made* form. The coverage is identical under both policies. The difference is the triggering of coverage. The claims-made form covers only those claims which take place first and are reported during the policy term. The occurrence form of the commercial general liability policy applies in the same manner as the old comprehensive general liability; it covers only claims for injury or property damage which occur during the policy period irrespective of when the claim is reported, subject to statutes of limitations and the policy's claims reporting provisions. Claims-made coverage should be avoided by contractors. Certain construction operations and firms such as remediation contractors currently have no choice but to accept claims-made coverage for their operations. The proper coverage for such remediation work is contractors pollution liability which is fully discussed in a later chapter of this book.

6.3 Coverages

The coverage granted by the commercial general liability policy is actually broken down to three distinct areas:

1. Bodily injury and property damage liability (coverage A)
2. Personal and advertising injury liability (coverage B)
3. Medical payments (coverage C)

Each of these coverage parts is defined by insuring agreements and exclusions. These forms are so common in insurance that many insurance professionals assume that the insurance lawyer is familiar with their meanings. However, it is important to understand what is contained within each of these:

1. *Insuring agreements* within the commercial general liability policy all begin using the same terms: "We will pay. . . ." The *we* is the insurance company. Therefore, it is safe to assume that the remainder of the clause states what it is that the insurance company will pay for.

2. *Exclusions* are, obviously, those circumstances which the insurance company will not pay for. Because the three coverages granted by the policy differ, the exclusions are not the same for each.

With these two key areas defined, the coverages granted will now be addressed.

6.3.1 Coverage A: Bodily injury and property damage liability

The insuring agreement. The commercial general liability form and almost all general liability policies have similar wording: "We will pay those sums that the insured becomes legally obligated to pay as damages because of 'bodily injury' or 'property damage' to which this insurance applies."

It continues by stating that the insured will be defended if a suit is brought seeking damages. In essence, reading this portion of the insurance agreement by itself leads the reader to conclude that the coverage grant is very broad, and in fact, it is. Without further refinement of what is meant by this agreement, an insured could rightfully believe that almost *anything* could be construed to be covered by this policy as long as bodily injury or property damage occurs.

At this point, the coverage A insuring agreement begins to narrow itself to define exactly what the insurance company intends to actually provide coverage for. First, it states that any claim or suit may be settled or investigated at the insurance company's discretion. The policy essentially says that the insured, while not excluded from participating in investigating or settling a case, agrees to allow the insurance company the final word in how a case is handled or settled. Second, the insuring agreement sets a limit of the total amount the insurer will pay as damages and also states that once that limit is actually paid, the insurance company no longer has the obligation to continue to defend the insured.

This clause provides an interesting provision when it mentions limits on *damages* actually paid. By implication this provides the insured with significant protection for critical and expensive costs: legal fees. In reading the policy in its entirety, it becomes clear that the insured will be defended even if the cost of the defense exceeds the stated limit of liability as long as the damages actually paid do not reach the limit. In addition, legal and other defense fees and costs are not considered damages and, therefore, do not reduce the amount of insurance available to actually pay losses. This again is a very significant coverage given by the policy, especially in light of the present litigious nature of the United States. This point is further borne out in another portion of the policy (supplementary payments) which specifically states that the insurance company will pay as respects any claim or suit defended by the company: "All expenses we incur."

The insuring agreement continues by stating that the policy applies to bodily injury or property damage only if it occurs in the coverage territory and is caused by an occurrence. Both of these terms (as well as all terms in parentheses in the policy form) are defined in a glossary referred to as Section V—Definitions at the rear of the policy. In this case, *coverage territory* means the U.S., its territories and possessions, Puerto Rico, Canada, and international waters or airspace while traveling between these areas. The policy also gives worldwide coverage if a product is made in these territories but sold elsewhere and when a person who normally lives in these territories is temporarily outside of them. In both these latter instances, the suit or claim must be brought in the U.S., its territories, etc.

An *occurrence* is defined as "an accident including continuous or repeated exposure to substantially the same general, harmful conditions." This definition is deceptively straightforward on first reading; however, the definition of exactly what it means has been subject to numerous and conflicting court interpretations. A full discussion of the key issues which have been (and continue to be) disputed is beyond the scope of this book. For the sake of this review, definitions will be read literally. One important effect the definition has is on the limits of liability owing to the fact that the policy has a limit on all payments arising out of a single occurrence.

Finally, the insuring agreement for coverage A concludes by extending the coverage for bodily injury to "care, loss of services, or death" resulting from the bodily injury.

While the coverage A insuring agreement may seem somewhat complicated upon first reading, it points out one characteristic which is true throughout the policy. Many of the clauses contained within the form have exceptions limitations and definitions both in the specific clause and in other sections of the policy which must be considered to fully understand the intent of each clause. This characteristic is typical of all contracts and reinforces the need to consider the entire contract in any interpretation of a specific clause.

One of the most critical sections of the policy, *exclusions,* is directly after each insuring agreement.

Exclusions.   The exclusions for coverage A under the commercial general liability policy are fairly extensive and can be classified into the general categories of:

- Uninsurable exposures—war, workmanship, product deficiencies
- Exposures more appropriately covered elsewhere—pollution, property damage, work under contract or in the care of the insured, employee injuries, mobile equipment, etc.
- Hazardous activities outside of normal operations racing events

This section will review the exclusions for coverage A as they appear in the insurance service office (ISO) form CG 0001 1093 and briefly review the intent of each.

1. *Expected or intended injury:* Obviously, the insurance company is not interested in offering coverage in those instances where the insured expects a loss to arise from a specific act. These types of incidents would not be considered truly accidental. The exclusion does cover "reasonable force" to protect persons or property which results in bodily injury. An example of this would be a person who attempts to enter a jobsite and has to be physically restrained, which causes an injury. If the person then alleges that the insured was negligent, the policy will respond.
2. *Contractual liability:* This clause excludes contractually assumed liability unless:
   a. The contract is an insured contract, which is defined and discussed in greater detail later in this chapter.
   b. The liability would have been the insured's even in the absence of a contract.
   Also, the exclusion stipulates that the contract must have been agreed to prior to the occurrence causing the bodily injury or property damage.
3. *Liquor liability:* Although this would seem to exclude such events as parties, receptions, or dinners sponsored by the insured, the clause is very specific in stating that it applies only to those actually in the business of making, selling, distributing, or serving alcohol. This clarification clearly gives a contractor protection if a loss is alleged to have arisen out of a holiday party where alcohol had been served.
4. *Workers' compensation or similar laws:* This clarifies that there is no coverage for what is specifically addressed by workers' compensation insurance.
5. *Employer's liability:* Again, this excludes employee-related injuries as coverage is granted by the workers' compensation policy. It also excludes injuries to the employee's immediate family as a result of the employee's injuries. There is more to this exclusion than meets the eye due to a one-sentence exemption at the end of it.

This exception specifically states that the exclusion does not apply to liability assumed by the insured under the insured contract.

This is an exception of great importance to contractors as it grants coverage for those instances where an injured employee of the contractor brings suit against the job owner or general contractor who then passes the liability back to the contractor under the indemnification provision of the construction contract. This is referred to as an *action over* case within the insurance community and essentially causes contractors and insurers to address the same injury under two policies.

6. *Pollution:* This extensive exclusion removes most coverage for these types of incidents. There are two exceptions which are not specifically stated within the exclusion but are read into it by its wording.

   a. Coverage is not excluded for pollution arising out of completed operations or products liability.

   b. Coverage is not excluded for materials which a contractor may come into contact with at a jobsite if the materials were not brought onto the site by the contractor and the job itself was not to in any way handle pollutants.

   Under all case orders to clean up materials or governmental directives regarding pollutants are excluded.

   A much more detailed review of the environmental or governmental directives regarding pollutants is excluded.

7. *Aircraft, auto, or watercraft:* This exclusion is intended to direct exposures from these types of equipment to more specific policies which were written to more accurately cover them. Again, it does have exceptions. These include:

   a. Watercraft while at the insured's premises or land.

   b. Watercraft you may borrow or rent which is less than 26 ft in length.

   c. Parking an auto on or next to the insured's premises (as long as the auto is owned by the insured).

   d. Contractually assumed liability for aircraft or watercraft. This again refers to the insured contract which will be discussed later.

   e. Coverage does apply to the use of mobile equipment such as cranes, front end loaders, excavators, etc., while not being transported.

8. *Mobile equipment:* This exclusion does not exclude operations of mobile equipment even though that is the reference made by the exclusion. Rather, this excludes two specific cases of mobile equipment use:

   a. The transportation of it by an automobile being used by the insured.

   b. The use, preparation, or practicing for a racing, speed, demolition, or statement event.
9. *War:* This exclusion includes war, civil war, insurgence, rebellion, or revolution and relates to the insurer's inability to insure such an event.
10. *Damage to property:* This exclusion is both lengthy and subject to some interpretation as to its total meaning. The exclusion's intent is to exclude damage to property which is more specifically covered by various property insurance forms including builder's risk and installation forms. To do this, it specifically excludes property damage to:
    a. Property owned, rented, or occupied by the insured.
    b. Premises sold, given away, or abandoned if the property damage arose out of the premises.
    c. Property loaned to the insured.
    d. Personal property (property other than buildings) in the care, custody, or control of the insured.
    e. "That particular part of real property . . ." being worked on by the insured or at his or her direction (if the property damage arises out of those operations).
    f. That part of any property which needs to be repaired because the work performed on it was incorrect.
    The rest of the exclusion attempts to clarify under what circumstances each of the preceding applies. One exception of interest relates to exclusion 10b. The exception states that the exclusion does not apply if the premises are or were "your work" (which grants completed operations coverage) and "were never occupied, rented, or held for rental by you." This exception can cause difficulty for those general contractors who also are engaged in property management. An example of this would be an apartment building the insured builds, rents for a period of time, and then sells. If a property damage loss then arises, the exclusion precludes coverage to the insured. In contrast, a similar building the insured builds but sells immediately would have coverage for a property damage liability loss.
11. *Damage to your product:* This exclusion makes clear that there is no coverage if a product the insured produces is damaged.
12. *Damage to your work:* Similar to the previous exclusion in that it seeks to exclude damage to the work itself once it has been completed. However, this provides an exception that coverage does apply if the damage arose out of a subcontractor's operations. This exclusion is very broad in that it makes no distinction between the entire worksite and only that part of the site or work out of which a loss arises. Again, this is an important point and can lead to misunderstandings of coverage in the event of a loss.

13. *Damage to impaired property or property not physically injured:* The intent of this exclusion is to not cover defective products or work or delays in completing work in accordance to a contract.

     It does provide an exception that grants coverage for loss of use of property as the result of damage to the work itself once it has been put to its intended use.

14. *Recall of products, work or impaired property:* This exclusion is self-explanatory as it clearly reinforces that the intent of the policy is not to cover poor workmanship or a dangerous condition in the product which needs to be recalled or repaired.

Finally, the last paragraph of the exclusions for coverage A provides for an exception for fire damage to premises granted or occupied by the insured. This is commonly referred to as fire legal liability and in fact has its own limit of coverage on the declarations page of the policy. The exception applies to all of the exclusions except (1) and (2).

6.3.2   Coverage B: Personal and advertising injury liability

Insuring agreement. The coverage provided by personal injury and advertising injury liability has much of the same wording as coverage A except that it specifically applies to exposures related to specified injuries other than bodily injury. These coverages are defined as follows:

Personal injury	*Advertising injury*
False arrest or detention	Misappropriation of advertising ideas
Malicious prosecution	Copyright, title, or slogan infringement
Wrongful eviction, wrongful entry	Oral or written material that slanders or libels
Oral or written material that slanders or libels	Oral or written material that violates a person's right to privacy
Oral or written material that violates a person's right to privacy	

Exclusions. There are a few exclusions to coverage B in the ISO policy as the definitions are quite specific about what is covered. The exclusions are grouped into those that apply to personal and advertising liability and those that apply only to advertising injury.

Those that apply to both are:

1. Material which is spoken or written which the insured knew was false

2. Material that was published prior to the policy effective date

3. Material arising out of the willful violation of a penal statute which was committed by or with the consent of the insured

4. Contractually assumed liability except liability which would have been the insured's if no contract was signed

Those that apply only to advertising injury are:

1. Breach of contract.

2. Goods or services which do not perform as advertised.

3. Incorrect pricing in advertising.

4. Those firms in the business of advertising, broadcasting, publishing, or telecasting—these firms can obtain more specific coverage for their activities.

These exclusions seem quite straightforward upon first review and, for the most part, they are. One area of concern for a contractor to make note of is the contractual liability exclusion. Some contracts seek to transfer personal injury liability from the owner to the contractor; if this occurs, the contractor will not have coverage under the policy as written.

6.3.3 Coverage C: Medical payments

This coverage is an attempt by the policy drafters to provide a no-fault medical cost limit to give an injured party immediate medical attention to reduce the chance that the injury will become worse without treatment, and also to resolve injuries without litigation. The limit on the policy is usually much lower (usually $2000 to $5000) than the limits for either coverage A or B.

There are several exclusions under medical payments (in addition to all exclusions for bodily injury which apply in coverage A) which clarify that the coverage is not for the employees of the insured, anyone covered by workers' compensation, etc.

6.4 Additional Contract Provisions

The first section of the ISO commercial general liability policy explains *what* the policy is intended to cover. The policy then continues in sections to define:

Section II: *Who* is protected by the policy?

Section III: *How much* will be paid for each coverage provided?

Section IV: *What conditions* are there for complying with the policy terms?

Section V: *What* do all the insurance phrases mean?

While all of the provisions contained in each section are important and can affect the coverage provided by the policy, there are several critical points within each that need to be understood.

6.4.1   Section II: Who is an insured?

This section is very specific as to who is entitled to coverage under the policy. It is broken down into two subsections which address different organizational structures and other persons insured.

The first subsection is a listing of three different business entities and who within each is covered as an insured. The insured is designated on the declarations page as either an individual (or sole proprietor), partnership, or corporation.

- Individual: Both the named insured and his or her spouse are covered for business-related activities.

- Partnership or joint ventures: Partner members and their spouses are covered for business-related activities.

- Corporation or other type of organization: The named insured and executive officers and directors (with respect to their activities as such). Stockholders are insured as well for their activities as stockholders only.

Additionally, the following are insured as well:

- Employees for acts within the scope of their employment. However, they are *not* covered for:

  Bodily injury or property damage to:
  - Other insured or coemployees
  - Immediate family of coemployees
  - Any sharing of an obligation to pay damages for other coemployees and their immediate families
  - Professional health care service

  Property damage to property:
  - Owned, occupied, or used by the insured or any employee of the insured
  - In the care, custody, or control of the insured or any employee of the insured

- Real estate managers of the named insured's property.

- Temporary representatives and legal representatives of the named insured's property if the named insured dies.

- Permissive users of mobile equipment while transporting the equipment on public roads and those people responsible for the user's conduct.

- Newly acquired or formed entities are covered automatically for 90 days from the date of acquisition or until the end of the policy period. In either case, such changes should be reported immediately to the insurance carrier to avoid any coverage issue.

The "Who is an insured?" section of the policy is quite broad but has a key deficiency which any contractor needs to be aware of. Nowhere in this section is there a reference to those entities which are required to be *additional insured* by contract. Since the typical contract calls for such a provision, this is a significant issue. In a later part of this chapter, methods for amending this policy (and this clause in particular) will be discussed to address this problem.

6.4.2 Section III: Limits of insurance

As its name indicates, this section of the policy details how limits of liability are applied to the coverage provided. The policy form itself does not contain actual dollar limits as these are noted on the declarations page and vary from contractor to contractor.

The limits of insurance are broken down into sections as follows:

- *General aggregate:* The most the insurers will provide in total for coverage A, B, and C except for products or completed operations liability.

- *Products/completed operations aggregate:* A separate limit applies to all such claims under the policy.

- *Personal and advertising:* A limit on these types of losses applies to all losses arising from one person or organization. It should be noted that this is not an *occurrence* limit per _____, but rather an aggregate of all losses from a particular claimant.

- *Per occurrence limit:* Applies to the bodily injury, property damage and medical payments arising out of a *single* occurrence.

- *Fire damage limit:* Applies to fire damage to premises rented or temporarily occupied by the insured.

- *Medical payments:* A district sub-limit from the per occurrence limit above, this applies to any one injured person. It is usually a limit of between $2,000 and $10,000 per person.

The limits of insurance are specifically noted to be *annual* limits and also apply to any short-term extension of less than one year.

As with many other aspects of the commercial general liability policy, the limits section has issues which contractors should be aware of. Specifically, prior to 1985 there was no aggregate at all on bodily injury claims (other than products/completed operations). By making these claims subject to an aggregate, the policy writers significantly reduced the limits which previously had been the standards. To address this reduction of coverage, an endorsement was drafted which makes the general aggregate apply separately to each job site. It must be noted that this endorsement is not automatic but must be added to the policy as part of the marketing negotiations with the underwriter.

6.4.3 Section IV: Commercial general liability conditions

The conditions of the policy impose certain obligations on the insured as well as the insurance company in order to assure consistent application of the coverage terms. In addition, this section states how the policy responds if two or more policies apply to the same occurrence, spells out how the insured must present a claim, and tells what happens if the insurance company decides not to renew the coverage.

Key conditions within the contract include:

1. Duties in the event of occurrence, offense, claim, or suit—this condition basically states that:
   a. The insurance company will be notified as soon as practicable in the event of an occurrence or if a claim or suit is brought against the insured.
   b. The insured will gather as many details as possible about what happened and who witnessed the accident.
   c. The insured will not make any payments without the insurance company's consent.
2. Other insurance—this innocuous-sounding condition states simply how this insurance applies in the event two or more policies apply to a claim. Basically, it says that:
   a. The policy is primary unless the other policy also is primary. This is not uncommon as this same form applies to the majority of insured. Or,
   b. The policy is excess over:
      (1) Property or related coverages which apply to the insured's work.
      (2) Property/fire insurance for premises rented by the insured.
      (3) Any insurance which addresses auto, aircraft, or watercraft, to the extent these exposures are not already excluded by the policy.

c. If the policy is neither *a* nor *b,* then the policy shares the loss with the other policies on an equal shares basis, if the other policies allow this. This means that the carriers share the loss dollar for dollar until the loss is paid or the limit of the policy is reached.

If the other policy does not allow equal shares then the loss is split by the percentage of the limits on this policy compared to the total limits of liability available.

3. Premium audit—the familiar provision allows the insurance carrier to review the records of the insured and decide what the actual charges are after the policy expires and at any time the carrier feels appropriate. In essence, charges are subject to interpretation. The implications of this clause are that the insured must be careful on establishing his or her business classification and have a realistic view of the exposures (sales, payrolls, etc.) for the coming year to avoid unpleasant surprises at year-end.

4. Separation of insured—this clause states that each insured on the policy is treated separately even if multiple claims are brought. The condition specifically states that the limits of insurance are not increased by this clause. This clause again can affect a contractor as it addresses insured, including additional insured which are often added as a result of construction contrasts.

5. Transfer of rights of recovery against others to us—in common terms, this clause is known as the *Subrogation* clause. It states simply that if a payment is made by the insurance company, it takes from the insured all rights the insured has to seek recovery of these sums by others.

Again, this clause may seem quite clear on the surface; however, the drafters of the policy added the following sentence which has created some confusion. "The insured must do nothing after loss to impair them." "Them" refers to the rights of recovery the insured has. By implication, this sentence would seem to allow an insured to waive such rights if done so *before* a loss occurs.

In practical terms, a standard practice is to notify the insurance company if a waiver of rights of recovery endorsement is needed by contrast to assure that there will be no issues of intent once a loss occurs.

6.4.4 Section V: Definitions

The last section of the policy provides a glossary of the terms used throughout the policy. These definitions seek to answer the question, "What does the insurance company mean by _____?" The definitions are quite specific yet have been continually challenged in various legal actions as claimants seek to expand the meanings to cover additional occurrences.

During the discussions of the previous four sections of the policy, several definitions have been reviewed to clarify the intent of specific coverage issues, yet several critical definitions have yet to be addressed. These include:

Insured contract. Obviously, a key consideration for any construction firm is how the insurance policy responds to contracts the company may enter into in the normal course of doing business. The policy is quite clear on this point and in fact provides coverage for most contracts entered into by the insured as respects legal liability assumed under the contracts.

One key point needs to be made before reviewing the specific contracts covered. The fact that the insured enters into a contract with another party does not amend any of the coverage terms under the commercial general liability policy form. This point is often overlooked in discussions of contracts and contractually assumed liability. The policy is a contract itself between the insurance company and the insured and no other part can affect its terms unless both the carrier and the insured agree. With this in mind the contracts included within the definition of insured contract are:

1. Leases of premises.
2. A side track agreement.
3. Any easement on license agreement except for work within 50 ft of a railroad.
4. Obligations required by ordinances to indemnify a municipality. It does not include work performed *for* a municipality.
5. Elevator maintenance agreements.
6. "That part of any other contract or agreement pertaining to your business" where the insured agrees to assume the tort liability of another to pay for bodily injury or property damage to a third party.
    Tort liability is defined as liability which would be imposed by law even if there was no contract or agreement.
    This broad definition of any other contract or agreement specifically excludes the following contracts:
    a. Railroads when the work is within 50 ft of nail property and affects bridges, trestles, tracks, roadbeds, tunnels, underpasses, or crossings. The intent of this exclusion is to have a railroad protective policy issued if such exposures exist.
    b. Agreements to indemnify architects, engineers, or surveyors for their professional acts.
    c. If the insured is an architect, engineer, or surveyor, coverage will not apply for contracts he or she signs to indemnify others for professional acts.

The professional exclusions within the definition of insured contract can be troubling if a contractor is being required to add an engineer who represents an owner as an additional insured by contract. This exclusion clarifies that the coverage for indemnifying an engineer does not extend to any professional acts (including supervision) of the engineer.

The insured contract definitions are the last that will be examined in detail here. There are other important definitions relating to *mobile equipment, suit, your product,* and *your work* which will not be examined but should be understood by contractors and reviewed prior to purchasing a commercial general liability policy.

6.5 Key Coverage Amendments

The previous section of this text dealt with the standard insurance policy for general liability used as the base coverage contract by the vast majority of the insurance industry. As has been noted, on a stand-alone basis the ISO policy has several clauses which need to be addressed to broaden the coverage and make it more applicable to construction accounts.

These amendments fall into several categories:

1. *Who is an insured?* A significant number of construction contractors call for the contractor to add the owner and/or the next higher tier contractor of the project as an additional insured to the contractor's liability policy. In order to bring the insurance contract into compliance with the construction contract, the policy must be endorsed to reflect the additional insured requirement. The recommended approach is to have the endorsement apply on a blanket basis to all situations where the contractor is required to name the owner or other contractor.

Other required changes to who is an insured include architects and/or engineers, property owners through leases, etc.

It is imperative that the policy reflect this endorsement as the contractual coverage provisions do not extend the insured status to third parties except when endorsed to specifically do so.

2. *Waiver of subrogation.* The insurance policy has a clause in it which speaks directly to the insurance company's right to recover losses it pays by recovering from the responsible party. Construction contracts often contain clauses which require that this provision be waived against the contract drafter. An endorsement to either do this individually by contract or on a blanket basis is needed to accomplish this.

3. *Pollution liability.* As we discussed earlier, most pollution incidents have been effectively excluded by the general liability basic pol-

icy. While there is some movement to selectively grant broader coverage, for the most part such coverage is only available on a broad basis through environmental impairment liability policies such as the *contractor pollution liability* (CPL) policy. These types of policies are discussed in greater detail elsewhere in this book.

6.6   Limits of Insurance and How They Apply

As previously noted, a critical endorsement for contractors is the *per project general aggregate*. Again, construction contractors can cause issues with this provision. Specifically, many contracts call for the contractor's insurance to apply on a *primary basis*. In order to accomplish this on the insurance policy, the contractor needs to have an endorsement to reflect the fact that the limits will apply before any other available coverages. If such an endorsement is not on the policy, the contractor's insurance may only pay a portion of the loss as it seeks contribution by other policies (such as the owner's) which may apply to the loss. In addition, this would technically cause a contract default by the contractor.

6.6.1   Partnership and joint ventures

These types of arrangements are excluded from general liability policies unless specifically endorsed onto the policy. This can cause a problem especially with old joint ventures which may not be named as *insured's* on the policy. The safest approach to covering this gap is to name all joint ventures the contractor ever was involved in to assure completed operations coverage.

6.6.2   Personal injury

Again, this coverage is one which can frequently be affected by a construction contract. Often, owners will require that they be protected contractually for personal injury occurrences. These types of losses are specifically excluded by the policy and an endorsement is imperative to cover this potential gap.

6.6.3   Other endorsements

The extent a commercial general liability policy can be amended is very much subject to negotiation with the insurance carrier. The marketplace of the early 1990s has made the process somewhat easier as competition for this industry has increased. Also, the need for coverage amendments has dramatically increased as construction contracts became more specific about insurance requirements. This trend will continue as case law evolves. The need for continual review of the insurance contract as compared to the construction contract's requirements is necessary to avoid unpleasant surprises such as no coverage.

6.6.4 Excess umbrella liability

These policies provide additional limits of liability protection which would exceed those provided by policies such as commercial general liability, business auto liability, and employer's liability.

Key issues for this coverage include:

■ No coverage standard exists in the insurance industry. Each insurance company's form varies, making negotiations with the carriers critical to assure the broadest coverage.

■ Forms tend to be somewhat restrictive and typically unfriendly to contractor needs. Hence, these policies typically contain significant endorsement activity to make them respond as closely as possible to underlying policies.

■ Limits of liability are typically purchased in increments of several million. Often $5 million or $10 million layers are purchased and frequently several carriers cover various layers of the total amount carried.

The umbrella and excess liability areas need to be carefully purchased. Often, contractors make decisions purely on a cost basis when, in fact, coverage is by far the more critical factor. These coverages are the true catastrophic protection a contractor purchases—often the difference between being in business and not being in business. Buying by price alone is a sure way to end up experiencing potentially disastrous results when a large claim occurs.

A full review of this coverage by a competent insurance professional is the only way to assure that these policies will apply as intended.

6.7 Summary

Commercial general liability combined with umbrella and excess coverage can form the backbone of a contractor's insurance program. With construction contracts demanding broader coverages all the time, it usually falls on the general liability and umbrella programs to address the demanded contract protection.

This being said, these policies on a stand-alone basis do not adequately address the typical construction contract without being amended substantially. The difference between a properly worded and placed liability program and an inadequate placement can be the assets of the contractor.

Contractors should protect themselves to the greatest extent by choosing competent construction agents or brokers and insurers with specialized knowledge of the industry. These professionals will be able to address the key issues readily as they deal with them on a day-to-day basis.

Builder's Risk Insurance

James H. Costner, CPCU, ARM
Willis Corroon Corporation

7.1 Introduction

The purpose of builder's risk insurance is to assure that fortuitous loss will not prevent completion of a construction project.

The project documents published by the American Institute of Architects (AIA document A 201), the Associated General Contractors (AGC document 415), and the Engineers Joint Contracts Committee (EJCDC document 1910-8) all require project owners to furnish builder's risk insurance. The insurance required must:

- Cover all risks not specifically excluded
- Cover the replacement cost of the project
- Cover the interests of the owner, contractor, subcontractor, sub-subcontractor, engineer, engineer's consultant and any other party with an interest in the project listed in the project documents
- Allow waivers of subrogation against all insureds
- Cover until final payment is made
- Cover materials stored offsite and while in transit

See Table 7.1 for a comparison of the insurance requirements specified in the project documents listed previously.

Project lenders almost always impose insurance requirements as a condition of the construction loan.

If the project owner is an operating business with a commercial property insurance policy in effect, the owner may decide to add the project to the list of covered locations on that policy. Adding clauses covering all the other interests, the subrogation waivers, and materials stored offsite or in transit would then be necessary.

TABLE 7.1   Insurance Requirements in Contract Documents

	EJCDC	AIA	AGC
Owner buys the insurance	X	X	X
Covers full replacement cost	X	X	X
Covers the interest of the owner	X	X	X
Covers the interest of the contractor	X	X	X
Covers the interest of the subcontractor	X	X	X
Covers the interest of the sub-subcontractor		X	X
Covers the interest of the engineer	X		
Covers the interest of the engineers' consultant	X		
Builder's risk all-risk form	X	X	X
Prohibits excluding fire	X	X	X
Prohibits excluding lightning	X		
Prohibits excluding extended coverage	X	X	X
Prohibits excluding theft	X	X	X
Prohibits excluding vandalism	X	X	X
Prohibits excluding malicious mischief	X	X	X
Prohibits excluding earthquake	X		X
Prohibits excluding collapse	X	X	X
Prohibits excluding debris removal	X	X	
Prohibits excluding law or ordinance demolition	X	X	
Prohibits excluding flood/water damage	X		X
Prohibits excluding other as required	X		
Prohibits excluding testing			X
Covers loss or damage to the work	X	X	X
Covers temporary buildings	X	X	X
Covers falsework	X	X	
Covers work in transit	X		X
Covers expenses to repair or replace insured property	X	X	
Covers engineers' fees	X		
Covers architects' fees	X	X	
Covers materials and equipment stored onsite or offsite provided such is included in an application for payment recommended by engineer	X	X	
Maintained in effect until final payment is made	X	X	
Owner to purchase boiler land machinery insurance as may be required	X	X	X
30 days' notice of cancellation	X	X	
The insurers have no rights of recovery against any insured or any additional insured	X	X	X
Permission to occupy		X	X

Or, the owner may decide to satisfy the insurance requirements by purchasing a builder's risk policy from an insurance company that specializes in covering construction projects. One reason for choosing this option is to purchase testing coverage if the owner's regular insurer can't or won't cover testing. Another reason is that the coverage required on the project may be broader than the coverage provided by the policy on the business.

Even if the owner buys the special builder's risk policy, it may still be prudent to cover business interruption and extra expense on the business policy.

Many contractors, especially the larger contractors, write their own contract documents, and they change the insurance specifications to require them to furnish the builder's risk insurance. They believe they know more about builder's risk coverages; they may have a blanket policy providing broad coverage on all their projects; in the event of a large complex claim, they prefer to deal with their own insurance company; they believe some owners will attempt to control project costs by buying the cheapest insurance available. The cheapest insurance may leave important risks (perils), property, or interests uncovered or covered inadequately. The cheapest insurance may be written by a company not enjoying a reputation for promptly and fairly paying its claims. Whatever their reasons, many contractors want to control the insurance on their projects.

The Willis Corroon Corporation recently presented a study on builder's risk insurance at the 10th-annual Construction Insurance conference. The findings were:

- The insurance policies examined for the study did not comply with the insurance requirements in the project documents.

- Property owners and operators, in general, were found to be better served by placing insurance on their construction risks with their normal property insurance company.

- Contractors, in general, were found to be better served by special builder's risk policies.

- In general, the policies did not cover hot testing, boiler explosion, machinery breakdown, or electrical injury.

- In general, stand-alone builder's risk policies provided little, if any, coverage on indirect loss (business interruption, extra expense, delayed opening, and soft costs).

- In general, stand-alone builder's risk policies contained fewer restrictions on perils covered and property covered.

- In general, the brokers and underwriters preparing the insurance policies had not seen the project documents.

Interviews with members of the EJCDC and the AGC showed that their members frequently experience a delay collecting losses whether the insurance policies were provided by project owners or contractors. These members feel they are often ambushed at claim adjustment time by unexpected exclusions or restrictions. Further, there are unexpected demands for reimbursement for claims paid because of negligent damage to work in progress. Until confronted with a demand for reimbursement of a claim, it is unlikely subcontractors and sub-subcontractors know they may be liable for damage to the project, other than their own work, caused by their negligence.

When a claim occurs, builder's risk insurance written on the broadest possible terms by brokers and underwriters familiar with the insurance requirements in the project documents is most likely to deliver complete satisfaction to everybody involved. This is even more true when the insurance company and the broker have staff members highly skilled and experienced at handling construction projects.

The broker and the underwriter must receive complete and legible copies of all project documents including the loan covenants. All the insurance specifications must be accumulated and displayed on a spreadsheet so the insurance policy can be designed in conformity to them.

In addition, prudence may require the purchase of insurance not listed in any of the project documents. The use of a checklist is recommended to identify these additional needs. A sample checklist is included in Fig. 7.1.

Some people will be concerned that the broadest insurance will cost more. At first, this may be true. A sampling of builder's risk policies on large projects showed an average rate of $.08 per $100 or $800 per million per year. The broader coverage might cost as much as 50 percent more, $1200 per million per year. On a $100 million dollar project, the broader coverage would cost $40,000 more per year. But the bottom-line project cost need not increase by that amount.

The broadest coverage will not permit the builder's risk insurer to subrogate against any interest for negligent damage to the project work, the builder's own work or the work of others. Negligent damage would normally be covered under *general liability* (*GL*) property damage insurance. When this exposure is insured on a first-party basis in a broad builder's risk policy, the GL property damage exposure is reduced to the extent that the GL underwriter can be asked for a price reduction.

Moving the exposure from the GL policy to the property insurance virtually eliminates litigation expense associated with liability claims. It eliminates duplicate claims investigation and adjusting expenses. And it should provide for faster claims payments because first-party claims are usually easier to adjust. The combined result is an actual reduction in the project's cost.

Since there is no standard builder's risk policy form, every policy must be audited to determine whether or not it complies with the spec-

Builder's Risk Insurance Checklist

		YES	NO
NAME OF OWNER/PRINCIPAL:			
NAME OF GENERAL CONTRACTOR:			
PROJECT NAME:			
PROJECT LOCATION:			
ATTACH COPIES OF ALL CONTRACT DOCUMENTS INCLUDING LOAN AGREEMENT			
Whose interest is to be covered?		YES	NO
Owners/Principals			
General Contractors			
Subcontractors			
Designers			
Architects			
Engineers			
Lenders			
What is the limit of liability to be covered?			
List all sublimits and aggregates that are to apply			
What are the deductible amounts?			
What is the policy period?			
What is the period of the construction contract?			
List all locations to be covered by the Builder's Risk Policy:			
What property is to be covered?		YES	NO
Foundations			
Footings			
Underground pipes and wiring			
Fixtures, machinery, and equipment used to service the building			
Building materials and supplies			

Figure 7.1   Builder's risk insurance checklist.

ifications in the contract documents. Compliance with several of the specifications will require negotiated improvements in policy language.

7.2 Document Specifications Requiring Special Negotiations

The next section will identify several project documents specifications that will, almost always, require special negotiations between the broker and underwriter.

What property is to be covered?	YES	NO
Building materials and supplies owned by the general contractor and the owners		
Building materials and supplies owned by the subcontractors		
Temporary structures on the construction site		
Scaffolding, cribbing and forms		
Contractors' and subcontractors' tools, equipment and vehicles		
Property in transit		
Property in Blue Water Transit		
Property in storage at a site away from the project		
Pre-existing structure at the job site		
Site lighting		
Fences		
Retaining walls		
Land		
Excavations, grading		
Backlifting or filling		
Signs		
Plane and blueprints		
False work		
Fine Art		
Precious Metals		
Landscaping		
Patios		
Paved Surfaces		
Bulkheads, docks, piers and wharves		
Paved surfaces and roadways		
Bridges		
Property lent to others		
Water		
Antenna		
Boilers and pressure vessels		
Motor Vehicles and motorized equipment		
Valuable papers		
Computers and data		
Property in the open at the site		
Glass		

Figure 7.1 *(Continued)*

7.2.1   All-risk coverage

All-risk coverage is required by all three project documents (EJCDC, AIA, and AGC). Two of the most popular policy forms cover damage to or destruction of covered property by *named perils*. The named perils include fire, wind, hail, explosion, and the like. Irrespective of how long

What perils to be excluded?	Not Excluded	Excluded	Excluded but Resulting Damage Covered
Breakage of glass			
Collapse			
Bulging			
Contamination			
Dampness or dryness			
Frost or freezing			
Delay, loss of market or loss of contract			
Destruction of property by order of public authority			
Employee dishonesty or fraud			
Dishonesty or fraud by any insured			
Employee negligence			
Defective design			
Defective work or materials			
Latent defect			
Inherent vice			
Earth movement			
Subsidence			
Electrical injury			
Boiler explosion			
Machinery breakdown or derangement			
Extremes of temperature or humidity			
Flood, surface water or sewer backup			
Guarantee or warranty			
Interruption of utility service			
Marring or scratching			
Mysterious disappearance			
Neglect of the insured to save and preserve the property at the time of and after a loss			
Settling, shrinkage, or expansion			
Nuclear reaction, radiation or nuclear contamination			
War			
Acts of terrorists			
Operation of building laws or codes			
Pollution			
Penalties or fines			
Seizure by public authority			

Figure 7.1   *(Continued)*

the list of named perils is, a named perils policy is not in conformity with the requirements.

An all-risk policy covers any fortuitous damage or destruction to covered property so long as the cause (the peril) is not excluded by name in a policy section called "peril exclusions" or "risk exclusions."

What perils to be excluded?	Not Excluded	Excluded	Excluded but Resulting Damage Covered
Rain, snow, sleet, ice, sand or dust			
Smog, smoke, fumes, vapors			
Weight of snow or ice			
Testing			
Theft			
Wear, tear, normal making good, corrosion, erosion			

Does the policy cover indirect damage or soft costs?	YES	NO
Accounting fees		
Additional commission expenses		
Additional cost of construction labor and materials		
Additional design costs		
Additional expenses to expedite repairs, reopening		
Additional interest on loans to finance construction or repair		
Additional leasing expenses resulting from opening delay		
Additional realty taxes and other assessments		
Architects', engineers', and consultants' fees		
Bond interest		
Closing costs		
Construction loan fees for rearranging financing		
Contingencies		
Debt service interest and principal		
Delayed opening because of damage to property in transit		
Expenses incurred to accumulate data and to prepare a proof of loss		
Expenses incurred for the sole purpose of reducing or avoiding an increase in the cost of repairs and re-opening		
Expediting expenses		
Fees for licenses and permits		
Founders fees refunds		
General overhead of developer		
Increased mortgage financing expenses		
Insurance premiums for Builder's Risk, WC and GL		
Interdependencies		
Lag time; loss of leases or contract		
Legal and professional fees		
Letters of credit		
Loan commitment fees		
Loss of licenses or permit		

Figure 7.1   (Continued)

Does the policy cover indirect damage or soft costs?	YES	NO
Loss of rental income including escape clauses in leases		
Loss of tax credit under IRS section 48(b)		
Loss resulting from denial of access to the project because of damage to other property in the same vicinity		
Normal operating expenses that continue even when the project is not operating		
Rental income or value		
Operation of building laws; demolition, increased cost of construction		
Overtime paid to architects, engineers and consultants to expedite their work		
Plans and records		
Refinancing charges		
Rental or lease expense of construction equipment		
Sue and labor charges		
Wheeling charges to an electric utility		
When does builder's risk coverage terminate?	YES	NO
When the policy expires or is canceled		
When the property insured is abandoned		
When any building described in the declarations is put to its intended use		
When the property is formally accepted by the owner as complete in accordance with contract conditions		
Occupancy (practical and substantial use).		
30 days after commencement of testing		
Expiration of the operational testing period		
90 days following substantial completion of the project		
This insurance covers while at the risk of the insured during transit, while awaiting and during construction and after construction		
Until the interest of the insured ceases		
Until the expiration date of this policy		

Figure 7.1 *(Continued)*

7.2.2 Peril exclusions

The contract documents prohibit certain peril exclusions:

- Fire (by all three documents)
- Lightning (EJCDC)
- Extended coverage, which means, "wind, hail, falling aircraft, risk and riot attending a strike, vehicle damage, but not by an owner's vehicle, explosion, and smoke" (all three)
- Theft (all three)
- Vandalism (all three)
- Malicious mischief (all three)
- Earthquake (EJCDC and AGC): There is a severe shortage of earthquake insurance and prices are skyrocketing. The most severe shortages are in the greater Los Angeles and San Francisco areas. For a

large project, it may not be possible to purchase earthquake insurance in amounts equal to the full replacement cost. Or, if available, the cost may be astronomical. A supplemental amendment to the documents may be required.

- Collapse (all three): Most insurers no longer consider collapse a peril but a result of peril. In other words, collapse must be caused by something. Coverage under an all-risk policy will depend upon whether the cause is or is not an excluded peril.

  Other insurers put an absolute collapse exclusion in their policies. Then they attach an endorsement adding back collapse coverage but at very low limits (usually $10,000). Higher limits can be negotiated.

  Still another group of insurers excludes all collapse. Then they add back coverage on collapse caused by perils that are listed. Design error or faulty construction are never listed! This approach does not satisfy the requirements.

  Collapse coverage without restrictions and for the full replacement cost of the project will, almost always, require special negotiations.

- Debris removal (EJCDC and AIA)

- The expense of demolition of the remaining portion of a damaged project when the demolition is required by the enforcement of a law or ordinance regulating construction or reconstruction. (EJCDC and AIA)

  Local building codes may require any building partially destroyed more than a certain percentage (i.e., 50 percent) to be demolished and the site cleared before reconstruction can begin.

  Without special endorsements, the vast majority of policies cover only the cost to repair actual damages. The remainder of the building, the demolition expense, and the debris removal expenses aren't covered.

  Not required by any of the contract documents is coverage for the increased cost of construction caused by the enforcement of building codes, laws, or ordinances regulating construction. If a code requirement changes after a project is begun, any repairs or reconstruction will be required to conform to the new code. Code changes could require, for example, automatic sprinklers, handicapped access, or compliance with revised earthquake construction standards.

  Law or ordinance coverage is available to cover both these exposures. It is almost always sold with a low limit on each claim or occurrence. Coverage on a full replacement cost basis may be difficult to get.

- Flood (and water) damage (EJCDC and AGC): If the project is located in a flood-prone area, commercial flood insurance may not be available. Away from flood-prone areas, coverage is almost always available via a special endorsement to the builder's risk policy.

In flood-prone areas, flood insurance from the national flood insurance plan may be available.

- Testing (EJCDC): Some people hold the mistaken belief they don't need testing coverage because new equipment being installed is covered by the manufacturer's warranty. Unfortunately, bad things can happen to equipment during installation caused by faulty installation or an installer's error. Warranties exclude damages caused by installers. Evidently, an installer's mistake is the most frequent cause of loss to equipment being installed.

There are but a handful of insurers willing to write testing coverage. These include Hartford Steam Boiler, CIGNA, Royal, ENCON, Zurich, Alliance, and perhaps a few others. Nobody writes testing without the rest of the builder's risk policy. Compliance with the testing requirement will restrict competition on builder's risk policies to the handful of insurers writing testing.

7.3 Conclusion

Although none of the policies included in the Willis Corroon study satisfied the document requirements, it is possible to get builder's risk insurance that will conform to the requirements and satisfy all the interests insured on the policy.

8

Workers' Compensation

Carl H. Groth III, ARM and Jeffrey A. Segall, CPCU, ARM
Willis Corroon Corporation

Ronald D. O'Nan, CPCU, ARM
Independent Insurance Consultant

8.1 History

The industrialization of the American economy in the early part of the twentieth century brought a large workforce into some very new environments. With the change in job types and the increased development of automation, employee injuries increased, as did the realization that the costs to society of financing the burden of the newly disabled would become staggering.

Workers' compensation insurance was developed to transfer the cost associated with workplace injuries to third parties. Today, states have enacted laws that stipulate the types and amounts of coverage which employers must have in order to employ workers. It is important to note that these are state-specific laws and differ to some degree. There are also statutes that the federal government has established to take federal and maritime employment into consideration.

The origin of today's workers' compensation laws comes from early common laws. These laws outlined the responsibilities employers had for their employees for work-related injuries and disease. Essentially, the employee had to be able to prove that the employer was responsible for his or her injuries due to negligence on the part of the employer, and that the negligence was the cause of the accident.

The employer had three primary defenses for these allegations. The first was the concept of *contributory negligence*. Here, the employer had only to prove that the injury was caused in part by the employee. If that was the case, the contributing factor could absolve the employer from liability.

Another defense was the *assumption of risk* rule. Using this rule, the employer only needed to prove that the employee knew of the conditions of the job that later resulted in the injury. Since the employee knew the risks associated with the employment and continued to work there anyway, the injury was not the employer's responsibility.

The third defense available to employers was the *fellow servant* rule. If the injury resulted from the contributing actions of another employee, the employer was not considered liable for the injuries. These defenses were powerful and employers were able to avoid paying for the injuries.

The Federal Employer's Act of 1906 attempted to create an environment in which the employees would have a more equitable process for collecting for their injuries. This act was declared unconstitutional at the time, but later modifications established a workplace where these defenses were removed. Employees still had to prove negligence on the part of the employer.

New York became the first state to establish a mandatory comprehensive workers' compensation law in 1910. It underwent many challenges and changes. It was not until 1914 that it was modified to the point where it was enacted. In the meantime, Wisconsin passed an elective law in 1911. Wisconsin's law allowed an employer to either adopt the new law or lose the three defenses previously mentioned. It was deemed constitutional because the employers had a choice of coming under its purview and generally opted to out of concern for the uncertainty of jury awards.

Each state now has some type of workers' compensation legislation. Some states differ in the number of employees an employer must have in order to be affected by it and some still allow certain types of employees to opt out. The single most important concept of these laws is that it eliminates an employee's right to sue and creates an automatic liability on the part of the employer. It is a no-fault system. As long as the injury or illness arises out of employment, the employer is financially responsible.

Since workers' compensation laws are enacted by state, it is important to understand how these laws may differ. Following are some important areas which should be considered in comparing various state statutes.

Benefits.   Each of the states requires the employer to be responsible for the cost of medical treatment, disability payments, and a death benefit. As laws are amended, more states are including costs for rehabilitation. The levels of the benefits are the key distinction. Some states stipulate a schedule of benefits. Disability payments may be reflected as a number of weeks of pay or an amount of benefit based on a doctor's analysis of the percentage of disability on a wage loss concept. Medical payments are usually limited only by a stipulated fee schedule or an amount considered usual and customary for the procedures involved.

Compulsory or elective. As mentioned previously, the development of workers' compensation laws was inhibited in part by the concern over constitutional issues relative to the deprivation of due process for employers. The first successful law allowed for the employer to make the election to come under the statute. Today, only three states—New Jersey, South Carolina, and Texas—still offer this option.

Disease. All existing state workers' compensation laws include a provision for benefits to be paid for "occupational disease." All states do not agree on the definition of these occupational diseases. Some restrict the list of diseases that are covered by actually naming them in the law. Other states may broadly provide coverage for any illness that is peculiar to the employee's occupation.

Employees covered. Domestic employees and farm laborers are usually excluded from all workers' compensation laws. There are other types of employment which, because of the associated hazards, are treated differently. Variances are evident in the number of employees an employer has before the law becomes mandatory. Casual and part-time employment are also treated differently. Volunteer workers may not be automatically included in the laws either; some states have provisions for including these types of workers and others do not.

Extraterritorial provisions. The workers' compensation statutes of most states have provisions for covering employees for accidents regardless of where they arise. In this way, the state law in effect in the state of hire (as long as some work is performed in that state) will have jurisdiction. The statutory differences are in the period of time that the employee is permitted to work in another state. The time may be from 90 days to a year. After the appropriate time has passed, the laws of the state of employment will prevail.

Type of insurance available. In most cases, employers are required to purchase insurance to satisfy the workers' compensation statute. A discussion of this requirement follows. For now, it is important to note that the various states allow for this type of insurance through very regulated channels. The majority of statutes permit private insurance through authorized carriers. Other states require the coverage to be sold exclusively through monopolistic state funds. States with these monopolistic funds are: Nevada, North Dakota, Ohio, Washington, West Virginia, and Wyoming. Still others permit private insurers to compete against state funds.

In most states (except some of the monopolistic states), an employer that meets certain size and financial requirements may qualify to self-insure or operate without the purchase of an insurance policy. The objective of requiring insurance is to be certain the employer will be able to meet the financial obligations of the workers' compensation law.

In addition to the state laws that stipulate the employer's responsibility to the employees, there are federal statutes which were passed to protect workers in industries involved with railroad and maritime employment. These are industries of commerce that are regulated by the federal government and therefore require federal legislation. Public work contracts performed for the U.S. government outside the United States and employment of civilian workers by the U.S. government also require unique treatment. At the time Congress passed the various statutes, interstate trucking was not an industry and today remains under the jurisdiction of the various states.

Depending on the nature of the employment, these employees are protected under either admiralty laws, the U.S. Longshore and Harbor Workers' Compensation Act, the Defense Base Act, the Federal Employees Compensation Act, or the Federal Employer's Liability Act. By virtue of the fact that these are federal laws, they take precedence over the state workers' compensation laws.

It is not possible to do justice to each of these statutes in this chapter. Suffice it to say that on analysis, they share a great deal in common with state statutes. These laws should be compared using the same basic criteria. The major issue is the type of employment these laws were designed to cover.

8.2   The Workers' Compensation Policy

Statistical information gathered from the *Social Security Bulletin* estimates that over $57 billion was spent in 1993 to insure or self-insure workers' compensation exposures. Of this, approximately 75 percent was paid out to insured workers or their dependents. The average cost of workers' compensation in the U.S. for 1993 was $2.30 per $100 of payroll.

8.2.1   Coverage

The workers' compensation policy is actually a package of two coverages. Coverage A provides coverage to protect the insured's liability under workers' compensation laws. It is written broadly enough that the same insuring agreement accommodates multiple state laws. The other coverage, coverage B, provides employer's liability protection.

Employer's liability covers suits brought as a result of injuries arising out of employment that is not covered by workers' compensation laws. It is similar to the type of coverage provided by the commercial general liability policy and would not be needed if not for an exclusion pertaining to claims by employees in that policy. An example of this may be a claim made by an employee arising from a disease alleged to have been contributed to by his or her employment which does not fit the definition of occupational disease in the workers' compensation law. Other more typical claims for which this coverage would apply are referred to as *action over* claims.

An action over claim (see Fig. 8.1) typically involves product liability-type claims. An injured employee sues the manufacturer of the apparatus that caused the injury, then the manufacturer sues the employer for poor maintenance. In the construction industry, an action over claim comes via an employee suing a property owner for an unsafe condition, who then sues the employer/contractor for negligence.

Coverage A has no limit stated on the policy. This is because the actual amount of payment is stipulated in the applicable state law. Coverage B does have a limit. The limit is most often represented in three ways: the most the policy will pay for claims arising from an accident, the maximum the policy will pay for any one disease, and a limit per person per disease. Typical employer's liability limits are stated at $100,000 per accident, $500,000 per disease, and $100,000 per person per disease.

8.2.2 Exclusions

The structure of workers' compensation is similar to most policies. In addition to the insuring agreements, the policy contains exclusions, definitions, and conditions. There are relatively few exclusions in this policy. Only two affect coverage A. The first avoids duplicate coverage by

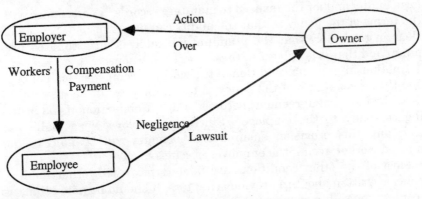

Figure 8.1 Action over claim.

excluding claims where other coverage is in force. The second exclusion is for domestic and farm labor; however, this exclusion does not apply where state law requires coverage for these types of employment.

The other four exclusions all pertain to coverage B, employer's liability. Excluded are claims assumed under contract or claims filed more than three years after the policy expires. Also excluded are claims made by employees who have been hired with the insured's knowledge that they were employed in violation of the law. The last exclusion applies to claims for state disability and unemployment laws.

8.2.3 Definitions

As with exclusions, there are few definitions found in the standard workers' compensation policy. The definitions that do exist are designed to tie the coverage to the specific workers' compensation laws of the states the policy was intended to cover. One policy defines workers' compensation laws to not include federal laws unless specifically endorsed. Another defines *state* to include U.S. territories and the District of Columbia.

Definitions are also used to clarify coverage for assault and battery but only when not committed by the employer. Bodily injury by accident and bodily injury by disease are defined to be certain that no single injury could be construed to apply to both limits and allow a claim to be paid twice.

8.2.4 Conditions

Premiums are determined, as you will see later in this chapter, in direct relation to payroll, or remuneration. The first condition of the policy deals with this premium determination. Other conditions relative to premiums include a provision for the insurance carrier to perform audits and for the insured to maintain records in accordance with the terms of the policy so that the audits can be performed. It is made clear in the policy that the premium indicated is an estimated premium and that it is subject to these audit conditions.

Another interesting condition of the workers' compensation policy is one that gives rights of the policy to the employees of the insured. In fact, the insuring agreement becomes a direct obligation of the insuring company to the employee. Since the employee is actually the claimant, this provision ensures that benefits will be paid to the injured worker even if the employer becomes insolvent.

Most of the other conditions are fairly typical with other liability policies written about in previous chapters. It should be noted that, as respects the rights of subrogation, the insured can do nothing to waive the rights of the insurance carrier without the carrier's written consent. Most other policies allow waivers of subrogation if made prior to a loss. With this condition, insurance companies can withhold giving up their rights or they may charge for them. While there is no standard premium charge, it may approximate 20 percent of the premium applicable for the work performed for the requesting party.

8.2.5   Endorsements

There are not many standard endorsements found on workers' compensation policies. One of the most important for insureds who may occasionally work in states that provide workers' compensation insurance through monopolistic state funds is stopgap liability coverage. For businesses doing business exclusively in states where monopolistic state funds operate and a workers' compensation policy with a private insurer is not purchased, this endorsement may be attached to a commercial general liability policy.

The stopgap endorsement is intended to offer coverage B, employer's liability coverage, on a stand-alone basis. It is needed for workers' compensation laws in monopolistic states. An unendorsed workers' compensation policy will exclude claims which come from operations performed in states not listed in the policy, i.e., the monopolistic states.

8.2.6   U.S. Longshore and Harbor Workers' Endorsement (USL&H)

As pointed out earlier, the unendorsed workers' compensation policy provides no coverage for any federal acts, only state acts for those states listed in the policy. This endorsement amends the definition of workers' compensation law to include the USL&H act. The interpretation of when an employee comes under the provisions of the act are left to the courts. Even if an employer has no employees typically considered to be covered by this law, it is important to remember that the inclusion of a particular employee is not only by job but also by the location of the job. Work on a container loading dock or ship passenger terminal may be interpreted as furthering commerce on navigable waters and, therefore, bring all employees at the jobsite within the scope of the law. To avoid the possibility of being unprotected for this type of loss, attachment of this endorsement is recommended. Failure to obtain USL&H coverage is punishable by fine and/or imprisonment.

8.2.7   Jones Act

The Jones Act is legislation which pertains to employee injuries for masters and members of vessel crews. Some contractors maintain pushboats or tugs for work around bridges and docks. Coverage for these individuals is typically provided by *protection and indemnity* (P&I) policies written on the vessel to which these employees are assigned. Another way to cover these people is to endorse the workers' compensation policy and attach the maritime endorsement. This modifies coverage B, employer's liability, to include liability claims made under the Jones Act.

When work is done outside the United States, one of two endorsements should be considered. For work performed on U.S. government property the Defense Base Act endorsement is appropriate. Otherwise, including the foreign voluntary compensation may be wise. Coverage for repatriation expenses and the cost of returning the injured worker to the United States for medical attention or burial is provided for by this endorsement. It also covers endemic diseases that would otherwise not be included in the definition of occupational disease. It pays the difference in benefits between the local workers' compensation coverage (if any) and that of the home state. It is probable that in addition to making the appropriate endorsements to your workers' compensation policy, you will also need to purchase workers' compensation in the jurisdiction the work is being performed in.

8.2.8 Costs

Workers' compensation costs fall into two categories: losses paid to injured employees and related expenses. Losses paid to injured employees include both medical expenses and benefits for lost wages. Expenses related to workers' compensation comprise a wide range of items such as claims handling costs and costs associated with underwriting, policy production, sales and marketing, administration, and insurance company profit. Workers' compensation premium is designed to cover all of these costs.

Workers' compensation premium is determined on a state-by-state basis. In each state, manual premium is based upon the amount of exposure, expressed as one hundred dollars of payroll for a given class of employee multiplied by the rate for that class. Employees are classified according to the relative hazards involved with a given type of business such as asphalt works and drivers, etc. Manual premium rates represent the loss experience of the average risk in a class. The manual premiums for each classification are added together to determine total manual premium for that state.

An example of a manual premium calculation for a particular state is represented as follows:

Employee description	Class code	Rate	Payroll	Manual premium
Accountant—Traveling	8803	0.38	$195,312	$ 742
Clerical	8810	0.68	113,550	772
Salesperson	0951	1.01	238,489	2409
				$3923

In this example, $3923 is the manual premium subject to experience rating. The manual premium reflects the cost of this particular workers' compensation risk based on the loss experience of the average risks in these classifications. Experience rating modifies the manual premium to reflect the loss experience of the individual insured to the average insured. Experience rating is mandatory for insureds who meet minimum workers' compensation premium requirements.

Manual premium is modified by the application of an experience modification factor commonly referred to as an *experience modifier* or *mod*. The experience modifier is determined by comparing the insured's actual loss experience to expected losses. The modifier is multiplied by manual premium to create a modified premium that reflects the differences between an insured's unique experience and the average insured.

Application of the experience modifier will result in either a premium increase or a decrease depending upon the relationship of the individual's loss experience to that of the average insured. An experience modifier greater than one indicates that the individual insured experience is worse than the average and modified premium will be greater than manual premium. An experience modifier less than one indicates that the individual insured's experience is better than the average, and modified premium will be less than the manual premium. Using the previous example, the application of an experience modifier (less than one in this case) produces a new premium as follows.

Employee description	Class code	Rate	Payroll	Manual premium
Accountant—Traveling	8803	0.38	$195,312	$ 742
Clerical	8810	0.68	113,550	772
Salesperson	0951	1.01	238,489	2409
				3923
Experience modification		0.89		(432)
State total				$3491

The modified premium, also known as modified standard premium, in this example is $3491. This premium reflects the relative experience of this individual insured to the average insured through the application of the experience modifier of 0.89. For this particular insured, the reward for a favorable loss experience is a premium credit of 11 percent or $432.

In each state, the rates for employee classifications and experience modifiers are calculated by rating the service bureaus. Most states subscribe to the *National Council on Compensation Insurance* (NCCI) to perform certain functions. The role of the NCCI is to determine adequate rates commensurate with the cost of workers' compensation in various states, determine individual insureds' experience modifiers, and work with states' legislatures to facilitate workers' compensation reform. States that do not subscribe to the NCCI have their own rating bureau to perform these functions.

8.3   Risk Financing Plans

The rising cost of workers' compensation in recent years has emphasized the importance of risk management policy and alternate financing programs to reduce costs. Selecting the most effective financing program for a contractor requires consideration of many factors. Among the factors impacting the risk financing decision are the contractor's expected losses, the financial ability of the contractor to assume losses, available cash and credit, effective tax rate, the rate at which losses are paid over time, the cost of borrowing, the rate of growth, capacity for handling administrative responsibilities, and the after-tax discounted cost of the financing program.

The range of alternative risk financing plans can broadly be characterized as offering various degrees of trade-offs between the opportunity for cost savings and risk. The greater the opportunity for cost savings, the greater the risk required to be absorbed by the insured. Figure 8.2 illustrates this point and presents the major alternative risk financing plans available for workers' compensation.

8.3.1   Guaranteed cost plan

A guaranteed cost plan is a premium rating plan that establishes a fixed premium prior to the policy's inception date for the policy period. Under this plan, the premium is fixed regardless of the insured's loss experience during the policy period, although it is usually adjusted to reflect an audit of actual payroll exposure versus the expected payroll estimated prior to policy inception. Guaranteed cost premium is determined by multiplying manual rates by payroll, then applying the risk's unique experience modifier as described in Sec. 8.2.8. Guaranteed cost premium can be further modified by premium discounts to reflect the economics of risk size and by application of scheduled credits based upon loss experience and underwriting judgment.

Premium payments for a guaranteed cost plan can be made in full at the beginning of the policy period or on an installment basis over the policy period. The flexibility of premium payment is usually a function

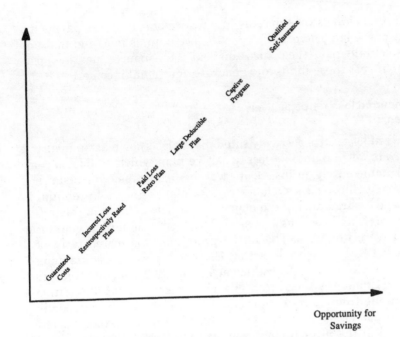

Figure 8.2   Alternative funding mechanism risk-reward trade-offs.

of premium size. Contractors with large premiums may have greater access to flexible premium payment schedules.

The primary advantage of a guaranteed cost plan is that it eliminates much uncertainty. Within any given policy year, the cost of workers' compensation insurance is fixed regardless of loss experience, and therefore is easily budgetable. This benefit is very important to small and medium size contractors whose loss experience cannot be predicted with an acceptable level of confidence. This plan may also be useful for contractors in reducing the chance that insurance costs would adversely affect the profitability of a job. Using a guaranteed cost plan, these companies can avoid sudden unexpected changes in costs associated with their actual loss experience.

The primary disadvantage of a guaranteed cost plan is the absence of any opportunity for immediate cost savings if loss experience is good. Although the experience modifier would ultimately reflect good loss experience, the benefit of premium credits comes in future years instead of immediately after the period of good experience. The magnitude of cost savings is also limited because it becomes increasingly difficult to keep reducing the experience modification factor with incremental improvements in loss experience.

Another drawback of the guaranteed cost plan is the negative impact on cash flow. Under this plan, through payment of the premium, the insured funds losses considerably in advance of actual loss payments. Because of the lag between the time the premium is paid and the time a loss is actually paid, the insurer, instead of the insured, has an opportunity to invest these funds and generate additional income.

8.3.2   Incurred loss retrospectively rated plan

An incurred loss retrospectively rated plan, also known as an incurred loss retro or retro plan, is a loss-sensitive plan which enables a company to immediately influence its workers' compensation insurance costs by controlling losses. Under a retro plan, an initial premium is paid by the insured during the policy period. The amount of the initial premium payment is usually based either on standard premium or expected retro premium. The initial premium is adjusted retrospectively based on the insured's actual loss experience. Premium adjustments are made using losses valued at 18 months after policy inception and every 12 months thereafter. The adjustments result in additional premium due from or return premium due to the insured depending upon whether losses are greater or less than initial expectations or losses used in the prior premium adjustment. The retro plan usually remains open until all losses are paid or until the insured and insurer can mutually agree to close the plan.

Although retro premium is based on the insured's actual loss experience, premium is usually subject to a minimum and maximum adjustment limitation. The minimum premium places a limit on the savings an insured may receive as a result of a favorable loss experience while the maximum premium places a limit on the amount of additional premium an insured may pay in the event loss experience is much greater than expected. Additional protection can also be provided through the retro plan by limiting the amount of loss arising from any single occurrence, which is included in the premium calculation.

There are four elements comprising the retro premium formula. These elements are described as follows:

- *Basic premium:* covers the insurer's expenses, profit and contingencies, and the maximum premium charge.
- *Converted losses:* the insured's total losses subject to any applicable loss limitation, multiplied by the loss conversion factor. The loss conversion factor, expressed as a percentage plus one, covers the cost of unallocated and/or general loss adjustment expense.
- *Tax multiplier:* includes state premium taxes and miscellaneous charges for assessments from assigned risk pools, guaranty funds, second injury funds, and ratings bureaus.

- *Loss limitation premium:* the cost to limit the amount of loss from any single occurrence to be included in converted losses.

Although there are many versions of retro formulas used by insurers, a simplified retro premium formula can be expressed as follows:

$$\text{Retrospective premium} = [(\text{basic premium} + \text{converted losses}) \times \text{tax multiplier}] + \text{loss limitation premium}$$

The primary advantage of an incurred loss retro plan is the insured's ability to influence the program costs by controlling losses. Significant premium savings may be recognized by an insured who successfully implements effective loss prevention and loss control procedures.

One of the disadvantages of an incurred loss retro plan is the uncertainty of the final premium, thereby complicating the budgeting process. Since retrospective premium adjustments for a given policy period are made over several years until all claims are paid, the insured's financial statements in each subsequent year can be significantly impacted by changes in each prior period's losses. This impact on financial statements can be particularly significant if claims are initially underreserved and readjusted upward to appropriate levels in later periods.

Another disadvantage of the incurred loss retro plan is that retrospective premium, which is based upon incurred losses, is required to be paid to the insurance company. Incurred losses include unpaid reserves which typically take several years to be paid. Because of this time lag, the insurance company, instead of the insured, has the opportunity to invest these funds and earn investment income. This negative impact on cash flow may be significant if the claims are overreserved.

8.3.3 Paid loss retrospectively rated plan

A paid loss retrospectively rated plan, also known as a *paid loss retro,* is a loss-sensitive program offering the insured many of the benefits of self-insurance without the increased administrative responsibilities and filing requirements accompanying a self-insured program. The calculation of premium for a paid loss retro is determined in much the same manner as an incurred loss retro. The primary difference between the paid loss retro plan and an incurred loss retro plan is that losses subject to the loss limitation are paid to the insurer as the losses are paid to claimants.

Under a paid loss retro program, the insurer agrees to accept a substantially reduced initial premium payment. The initial premium payment is normally equal to the insurer's basic expense, loss limitation premium, and related premium taxes and assessments. Basic expense under a paid loss retro plan is usually greater than under an incurred loss retro to make up for a portion of the insurer's lost opportunity to earn investment income on unpaid losses. The difference between the initial policy year's payment and the estimated standard premium is collateralized with a letter of credit, surety bond, or other type of financial security and may be guaranteed by the insured with an indemnification agreement or promissory note. The financial security is adjusted at least annually to reflect the insured's current outstanding loss reserves under the program. The plan continues for several years following the policy's expiration until such time as all outstanding losses are paid or the premium payment agreement is terminated. When the premium payment agreement is terminated, the program converts to an incurred loss retro.

In addition to the initial premium payment previously described, the paid loss retro typically requires an advance paid loss deposit of two months estimated paid losses to establish a claims escrow fund. As claims become payable, the insurer makes payments from this fund and, in turn, the insured is billed for actual losses paid and reimburses the fund on a periodic basis, usually once every month. An increasingly popular alternative to the escrow fund is a zero balance account utilizing daily wire transfers of funds.

The primary advantage of the paid loss retro is the cash flow benefit to the insured resulting from deferring the payment of a substantial portion of the premium until it is required for loss payments. The longer the period of time over which payment extends, the more valuable this benefit becomes. At the same time, the insured retains the opportunity to realize cost savings through the retro formula by reducing losses.

The major drawback to the paid loss retro is the financial security requirements accompanying the program, particularly when collateral requirements for several policy periods overlap. Also, like an incurred loss retro, this plan is characterized by a degree of financial uncertainty. Budgeting difficulties may arise from both the unpredictable cost associated with the program as losses change over time, and the unpredictable amount of paid losses payable each period.

8.3.4   Large deductible plan

A large deductible plan allows an insured to retain liability up to a predetermined level and purchase excess insurance to cover losses above the deductible. In this example, large is defined as $250,000 or higher. Although large deductible plans for workers' compensation are not available in every state, most states have accepted them. In unapproved states, other funding mechanisms described in this section can be used.

As with the paid loss retro plan, a large deductible plan offers an insured many of the benefits of a qualified self-insurance program without the increased internal administrative expenses and the additional staffing and filing requirements that accompany a self-insurance program. Under a large deductible plan, the insurer typically provides all the administrative services necessary to maintain the program, including insurance certificates. However, a large deductible plan differs from a paid loss retro in that losses within the deductible are not included in the premium calculation, thus reducing much of the premium taxes and assessments levied by the insurer.

The cost components of a large deductible plan are the administrative charge, claims administration expense, security costs, and excess premium. The administrative charge covers the insurer's administrative expenses and profit loading. Claims administration services may be contracted with the insurer or with an outside vendor. In most instances, the insured is required to provide an advance deposit to the insurer or independent claims administrator who pays claims from this fund and, in turn, the insured reimburses the escrow fund on a monthly basis. The amount usually retained in the fund is approximately equal to two months of estimated paid claims. Upon termination of the plan, any amount remaining in the fund is returned to the insured.

A large deductible plan provides substantial cash flow benefits since the insured is able to delay the funding for losses until payment is required, thus avoiding the financing cost associated with the early funding of losses. The large deductible plan also eliminates many of the taxes and assessments because the basis for these assessments (in most states), the premium, is reduced under the plan.

On the negative side, the insurer regularly requires the insured to provide collateral for the amount of the estimated ultimate losses within the insured's deductible. These financial security requirements may prove burdensome, especially when the collateral requirements of several policy periods overlap. Aggregate stop loss protection may be more costly or may not be available under a large deductible plan. Additionally, as with the paid loss retro plan, costs and paid losses are not known with certainty in advance and therefore are difficult to budget.

8.3.5 Captive insurance company

A captive insurance company is an insurance company used to formalize a program of loss retention and prefunding of anticipated losses. A pure captive is a wholly owned subsidiary established for the purposes of insuring a portion of the parent company's risks. Captives can also be organized by multiple owners whose risks are collectively insured by the captive (group captive) or by an independent third party whose role is that of an investor (rent-a-captive).

State workers' compensation regulations require a licensed company to write workers' compensation coverage. Therefore, the methods for the captive to assume workers' compensation risk are limited to the following possibilities:

1. Reinsuring a licensed company (fronting company) for the desired amount of risk

2. Writing a loss reimbursement policy directly to the insured for retained losses under a large deductible or qualified self-insurance program

A fronting arrangement with a commercial carrier allows the captive to assume part of the parent's risks without being licensed and admitted in all states in which the parent has operations. Under this arrangement, the captive pays the fronting company a fee, typically based on a percentage of premium, for its policy issuance and administrative services. The fronting carrier is also reimbursed for state premium taxes and assessments incurred as a result of the policy issuance.

A captive program creates underwriting income from an effective risk management program and investment income from invested loss reserves. This income enables the captive to serve as a performance measurement vehicle for a company's risk management function.

A captive's surplus can be put to many uses. It can improve the flexibility of cost allocation systems where some profit centers receive cost allocations based on a smaller loss retention than the contractor's master retention. In this case, the captive surplus will fund any losses between a profit center's retention and the master retention in return for a fixed premium. Captive surplus can also be used to fund the risk management budget. During periods of cost cutting, this funding source can be particularly valuable.

Another benefit of a captive is its ability to realize underwriting and investment income from contractor-controlled programs. When a wrap-up project is needed, it may be possible for the contractor, instead of the owner, to buy insurance. In these cases, the contractor can exercise more control over the risk management process which may generate a low-cost insurance program. If premiums are reinsured into the captive, the resulting profits can be captured by the contractor.

The disadvantages of using a captive insurance company for funding retained losses are largely financial. Capital must be committed to the entity at the time it is established and throughout its operation, which, depending upon its domicile, can take the form of cash, securities, or letters of credit. Aside from the loss of use of funds associated with a capital commitment, the amount of capital required may prohibit a contractor from utilizing this type of funding mechanism. In addition, there is an implied finance charge or opportunity cost incurred by the parent to maintain the capital in the captive. Additionally, anticipated losses insured by the captive are prefunded by the payment of the premium to the captive. The financing costs of prefunding losses and contributing capital to the captive are usually only partially offset by investment income earned by the captive on its assets.

Other expenses incurred in the operation of a captive insurance company include accounting and auditing fees, legal fees, actuarial fees, management fees, filing and license fees in the domicile of incorporation, and cost of letters of credit to the fronting carrier, if any.

8.3.6 Qualified self-insurance program

Qualified self-insurance is a carefully designed plan for retaining liability, up to a predetermined limit, for those risks which a contractor decides it is financially prudent to assume. Excess insurance is usually purchased to cover losses above the desired retention level. If well structured, self-insurance can, in many cases, offer the most efficient and least expensive means of loss financing. It is particularly true if a company has a high cost of borrowing and a low or zero effective income tax rate.

The decision to establish a qualified self-insurance program is dependent upon the philosophy of management in an organization, financial capabilities and objectives, the nature of operations, and exposure to loss. Self-insurance programs can be suitable for workers' compensation coverage where the loss exposure is large but predictable and not catastrophic in nature. Excess insurance, covering those losses which are less predictable and therefore likely to impair a contractor's financial condition, can enhance the stability of a self-insurance program.

Self-insurance requirements for workers' compensation vary from state to state, as determined by state insurance departments and regulations. The areas of particular importance to a self-insurer in determining the total cost of a program include: filing and reporting requirements, state premium taxes and assessments, excess insurance, and financial security requirements.

To become a qualified self-insurer for workers' compensation, a company must file an application and pay a filing fee in each state where it intends to qualify. A recent annual report and other relevant financial information is generally required with each application, in addition to information detailing an insured's historical loss experience. Furthermore, some insurance departments require an applicant to submit incurred losses, paid losses, and claimant information on a periodic basis. Self-insurance taxes and assessments are typically calculated as a percentage of losses incurred or as a percentage of the premium that would have been paid had a commercial insurance policy been purchased. The amounts vary from state to state. Likewise, excess insurance requirements and security requirements vary from state to state. While some states do not require the purchase of excess insurance, it is generally recommended and in some instances may reduce a state's security deposit requirement. The most common forms of security used in conjunction with a self-insurance program are surety bonds, letters of credit, and certificates of deposit. Self-insurance surety bonds may not be easily obtained in the marketplace as states are imposing higher security requirements, thereby increasing the complexity and cost of these programs.

It should be noted that the market for excess coverage for self-insureds, especially for aggregate excess coverage, is limited. The availability of excess coverage at reasonable premiums, limits, and attachments is key to the viability of a qualified self-insurance plan, especially when occupational disease exposure exists in the workplace of the self-insured.

Under a qualified self-insurance program, the self-insurer is also responsible for providing loss control claims handling services, data processing, and other internal administrative services. Because such programs must be designed and implemented, data collected and analyzed, and claims settled and sometimes litigated, an internally administered self-insurance plan requires a substantial commitment on the part of company management. However, it is possible to contract with outside vendors for these services and thereby reduce the internal administration to a level more closely comparable to that of a commercial insurance program. Furthermore, a qualified self-insurance program typically leaves the insured and employees of the insured without access to insurance company guaranty funds in the event of insolvency.

Self-insurance may create several advantages. First, by assuming the responsibility for losses occurring with predictable frequency and of a severity level that is within the contractor's funding capabilities, the contractor can save the administrative and profit charges made by the commercial insurance marketplace.

Improved cash flow is another potential advantage. The self-insurer usually delays the funding for losses until payment is required, thus avoiding the financing cost associated with the early funding of losses.

A number of administrative benefits may also be made possible through self-insurance. Although virtually all of the administrative services provided by a commercial carrier must still be provided by the self-insured contractor, it may be possible for the self-insured to realize increased control over claim settlements and claim adjusting procedures. Also, employees may be less likely to file a compensation claim for nonsevere injury and may be more prompt in returning to work when it is their employer's name on the check as opposed to an insurance company. Self-insurance often results in an improvement in the attitude of employees at all levels in the corporation toward loss prevention and control.

One negative aspect of a qualified self-insurance plan is the commitment required. The costs of the increased administrative tasks will offset at least some savings achieved by the program. Because self-insurance regulations vary from state to state, administrative burdens are further increased. There is less stability and predictability to the assessments charged self-insured. Moreover, a qualified self-insurance program requires the filing of financial statements in the public domain, which represents a deterrent to privately held entities.

Workers' Compensation Insurance Cost Control

Richard Carris, CPCU, CLU
Ernst & Young LLP

The cost of workers' compensation insurance has become a problem for contractors of all trades and financial net worth. In all states, the sharp boost in benefits and resulting premiums, soaring health care costs, and a growing number of claims are among the chief factors fueling a system gone amok.

Even more out of control than health care costs, workers' compensation costs have risen steadily both in amounts and percentage of payroll. Restricted by state law that disallows many cost-control initiatives, many contractors and their professional associations, such as the Associated General Contractors, have been pushing legislators for reform but to little avail to date. Instead, insurance rates for various classification codes for workers' compensation have risen by almost 100 percent and sometimes more during the past five years. Table 9.1 illustrates typical rate increases for one state.

According to a Johnson & Higgins' 1993 construction industry survey of more than 500 contractors, the cost of workers' compensation insurance was the single common concern of contractors. Additionally,

TABLE 9.1 Construction Rate Increases—Selected Class Codes

Class code	NCCI code	1989	1990	1991	1992	1993	1994	Annual increase, %
Masonry	5022	$13.36	$15.16	$14.91	$20.08	$20.68	$23.62	12.1
Iron or steel	5057	24.06	25.67	26.94	37.99	39.13	47.44	14.5
Sheet metal	5538	9.75	15.87	16.29	20.22	20.83	25.55	21.3
Pile Drive.	6003	17.49	24.81	22.27	31.41	32.35	36.69	16.0
Clerical	8810	.35	.47	.55	.64	.69	.58	10.6

the survey noted that contractors have implemented structured safety programs but the smaller contractors have lagged in adopting safety procedures.

In many states, the contractor and insurer can't employ the full range of managed care options for workers' compensation that are common in health care insurance. For example, in many states the injured employee has full freedom to choose the treating physician.

Unsuccessful attempts at reforming the law are compounded further by other critical developments such as constraints posed by the Americans With Disabilities Act (ADA), the continuous emergence of new types of claims, and increasing difficulty insuring workers' compensation insurance in the commercial market.

Insurance, and in particular, workers' compensation insurance, is a significant cost item for any size contractor, yet some contractors are able to keep workers' compensation costs under control while others are mere victims of the system and are forced to pay whatever the insurer charges. Workers' compensation insurance costs alone can be 5 percent or more of a contractor's bid and usually represent the largest single cost item of insurance. And in recent years, the annualized rate of increase in workers' compensation costs began to double or triple the annualized rate of payroll increases.

9.1   Introduction to Workers' Compensation Insurance

Workers' compensation insurance is a form of no-fault insurance and is paid for by the contractor who provides replacement wage benefits and medical care for workers that are injured or become sick due to job-related injury. In many sites, a small waiting period applies from one to three days. Even this waiting period is sometimes waived, however, if the employee's injury causes the extent of disability to last beyond a certain period of time. Unlike health care insurance, with workers' compensation insurance there are no deductibles or coinsurance requirements to be incurred by the injured employee.

Every state except Texas, South Carolina, and New Jersey requires that contractors provide workers' compensation benefits for employees who are injured or become sick due to accidents or injuries within the scope of employment. Although the workers' compensation laws are not mandatory in the three states cited, due to financial requirements most contractors in these states, too, continue to purchase workers' compensation insurance.

As a type of no-fault insurance, workers' compensation entitles the employee to benefits regardless of fault or negligence. In exchange, however, the employee forgoes certain rights, the most significant being the ability to sue the contractor for negligence under tort law. The issues of no-fault and compensability are important to note because, in the absence of contrary evidence, any employee who claims to be injured within the scope of employment is entitled to workers' compensation benefits. These benefits include unlimited medical benefits, wage loss replacement, and, in most states, vocational rehabilitation.

According to the Alliance of American Insurers, a network of insurers that write workers' compensation insurance, insurers' costs have increased at an annual compounded rate of approximately 10 percent between 1977 and 1990, while wage inflation for this time period was approximately 5 percent. During the first five years of the 1990s, while the level of rate increases has begun to subside in some states, contractors in most states continue to experience out-of-control workers' compensation costs.

From a business perspective, contractors need to provide workers' compensation insurance to comply with state laws as well as to bid on contracts and ultimately provide certificates of insurance to owners and other parties.

Since workers' compensation insurance is regulated by the states, and not by the federal government, a complex array of laws has developed. Although the U.S. Constitution states that all people are created equal, when it comes to workers' compensation insurance apparently this is not always true. Not only do premiums vary considerably, but, in addition, benefit levels vary considerably throughout the country. The current system is inequitable as benefits to injured employees vary significantly depending upon the state in which the accident happened. For instance, a loss of a hand in one state is worth $146,060, but only $43,450 in another state! In addition, weekly income benefits vary significantly.

For an excellent compendium of these regulations and benefits state by state, contractors can obtain the *1994 Analysis of Worker's Compensation Laws* from U.S. Chamber of Commerce (1-800-638-6582). By understanding the workers' compensation system in a particular state, contractors may discover whether their insurance companies or claims administrators are taking a proactive role to control all aspects of claim costs.

In most states, the means by which workers' compensation insurers request rate increases is through the *National Council on Compensation Insurance* (NCCI). NCCI is a private corporation to which member insurance companies submit loss and expense information.

Unlike other lines of insurance, workers' compensation insurance is sold by private insurers or state funds. And six states don't allow private insurers to even do business (Nevada, North Dakota, Wyoming, West Virginia, Washington, and Ohio). These states are known most appropriately as the monopolistic state funds. Contractors in these states must either purchase workers' compensation from a state fund or become qualified self-insurers, if permitted.

Thirteen other states have so-called competitive state funds. The problem is that the only competition is from private insurers. A contractor buys this mandated line of insurance either from the state or a private insurer and that's it. And for many small contractors, the private insurers won't even quote unless they are given other lines of business to insure. The states with so-called competitive state funds include New York, California, Arizona, Colorado, Idaho, Maryland, Michigan, Minnesota, Montana, Oklahoma, Oregon, Pennsylvania, and Utah.

Finally, in the remaining states the contractor has only private insurance companies from which to purchase insurance. If a contractor can't obtain private insurance, since this is a mandated type of insurance just like assigned risk automobile insurance, the contractor is dumped into an expensive assigned risk program. At all costs, most of the states' assigned risk programs should be avoided.

9.1.1   Safety is the key to cost control

The most important cost-control measure for a contractor is to run a safe jobsite. This is because the single largest cost of workers' compensation insurance is losses. Workers' compensation insurance costs are a function of loss experience more than any other line of insurance. This is because most contractors have an experience modification factor (*mod*) that tracks loss experience. As will be noted, loss experience stays with a contractor for many years through the experience modification factor. If a contractor does not have good loss experience, the options to control workers' compensation costs are rather limited.

Worse than this is basic business survival. Many owners will not allow a contractor to bid on work that has a high experience modification factor. Unless a contractor has efficiencies in other cost areas, a high experience modification factor translates into high workers' compensation costs that take away the competitive advantage when bidding on projects.

Regardless of the inconsistencies and the patchwork of workers' compensation insurance, the following tips are valuable to all contractors in the United States.

9.2   Tips to Reduce Workers' Compensation Costs

The effort expended in reviewing the workers' compensation program will be financially rewarding. For the tips that are technical in nature, a contractor should have his or her insurance agent or broker undertake action on the specific recommendation or tip.

9.2.1 Review experience modification factor (mod) for accuracy

Most contractors have been assigned what the insurance industry calls an experience modification factor. Through a complex formula this factor rewards or penalizes contractors for their loss experience. The mod is calculated by the NCCI and state bureaus by comparing three years of losses with three years of payroll and specific classification codes of employees. For example, assume a contractor's insurance policy is for the one-year term of January 1, 1995, to December 31, 1995. The insurance company will evaluate losses on all its claim files for the particular contractor valued as of December 31, 1994. Losses that have occurred during calendar years 1991, 1992, and 1993 are then sent to a rating board to be loaded into the formula that produces the mod.

Note, insurance companies don't consider just paid losses as losses. Any reserve (an amount set aside to pay future claims) can count just as much as if the insurance company had paid the claim! At renewal, contractors need to make sure that the insurance agent has confirmed that only the insured's losses are being loaded into the experience modification formula. The higher the losses, the higher the mod, which translates into higher premiums. The contractor's agent or broker should review high insurance company reserves and noncompensable and other open claims which should be CNP (closed no payment), as these reserves will cause the mod factor to increase.

Contractors need to question any unusual increase in the mod. Mods don't usually change significantly from year to year. If the experience mod changes by more than a 25-point swing in one year, something may be wrong.

9.2.2 Review rating classifications

Annually, the insurance agent should undertake an audit of workers' compensation policies to confirm proper rating classifications. Employees are assigned classifications from a complex classification scheme with over 500 classification codes. Contractors' workers' compensation premiums are a function of several variables. To develop a contractor's premium, the insurer begins by assigning every employee into a classification code. Then payroll attributed to that code is divided by $100 and multiplied by the applicable rate and the applicable experience modification factor. Fixed costs such as expense constants and fees to the state are applied also. Table 9.2 is an example of how the workers' compensation premium is determined by an insurer.

TABLE 9.2   Development of a Typical Contractor's Workers' Compensation Premium

Class (a)	NCCI code (b)	1994 NCCI rates* (c)	Payroll (d)	Payroll/ $100 (e)	Estimated premium class rate (f): (c) × (e)
Masonry	5022	$23.62	$15,000	$150	$3,543
Iron or sheet	5057	$47.44	$8,000	$80	$3,795
Sheet metal	5538	$25.55	$30,000	$300	$7,665
Pile drive.	6003	$36.69	$10,000	$100	$3,669
Clerical	8810	$.58	$50,000	$500	$290
Premium subject to experience modification					$18,962

Development of premium	Calculation
Experience modification (i.e., 1.15)	× 1.15
Standard or modified premium	= $21,806
State assessment (i.e., New York State—13.5 percent)	× 1.135
Rating board premium	= $24,750
Premium discount (5 percent of modified premium)	−`(1090)
Discounted premium	= $23,660
Expense constant	+ $160
Annual premium prior to payroll audit	= $23,820

* Rates per $100 of contractor's payroll

The preceding illustration shows the typical mechanics in the development of a workers' compensation premium. Thus, a contractor will note that reviewing classification codes is important. For instance, for one state as of 1994, classification code 5190 for electrical wiring was an $8.61 rate, code 3737 for electrical surface work without installation was a $10.58 rate, and code 3724 for electrical apparatus installation was a $16.46 rate. A search of the workers' compensation manual could lead to a possible lower classification. Many times it is not exactly clear what classification an employee belongs in. As illustrated in Table 9.2, the proper assignment of the employee is critical because there is a multiplier effect due to the application of the experience modification factor.

9.2.3 Have one workers' compensation policy

Contractors may have subsidiaries that are independent corporations each with their own workers' compensation policy. By combining policies, the charges for loss constants and expense can be eliminated. Most important, with a broader premium base, a contractor may be eligible for retrospective rating or other cash flow programs such as a large deductible program.

9.2.4   Check payroll limitations and other mathematical calculations

All contractors need to determine that proper payroll limitations are applied. Not included in payroll calculations are the portion of payroll attributed to overtime (just straight time applies), a portion of executive salaries, and the payroll for officers or directors that may not be covered by workers' compensation insurance.

Insurance companies can and do make mistakes. These mistakes can either be advantageous or costly to the contractor. As discussed, the premium is developed by multiplying every $100 of payroll *times* the rate for the classification code *times* the experience modifier. There are often other charges, including the premium discount, which are constants and are applied to the modified premium to develop the final premium. Check the math!

Paid losses, reserves, and payrolls all have an effect on the modification factor. The contractors need to confirm that the correct numbers are being used. In addition, they need to check to determine if proper classifications, rates, and payrolls were employed by the insurance company.

9.2.5   Simulate calculation of the experience mod

The insurance agent or broker needs to calculate the contractor's experience mod for accuracy, although this is complex for most contractors. The insurance agent should already have all the underlying data (payrolls, classification codes, and losses) necessary to simulate the mod.

Note, experience mods are a function of loss frequency more than loss severity. For instance, it is more favorable to have one loss of $50,000 rather than ten losses of $5000 each. The system is designed this way since any one of the ten $5000 losses can result in a more severe claim. In addition, a contractor with a frequency of losses probably does not have an effective safety program. In contrast, a contractor without frequency of loss but one severe loss could have a very good safety program. Accidents can happen and the workers' compensation system in the United States is designed not to significantly penalize contractors who have good safety programs.

9.2.6   Consider retrospective-rated insurance

As a contractor's business grows, a time will come when retrospective insurance should be considered. *Retrospective insurance is a cost-plus insurance contract.* Under retrospective-rated insurance the premium is paid at policy expiration rather than at policy inception and is a function of the contractor's own loss experience. Total insurance costs are divided between fixed and variable costs. Fixed costs include a minimum amount to the insurer known as the basic premium, a charge to limit losses to a certain amount so that a catastrophic loss is capped, and a charge to cap the total costs of the program. The variable costs component is that of losses—the greater the losses, the higher the premium, until the maximum premium is reached.

9.2.7 Evaluate state funds

Many times workers' compensation insurance from the state funds costs less than insurance from the private carriers. This is because the expense loadings are not as great and the state funds usually pay no premium taxes. Contractors in New York state, for example, have two choices when purchasing workers' compensation insurance. Contractors can either purchase workers' compensation insurance from private insurers or from the state facility called the *State Insurance Fund.* In New York, as with many other states, there is usually more than just the guaranteed cost insurance policy. For example, contractors in New York state have the following options in addition to the private insurance options:

- *Safety Groups.* Specifically, groups Number 469 (general contractors), Number 135 (painters), Number 420 (steel contractors), and Number 411 (sheet metal contractors) are examples. One way to reduce costs for the small contractor is to enjoy the economies of scale that the State Insurance Fund offers larger contractors. There are few reasons to be alone in the workers' compensation system when a contractor can be in a safety group that has a professional to manage the entire process. Many of the tips for controlling workers' compensation costs can be achieved via a safety group professional with aggressive claims handling.

    As one example, in New York, Allied Safety Management, Inc. administers several safety groups for the State Insurance Fund. One such group, Safety Group Number 432, incorporates several trades. Allied Safety Management has also begun a safety group in New Jersey as well.

    Contractors in states with state insurance need to determine if safety groups are available in their states as well. Several states are studying the success of the New York State Insurance Fund. Assuming that state funds can effectively manage medical costs and undertake investigations to deter fraud or to identify subrogation possibilities, safety groups remain an excellent way to purchase workers' compensation insurance.

- *Group Self-Insurance.*   Many states allow groups of a homogeneous nature to self-insure their exposures. For states that do not have state insurance funds, this is an alternative that needs to be evaluated by the contractor.

- *Incurred Loss Retro Programs.*   The New York State Insurance Fund has a retro program with a low basic premium. In addition, no loss conversion (claim handling charge) and no tax factor apply. Obviously, commercial insurers in New York can't match the retros that are available from this state fund. Contractors who are in states with either private insurers or the state funds need to obtain a quote from both.

9.2.8   Undertake proper preemployment screening, subject to ADA rules

In compliance with the American With Disabilities Act (ADA), now effective for contractors with 15 or more employees, a contractor needs to make sure the hired employee is the right person for the task. Contractors cannot discriminate in the hiring process because of a prospective employee's preexisting medical condition or previous workers' compensation claims.

9.2.9   Undertake a competitive bid and compare "premiums" and coverage

Contractors should have their insurance, including workers' compensation, be bid on every three to four years. In managing the bid process, the contractor should give two or three other insurance agents the opportunity to compete for the business. In addition, the contractor should have a direct writer that does not use the independent agency system to bid on the contractor's program. When obtaining a premium quote from another agent or broker, the contractor should ask what the bottom-line cost will be. Contractors need not let an insurance salesperson sell insurance based upon a so-called lower premium. This is because in workers' compensation insurance a premium can mean any of the following:

- Standard premium
- Manual premium
- Rating board premium
- Discounted premium
- Undiscounted premium
- Subject premium

- Nonsubject premium
- Modified premium
- Unmodified premium

And when obtaining quotes, the contractor should make sure that it is an apples-to-apples comparison. Besides the basic Part 1 (coverage A) protection, workers' compensation policies provide Part 2 (coverage B) for employers' liability insurance. Employers' liability insurance is coverage in the event an employee who is injured on the jobsite sues a third party. This third party can then claim that, due to the negligence of the contractor/employer, the employee was injured. A contractor's general liability policy will not respond because there is an exclusion for injuries occurring to employees.

When comparing coverage, one needs to make sure the limits of liability for Coverage B are the same. Additionally, other endorsements in the coverage comparison should include: other states' covers, Stopgap Employer's Liability, United States Longshore and Harbor Workers' Compensation Act, voluntary compensation, foreign voluntary compensation, and repatriation expense.

9.2.10   As allowed by law—pay small medical bills

Several states, including California and New York, specifically provide authority for all employers to absorb small medical bills that generally do not involve loss of time beyond the day or shift, or medical treatment beyond two visits. This is important because then the loss is not paid by the insurer which avoids its being counted toward the calculation of the experience modification factor.

9.2.11   Review loss-sensitive programs

For those contractors already purchasing workers' compensation insurance via a dividend program or a retrospective-rating program, are the additional or return premiums being calculated correctly? The contractor needs to check the math and be sure to check that the proper retro factors are being used.

9.2.12   Confirm that the insurer is assessing second injury funds

The purpose of states' second injury funds is to encourage contractors and other employers to hire the physically handicapped by protecting employers against a disproportionate liability in the event of a work-connected injury to the handicapped worker.

The law limits a contractor's exposure to 104 weeks of workers' compensation benefits and medical expenses. In all states except New York, the employer must have knowledge of the prior permanent physical impairment. Thus, once an employee is hired, documenting previous injuries is necessary.

Three prerequisites must be presented in order for a compensation case to qualify under second injury laws. They are:

1. The employee must have a preexisting permanent physical impairment which is, or is likely to be, a hindrance to future employment.

2. The worker must have, subsequent to being hired, a compensable injury or occupational disease at work which by itself results in a permanent disability.

3. The combined resulting disability must be substantially and materially greater than that caused by the second injury alone.

Effect on the contractor. Because each claim remains in an employer's experience rating for three years, the difference between a claim of $50,000 and $75,000 or $85,000 and $150,000 can result in substantial premium savings. Under the law, savings can be retroactive once a claim is established as a second injury claim.

9.2.13 Buy insurance from an insurer with a successful managed care program

In some states the insurance company and employer can demand that an injured employee seek treatment from a particular physician. In these states, contractors should make certain that the insurer is only paying claims when the injured employee sees a physician from within an authorized preselected list and not the employee's own physician.

In other states, an employee has full freedom of choice for whom the treating physician will be. In these states, it is illegal to direct the employee to a specific physician.

For all states, especially those in which the injured employee has freedom of choice, it is essential that the insurer has a managed care program. Even in states in which the employee has freedom of choice in terms of physicians, the insurer should still be undertaking efforts to control the cost of the claim. Contractors should review insurers that have utilization review services.

A utilization review program evaluates case records and bills related to treatment to determine medical necessity of proposed treatment, identify cost-effective alternatives, confirm work relatedness of injury, and facilitate the worker's timely return to work. Utilization review services include hospital review rectification for admissions, concurrent review, discharge planning, retrospective review, and specialty reviews for certain types of treatment such as chiropractic physical therapy and expensive diagnostic testing.

Medical case management and vocational rehabilitation services. A medical case management and vocational rehabilitation program provides access to nurse consultants, vocational counselors, rehabilitation specialists, and case management specialists dedicated to getting the employees back to work. Their medical knowledge and vocational expertise, coupled with the ability to assess cases and motivate recovery, result in faster closing of claims.

Preferred provider networks. Most insurers have established preferred provider networks of physicians and hospitals which combine discounts with adherence to utilization review protocols. Employers should avail themselves of the option of using these networks whenever possible. In addition, some providers will visit the worksites to more closely understand the needs of the employers.

Implement light duty and return-to-work programs. The highest dollar losses paid within workers' compensation are for the indemnity payments. Returning the employee to a work environment, even if it is not the employee's regular occupation, may then motivate the employee to get back to work.

Programs can reduce costs by decreasing injury frequency and severity, decreasing work delays and interruptions, reducing charges for permanent disability, decreasing charges for injuries, and reducing employee turnover.

9.2.14   Request surveillance checks if fraud is suspected

This activity is conducted when a red flag is cited. Cases usually involve extended disability on which the medical information does not coincide with the employee's reported activity, and wrongdoing is suspected. Since the employer and coemployees of an injured worker may have hands-on knowledge of the injured worker, if any fraud is suspected, the contractor should let his or her insurance broker or agent and the insurance claims representative know. The contractor needs to send a certified letter to the insurance company claims representative and a copy to the insurance agent or broker. States such as California that have had problems with workers' compensation fraud mandate that insurers have investigative units. Contractors interested in taking a proactive involvement with suspected workers' compensation fraud should contact the National Insurance Crime Bureau (NICB) at

(703) 430–2430 and ask for a free brochure entitled *Indicators of Workers' Compensation Fraud.* Last, any contractor that suspects fraud should call the NICB's 24-hour hotline at (800)TEL–NICB. And contractors should be sure to ask NICB for its wall poster which should be placed on a jobsite so that all employees are aware of the hotline. Some contractors hang this notice next to other required posting notices such as the workers' compensation notice of compliance.

9.2.15 Develop incentive programs

An incentive program designed to give cash rewards can be a big plus in controlling workers' compensation losses. Departments can compete with each other and be awarded based on exceeding a certain period of time without a lost time accident.

Many contractors that have improved their safety record, and thus controlled workers' compensation costs, have used these psychological motivational or behavioral modification techniques. The insurance industry has developed the experience modification factor to motivate all employers to improve performance. In this vein, contractors should attempt to apply their own motivator to modify employee behavior.

9.2.16 Be active in claims management— keep track of the claims

If an employee is injured, keep track of the claim and keep in contact with the insurance company. The contractor should request that the insurance agent or broker obtain the name of the claims examiner who will be paying the medical bills on the employee. Then, the contractor needs to communicate on a regular basis with this insurance company employee until the injured worker returns to work.

Encourage the insurer to make use of independent medical examinations. If wisely used, the *independent medical examination* (IME) remains one of the most effective strategies for managing workers' compensation claims. An independent medical exam is important, especially in those states in which employees are under the control of their own physicians. Often, after the results of an IME, an insurer can then request a formal workers' compensation hearing to discuss the medical findings and stop workers' compensation benefits. In practice, by this point many employees with subjective complaints only often go back to work.

Confirm that index checks are being done.   One step in the investigative process for insurers when employees submit claims is for the insurer to run an index check. The index system is a central database whereby all employees and third parties having claims with insurers are maintained.

Consider subrogation recoveries. In order to obtain work from owners, contractors frequently enter into hold-harmless agreements of some nature. Any recovery that an insurer realizes should be reflected in the mod. Contractors need to demand that a recalculation of the mod be undertaken. A subrogation recovery may be possible if an employee is injured not due to the sole negligence of the employee or employer. For example, suppose an employee at a jobsite is injured due to the negligence of another employer. A subrogation action exists which insurers don't always pursue due to informal agreements.

More common, however, is when an employee at a site is injured by a piece of equipment not working correctly. In these cases, the employee will collect compensation but sue in tort the negligent party. In such cases, a contractor needs to determine that the carrier correctly files the lien to recover monies paid out on the workers' compensation claim.

9.2.17 Get certificates of insurance for subcontractors

Workers' compensation laws in many states provide that contractors shall be responsible for payment of benefits to employees of uninsured subcontractors. Contractors need to make sure that any subs working the site have adequate insurance.

9.2.18 Consider participation in wrap-up insurance programs

At times contractors may be asked to bid "ex-insurance." One needs to consider participation in these programs. At some time, most contractors will participate in a construction wrap-up. If one was ever asked to bid without insurance, then one participated in a wrap-up.

In a wrap-up, the owner, construction manager, or general contractor and subs are the insureds. A contractor can also provide wrap-up insurance on a turn-key project. A wrap-up insurance program usually provides workers' compensation and general liability insurance for a construction project at one site. In a wrap-up, the owner, construction manager, general contractor, subcontractors, engineers, and architects are combined insureds under one jumbo insurance program.

The owner or general contractor pays the total premium for all the insureds and receives the benefits of any return premiums for favorable loss experience. Wrap-ups are usually not based on guaranteed costs insurance (premium at policy inception), but are retrospectively rated. In effect, it's like a cost-plus construction contract, whereby the premium paid by the owner is a function of the loss experience on the project.

Since insurance is being provided en masse, all bidders on the project are asked to bid on an ex-insurance basis, thus saving the owner the cost of insurance as one cost component of the overall bid. Some of the advantages of the wrap-up approach for small contractors are reduced insurance costs, known and provided coverage, loss control assistance, and less administration.

9.2.19 Large contractors need to evaluate self-insurance

Contractors paying more in insurance premiums than their cost of claims should consider self-insurance. The largest part of the total cost for a self-insured is the cost of claims. This should be estimated by a casualty actuary because estimating claim costs is very difficult. An incorrect estimate could lead to a wrong and costly decision about self-insurance.

The major cost components of self-insurance include catastrophic excess insurance, states' assessments, security, claim handling fees, and loss prevention services.

9.3   Reevaluate Safety Programs

The cornerstone in controlling workers' compensation costs is effective loss control. A comprehensive loss prevention and loss reduction program is the best method to control workers' compensation costs. A contractor needs to invest in pretested protective equipment, undertake a study of job hazards, and rotate employees. For example, a contractor should not let an employee stay on a jackhammer for extended periods of time. Contractors wanting additional information on safety devices for employees should contact the Safety Equipment Institute at (703) 525–3354 and ask for its free certified product list of safety equipment.

The ability to prevent injuries is the cornerstone of an effective safety program and the one guaranteed way to keep workers' compensation costs low. Many insurers have loss-control engineers that will help a contractor prevent losses. Areas to have the loss-control engineer evaluate include: review of the contractors' OSHA reports, employee training manuals, safety handbooks, posters, tools, and protective equipment.

Acknowledgment

The author thanks Dr. Greg Johnson of Ernst & Young's San Francisco location for providing technical review.

10

Environmental Impairment Liability Insurance for the General Contractor

David J. Dybdahl, CPCU, ARM
Willis Corroon Corporation

10.1 Introduction

Environmental liability poses a unique risk exposure to general contractors, in particular, general contractors performing environmental remediation work. Their liability may arise out of the common law principles which are based on custom and court decisions.

Environmental liability may also be imposed by statutory law created by legislative bodies. Environmental laws are being enacted at the federal, state, and local levels and include such acts as the *Resource Conservation and Recovery Act* (RCRA) and the *Comprehensive Environmental Response Compensation and Liability Act* (CERCLA). Since 1981, the volume of federal environmental regulations alone has increased nearly 50 percent. These laws may impose both civil and criminal liability on contractors and others for their past and present environmental activities. Under these laws, contractors may face retroactive, strict, and joint and several liability. The pollution events resulting in civil and criminal liability may result not only in the payment of fines and cleanup costs, but also in third-party claims for bodily injury or property damage.

10.2 The Risk Management Process

In the classical risk management framework, there are four steps that are used to effectively manage any risk. These are:

1. Risk Identification

2. Loss Prevention

3. Loss Control

4. Loss Financing

Risk identification can be defined as the process of identifying, analyzing, and measuring a particular loss exposure. Loss prevention can then be defined as methods to prevent a loss from occurring. Loss control addresses the need to contain and minimize the adverse affects of a loss once it has been incurred. Finally, loss financing addresses the funding of losses while maintaining the entity's long-term financial solvency. This chapter will devote most of its discussion to the financing of environmental losses from construction activities.

10.3 Past Environmental Liability

Environmental cleanup legislation can be a source of liability for contractors who may be classified as *potentially responsible parties* (PRPs) at superfund sites. PRPs include present and past owners or operators of the site, from those who arranged for hazardous materials to be transported to or disposed of at the site, to those who generated the materials deposited at the site. A contractor can be classified as a PRP by disposing of waste at the site during construction activities; a contractor may also be a generator of wastes if hazardous materials are brought onto the site by the contractor or they are moved from a contaminated to a clean portion of the site. Also contractors may be classified as PRPs by exercising a measure of control over activities that make them "operators" for purposes of CERCLA liability. Retroactive liability may hold PRPs responsible for the cleanup costs at a site where disposal occurred in the past despite the PRP's compliance with all laws and regulations at the time the waste was generated or disposed of on the site. PRPs are also subject to strict and joint and several liability standards which could result in disproportionate responsibility for the costs of a site cleanup.

To finance the environmental cleanup and potential bodily injury and property damage claims from prior activities, risk managers have few available options. A responsible party at a superfund cleanup cannot prevent that loss from being incurred, and it is not possible to purchase new insurance coverage for loss which has already occurred. Trying to obtain liability insurance to pay for the actual cleanup of a superfund site is analogous to insuring a burning building.

However, firms may be able to recover the costs of cleanup and third-party claims under old commercial general liability policies in effect prior to 1986. These older policy forms do not include the absolute pollution exclusion now found in most policies. To date, court decisions on coverage disputes have been mixed on allowing claims for pollution damages. Contractors who are named as PRPs should also consider that the contracts they were working under at the time they disposed of or generated waste might respond for the funding of their share of a superfund cleanup. In particular, federal government contracts have successfully claimed remediation costs as allowable costs of contract performance in cases where the costs of claims or cleanup are not the result of violation of any laws.

10.4 Present and Future Environmental Liability

For contractors, the environmental cleanup movement should be viewed as both a source of revenue and a source of potential liability. The opportunity to make money on cleanup contracts has resulted in the development of environmental remediation divisions within many large contracting companies as well as the formation of smaller firms to compete for numerous subcontracting opportunities. For these contractors, the profit potential must be carefully measured against the inherent risks related to environmental remediation activities. One of the major exposures these contractors face is liability arising from an accidental release of pollutants from the cleanup site. Under tort liability, third parties may sue contractors who are working on government projects. Under certain environmental laws, these contractors also face strict, joint, and several liability. Thus, an individual contractor may bear the full responsibility for contamination—regardless of the contractor's negligence or his or her percentage of contribution to the contamination.

For ongoing operations, the risk manager has many more proactive options available to effectively address the pollution liability exposure. Risk assessments should be used to identify and quantify potential liabilities that could be incurred from ongoing operations. Based on the analysis of the risk assessment, loss prevention strategies can be formulated to include risk avoidance as an option where the exposure is too great. Contingency planning for loss exposures can be an effective method of loss control. Hopefully, the ongoing operations of the contractor do not present a burning building scenario, allowing the risk manager access to the entire range of loss financing strategies.

10.5 Risk Financing

There are basically two broad classifications of risk financing tools available for payment of environmental losses. Risks can be financed either through risk retention or risk transfer. It is very common to see both approaches used to address a particular loss exposure as in the case of an insurance policy with a deductible.

Risk retention can be thought of as paying for losses out of pocket, but this should not suggest that this technique is unstructured or that expense associated with retentions cannot be predicted. Risk retention includes such tools as deductibles and planned self-insured retentions in insurance programs, or a conscious decision not to purchase insurance where losses are predictable or where the risk associated with an exposure is low. In theory, losses that have a high frequency and low severity can be retained without jeopardizing the financial strength of the entity. The reason for this is the increased predictability of these losses which allows the firm to better plan and budget for the losses.

In contrast, losses with a low frequency and potentially catastrophic severity are unpredictable and are often beyond the budget capacity of the firm and should, where possible, be transferred to another party. Indemnification by a third party and the purchase of insurance are the two most commonly used forms of risk transfer. Remediation contractors should be aware of the indemnification sources and insurance products that are being developed to meet the dynamic environmental exposures that they face. Contractual indemnification clauses can shift liability from the contractor to site owners, government bodies, subcontractors, and others involved in the site cleanup. The contractor should have all indemnity provisions in contracts reviewed by legal counsel, and a valuation of the financial ability of the indemnifying party should be made as a part of the overall consideration of the protection afforded by such provisions. Environmental contractors should pay careful attention to the indemnification and insurance provisions contained in cleanup contracts when designing their risk financing programs.

10.6 Contractor Pollution Liability Insurance Considerations

Developing a comprehensive contractors' pollution liability insurance program is a very complex task. The complexities are brought about by the fact that much of the insurance that covers environmental exposures is written by specialty underwriters providing only pieces of the necessary coverage. Turf battles for the various insurance products, even within the same company, sometimes create a situation where multiple insurance policies, each covering small pieces of the exposure, must be purchased to complete the insurance coverage matrix. For example, an insurance program written for a firm providing design-build or single-source environmental contracting, including asbestos abatement services, could include as many as eight basic insurance coverage parts. All of the coverages must be interfaced with each other to properly structure an integrated insurance program.

The following is an outline of the individual coverages necessary to build an insurance program for remediation firms. The insurance coverages described here are available in the marketplace today. Since the insurance market is expanding at an increasing rate, it is difficult to compose a picture reflecting current conditions that will still be totally accurate weeks or months later. The areas which are most susceptible to change include the available limits of liability and underwriting guidelines of the various markets. The basic coverages, however, will remain unchanged for a longer period of time.

It is also important to note that some of these policies are variations of the same basic coverage form, but exclusions in the policies may make the purchase of more than one policy necessary. Particular attention must be paid to the rating basis on all policies to avoid duplicating premiums. The premium cost for some of these coverages may exceed the normal operating margins of a contracting firm; consequently, failure to coordinate rates and revenue streams could cause severe cash flow problems for a small firm following a premium audit at the end of the policy period.

Nearly all liability policies written for contractors or engineers working in the environmental area are written on a claims-made basis, except for conventional commercial general liability and automobile liability policies that are available on an occurrence form. Careful attention must, therefore, be given to retroactive date and extended discovery provisions of the policies to assure continuity of coverage from one policy period to the next and to address the site owners' concerns about completed operations coverage.

10.6.1 Commercial general liability insurance

The *commercial general liability* (CGL) insurance policy is the third-party litigation insurance coverage relied on by most businesses. This policy provides coverage for bodily injury and property damage claims arising out of the insured's operations, premises, completed operations, and products, plus the defense of those claims.

Faced with increasing pollution liability losses under insurance policies issued in prior years, the insurance industry rewrote the CGL policy in 1986 and again modified the pollution exclusions in 1988. As part of these revisions, the pollution exclusion was modified to prevent insureds from recovering the costs of environmental claims from CGL underwriters. The new standard exclusion reads:

1988 ISO Pollution Exclusion

1. "Bodily injury" or "property damage" arising out of the actual, alleged or threatened discharge, dispersal, seepage, migration, release or escape of pollutants:

   a. At or from any premises, site, or location which is or was at any time owned or occupied by, or rented or loaned to, any insured;

   b. At or from any premises, site, or location which is or was at any time used by or for any insured or others for the handling, storage, disposal, processing, or treatment of waste;

   c. Which are or were at any time transported, handled, stored, treated, disposed of, or processed as waste by or for any insured or any person or organization for whom you may be legally responsible; or

   d. At or from any premises, site, or location on which any insured or any contractors or subcontractors working directly or indirectly on any insured's behalf are performing operations:

      (1) if the pollutants are brought on or to the premises, site, or location in connection with such operations by such insured, contractor, or subcontractor; or

      (2) if the operations are to test for, monitor, clean up, remove, contain, treat, detoxify or neutralize, or in any way respond to, or assess the effects of pollutants.

   Subparagraphs a and d (1) do not apply to "bodily injury" or "property damage" arising out of heat, smoke, or fumes from a hostile fire.

   As used in this exclusion, a hostile fire means one which becomes uncontrollable or breaks out from where it was intended to be.

2. Any loss, cost, or expense arising out of any:

   a. Request, demand, or order that any insured or others test for, monitor, clean up, remove, contain, treat, detoxify, or neutralize, or in any way respond to or assess the effects of pollutants; or

   b. Claim or suit by or on behalf of a governmental authority for damages because of testing for, monitoring, cleaning up, removing, containing, treating, detoxifying or neutralizing, or in any way responding to or assessing the effects of pollutants.

Pollutants means any solid, liquid, gaseous, or thermal irritant or contaminant, including smoke, vapor, soot, fumes, acids, alkalis, chemicals, and waste. Waste includes materials to be recycled, reconditioned or reclaimed.

Although there is supposedly limited pollution coverage for products and completed operations loss exposures within the CGL policy, in the final analysis, insureds with an exposure to pollution liability losses will almost always have no coverage in any of the standard policies. Pollution exclusions with a similar impact are also found in the automobile liability and professional errors and omissions liability policies. To fill the gap caused by this exclusion, the purchase of specialized pollution liability coverages is necessary.

10.6.2   Pollution legal liability insurance

Pollution legal liability (PLL) insurance was the first of the specialty environmental coverages and is designed to provide coverage for offsite third-party bodily injury and property damage caused by environmental impairment that results from either sudden or nonsudden pollution incidents taking place at named sites. Unless modified by endorsement, PLL insurance does not provide coverage for cleanup of an insured's site, liabilities associated with the use of nonowned disposal sites, and liabilities resulting from known environmental damages. In many instances PLL is mistakenly required under the terms of a contract or requested by a remediation contractor. What might be covered for the contractor is the payment of a third-party claim for damages to the property of a neighbor due to the release of pollutants from the contracting site onto the neighbor's property.

10.6.3   Contractor's pollution liability insurance

Contractor's pollution liability (CPL) coverage was developed in response to the pollution exclusion in commercial general liability policies. In that sense, it might be considered a stopgap coverage for general liability exposures. Unlike general liability policies, CPL insurance cannot be compared to all-risk protection. It provides specific pollution liability and defense coverage for bodily injury, property damage, and environmental cleanup costs for pollution conditions arising from the contractor's described operations. Coverage can be purchased on either a blanket/reported sites or project-specific basis.

CPL coverage has its roots in environmental impairment liability insurance which also was the basic form from which pollution legal liability policies were derived. In spite of the shared similarities, buyers should be aware that a standard PLL policy provides little or no insurance protection for contracting activities. A CPL policy, on the other hand, modifies the coverage to more closely reflect the exposures of a contracting firm.

Since the CPL is a stopgap coverage that addresses only pollution claims arising from insured contracting operations, it is necessary to purchase a commercial general liability policy for other contracting exposures. While there are a limited number of insurers that can write CGL and CPL coverages on a combined form, it is more common for contractors to purchase these policies from separate underwriters and in separate policies.

10.6.4 Architect's and engineer's errors and omissions insurance

For design professionals, the purchase of *errors and omissions* (E&O) coverage has historically been necessitated by a special exclusion typically added by endorsement to CGL policies written for such risks excluding coverage for claims of injury of damage arising out of design error. To fill this gap in coverage, the purchase of a professional E&O policy is necessary. The typical insuring clause of these policies agrees to pay on behalf of the insured for negligent acts, errors, and omissions arising out of the rendering of a described professional service.

The majority of E&O policies on the market today contain an exclusion for claims arising out of a pollution incident. It is now possible, however, to purchase E&O coverage from specialty markets that provides coverage for pollution claims as part of the traditional errors and omissions coverage.

Unlike the CPL policy, an architect's and engineer's errors and omissions policy is a practice form that provides coverage for all professional liability exposures, not just those related to pollution incidents. This form replaces other E&O policies of the architectural or engineering firm that need coverage for pollution exposures.

10.6.5 Environmental consultant's E&O insurance

A number of new custom-tailored errors and omissions policy forms have been introduced to accommodate firms that provide environmental remediation services. These policies usually take the form of a traditional errors and omissions policy, extending professional liability coverage to claims arising out of a pollution incident by amendment or elimination of the pollution exclusion. It is interesting to note that the new policy forms are often less expensive than the traditional forms with the pollution exclusion.

10.6.6 Asbestos abatement liability

Asbestos abatement liability insurance policies typically track the CGL policy form. Most provide coverage for asbestos abatement operations by amendment of the pollution exclusion in the standard CGL policy. Both the CGL and asbestos abatement liability policy use payroll and receipts as the rating basis to determine premium. Since these coverages are redundant for premium computation purposes, each of the carriers should be aware of the existence of the other policy and should issue endorsements that enable the insured to avoid paying double the necessary premium. Policy forms and insurer integrity vary a great deal within the asbestos abatement liability insurance market. The advice of an expert broker is highly recommended on this line of coverage.

10.6.7 Asbestos consultant's E&O insurance

Coverage for asbestos consultants is usually written under a conventional professional E&O liability insurance policy, with the pollution exclusion eliminated or amended to the extent necessary to provide coverage for professional errors, acts, or omissions arising out of the design of asbestos projects. Asbestos consultant's E&O coverage may be redundant for some risks, however. Such specific insurance would be unnecessary if the insured has an environmental consultant's professional pollution liability policy or some other form of errors and omissions coverage with sufficiently broad wording so as not to exclude the asbestos hazard.

10.6.8 Combined policy forms

Some underwriters are introducing specialty policy forms combining pollution coverage with either professional liability or general liability insurance. The principal advantage of these forms is that they are lowest in cost. Since all the coverages share one limit of liability, the underwriter has lower total exposed limits and can charge a lower premium than where separate policies are issued. Another advantage for the buyer is that a single policy covering pollution as well as other liability exposures should eliminate the problem of disputes between insurers over whose policy covers a particular loss.

A major disadvantage of combined forms is that they do not lend themselves to accommodating reinsurance or excess liability coverage. Since pollution coverage is written by a limited number of specialty markets, there are very few reinsurers or excess liability underwriters willing to insure pollution exposures. Therefore, the available limits of liability will remain relatively low, not only for the pollution coverage, but for the traditional general liability and professional liability coverages as well. Another disadvantage of combined forms stems from what may be characterized as aggressive marketing of these products by certain underwriters. These insurers often require buyers to disturb long-standing relationships with the incumbent underwriters on their conventional insurance programs, only to lose the combined coverage forms at the first sign of a hard insurance market. This can place buyers in a distressed position requiring them to find a replacement insurance program for a difficult class of business at a time when market conditions are least favorable.

10.6.9   Commercial automobile liability

General liability policies and professional liability insurance policies exclude claims arising out of the maintenance, use, or operation of a motor vehicle. Coverage for such claims is typically written under the standard Insurance Service Office *business auto policy* (BAP), which has a pollution exclusion. That exclusion is particularly significant for any firm which transports materials that could lead to environmental damage and consequent liability claims or government-mandated cleanup. It is also important to note that the MCS–90 endorsement, which is required under the Motor Carrier Act for certain insureds transporting hazardous substances, does not provide environmental insurance for the transporter. The MCS–90 endorsement is an agreement by the underwriters to pay the costs of environmental cleanup and third-party claims resulting from over-the-road transportation risks, but it requires the insured to reimburse the underwriters for any cost incurred. An additional endorsement to the automobile policy is required to transfer this risk to the insurer.

The rating of automobile liability insurance is based in part on the number of owned or leased vehicles. Consequently, its coordination with the other liability lines is not as critical from the standpoint of avoiding overlapping premiums.

10.7   Structuring the Program

Piecing together the coverages necessary to address the risks of environmental contracting firms has developed into something of an art form. Proper structuring of the insurance program assures insurance protection without gaps or overlaps in coverage and avoids unnecessary duplicate premiums for redundant coverages. As engineering and contracting businesses throughout the country see the flow of money toward environmental cleanup, many of these firms are exploring the ramifications of entering this field. As mentioned earlier, the astute among them recognize liability and insurance as among the more significant barriers to entry.

10.7.1   Subsidiary versus master plan approach

To deal with the insurance barrier, the creation of an environmental contracting subsidiary produces a vehicle that can be used to structure an insurance program for the environmental contracting work. If done properly, this approach allows the firm to develop completely separate insurance programs for the nonenvironmental work and the environmental work.

An alternative approach is a master insurance program combining the environmental contracting exposures with other exposures insured by the firm. This may require shifting the master program to an insurer that has the ability to handle the environmental coverages.

10.7.2 Blanket versus specific coverage

Another question that arises when setting up a remediation contractor's insurance program is whether to use blanket or project-specific insurance coverage for the environmental work. Under the blanket approach, the firm obtains coverage for all of its described operations. Depending on the insurance policy under review, the coverage can be either true blanket coverage or something fairly unique to remedial action contracting: blanket coverage for the described operations at specified sites named on the policy. The latter may provide some measure of automatic coverage on new sites for a period of time, pending the reporting to the insurance company for the specified site.

The true blanket approach almost always produces the lowest rate for insurance, which is then applied against the total revenues of the named insured to arrive at the premium. Firms doing a significant amount of environmental contracting will probably find it beneficial to utilize both blanket and project-specific insurance programs, depending on the situation. If the dual program approach is undertaken, the coordination of these policies becomes critical to avoid duplication of coverage and premiums.

10.7.3 Wrap-up plans

Another option exists to address the insurance requirements of remedial action contractors involved in environmental work. This is the project wrap-up approach, which was pioneered by Willis Corroon. Under a wrap-up program, general contractors provide the primary casualty insurance protection on the entire job for themselves as well as their subcontractors. Typically, such a program includes general liability, contractor's pollution liability, professional errors & omissions, and engineer's pollution liability insurance. When there is an asbestos exposure, asbestos abatement liability and asbestos consultant's errors and omissions coverages may also be purchased.

Despite the advantages of the wrap-up approach, it is generally feasible only for those firms on jobs exceeding $1 million in annual receipts. Many of the necessary coverages mentioned above carry minimum premiums of at least $25,000 each; consequently, annual receipts associated with the job must be sufficiently large to absorb the overall minimum premium charge.

10.7.4 Determining coverage requirements

The scope of operations and services provided by a firm determines which insurance coverages are necessary. For example, specialty contractors providing only engineering services may need to purchase only professional pollution errors and omissions, automobile liability, commercial general liability, and workers' compensation coverages.

At the other end of the spectrum, a firm providing design-build environmental contracting services, including asbestos abatement, should purchase commercial general liability, contractor's pollution liability, professional errors and omissions with pollution and asbestos liability, asbestos abatement liability, automobile liability, and workers' compensation coverages.

If an environmental contracting firm is a subsidiary covered under the parent corporation's insurance program, that program should include, in addition to the specific coverages for the environmental contracting operations, commercial general liability, professional liability, automobile liability, workers' compensation, and probably umbrella liability insurance policies.

Management of the environmental liability risks that contractors face requires a comprehensive approach. Contractors must first identify all sources of environmental liability which may arise out of their activities. Insurance for contractors' pollution liability exposures is generally available in the marketplace today. The available limits of liability are increasing rapidly, and rates continue to fall. Due to the diversity of underwriters providing various policy forms and the complexity of the market, remediation contractors' insurance programs are difficult to administer. For relief from some of that burden, a contractor should be able to rely on a qualified environmental risk management consultant or broker to access new and simplified packages and to coordinate coverage with other sources of risk financing.

10.8 Adverse Legal Climate for Contractors

Legal liability based on environmental damages arising out of releases or discharges of pollutants can be a threat to the general contractor on nearly any kind of job.

A contractor that inadvertently encounters hazardous waste at a jobsite can be liable under CERCLA (superfund) for either operating the facility where the hazardous wastes are located, for transporting the wastes to the site, or for generating wastes. In the case of *Kaiser Aluminum & Chemical Corporation v. Catellus Development Corp.*, 979 F.2d 1338 (9th Cir. 1992), a contractor that was excavating land for a new housing development was sued under the theories of CERCLA liability when all he did was move soil from one part to another part of the same construction site.

A contractor that encounters hazardous wastes at a site and subcontracts with someone else to remove it, treat it, or transport it, may be liable under superfund for arranging for transportation of the wastes according to the decision rendered in *Environmental Transportation Systems, Inc. v. Ensco, Inc.,* 763 F. Supp 384 (C.D. Ill. 1991).

The first line of defense for the contractor is a well-drafted contract that contains terms and conditions protecting him or her from environmental liability. Several key contract provisions which can be particularly critical to allocation of risk among the parties include the following:

10.8.1 Scope of work

A well-defined scope of work can be critical to avoiding disputes as to whether the contractor is required to address environmental problems that may be encountered on the worksite, such as asbestos, lead paint, or hazardous wastes. If the contractor does not intend to include such work within the contract scope, this should be clearly stated.

A separate contract provision may provide for what is to happen in the event that an unanticipated environmental problem is encountered. Is work to stop? Who is to be responsible for addressing the problem? Is a change order to be issued? If it is the owner's duty to issue a change order, will the contractor be entitled to an equitable adjustment for a suspension of work while the owner has another contractor deal with the situation? These issues should be addressed in the contract.

Some contracts include a change order clause that gives the owner the authority to issue a change order directing the contractor to perform additional work—even where this may constitute a significant change to the type of work involved, or to the method or manner of performance. Such a provision may be an invitation for trouble for the contractor since it can allow the scope of work to increase into areas in which the contractor lacks expertise and experience. It may even result in work for which the contractor lacks adequate insurance coverage.

If the scope of work states that the contract is not to include any hands-on remedial action work, a paragraph may be added to the scope of work clause stating:

> It is understood and agreed that Contractor is not, and has no responsibility as, a handler, generator, operator, treater or storer, transporter, arranger for transport, or disposal of hazardous or toxic substances found or identified at site, and that Contractor shall not be responsible to under-

take or arrange for the handling, removal, treatment, storage, transportation, and disposal of hazardous substances or constituents found or identified at a site.

10.8.2 Standard of care

If a contract is silent concerning the standard of care that will be exercised by the contractor, it is generally presumed that the contractor will be held to the ordinary negligence standard. That means the contractor is expected to perform his or her work using that degree of care and skill ordinarily exercised under similar conditions by a contractor practicing at the same time in the same or similar locality. When the contract contains a specific clause setting forth a standard of care, the contractor needs to be sure to closely follow the language of the clause.

In recent contracts where environmental design and remediation were involved, some owners have sought to increase the responsibilities of the consultant or contractor by raising the standard of care. An example of one such clause to be avoided is:

> Consultant/Contractor represents that the services will be performed in a manner consistent with the highest standard of care, diligence, and skill exercised by nationally recognized consulting firms for similar services.

Warranting to perform in accordance with this higher standard could subject the consultant or contractor to liability even though all work has been performed in accordance with generally accepted standards within the industry. Furthermore, this failure to meet the required standard of performance could constitute a contractual liability that may not be covered by insurance policies.

An appropriate standard of care clause is as follows:

> Contractor will perform its services using that degree of care and skill ordinarily exercised under similar condition by contractors performing with at the same time in the same or similar locality.

10.8.3 Warranties, guarantees, and representations

Avoid warranting that at the conclusion of the work the project site will be 100 percent cleaned up of all contaminants. It may be impossible to guarantee that the entire site will be cleaned up as a result of the contractor's work, even if the work is performed in accordance with the standard of care and other terms of the contract.

On the other hand, the owner might be satisfied by an agreement that any site remediation work that is done will meet specified federal or state requirements. Limit such agreements to apply only to the actual material that is treated or removed. Avoid agreeing that the site location from which the material was removed will be clean to federal and state standards after the cleanup operation is complete. The safest course of action in such situations is to limit the warranty to the representation that all work will be performed in accordance with the plans and specifications.

Do not warrant that work will conform to all EPA standards, policies, and guidance documents. There are too many guidance documents published by EPA for a contractor to read them, let alone be able to understand and apply them. Moreover, these items are constantly changing, and at any given time they are not entirely consistent with each other.

It may not ever be appropriate to warrant that the contractor will comply with the laws, regulations, and ordinances since a law or regulation could be inadvertently violated even with the exercise of the utmost good faith and diligence. Instead, it may be prudent to represent only that such laws and regulations will not be negligently violated. An example of such a clause is as follows:

> Contractor and Client will use reasonable care to comply with applicable laws in effect at the time the services are performed hereunder, which to the best of their knowledge, information, and belief, apply to their respective obligations under this Agreement.
>
> Client shall pay for any reasonable charges on written change orders from Contractor for services, modifications, or additions to facilities or equipment required on the part of the contractor to comply with laws or regulations that become effective after the execution of the Agreement, and any change order to this Agreement.

10.8.4 Site information provided by client

The site owner is in a better position than the contractor to know the site conditions, utilities, contaminants, wastes, and prior uses. The client should be required to provide all known site data and information to the contractor. A contract clause may be included that establishes an affirmative representation by the client that all known site information has been provided and that the contractor may rely upon it accordingly in performing the work.

10.8.5 Disposal of contaminated material

If, during the course of performing the work, the contractor encounters hazardous wastes that must be removed from the jobsite, the likelihood of incurring liability under the various environmental laws can be reduced by:

- Stopping the work and requiring the client or owner to take appropriate action to remove the wastes

▪ Where this is not feasible, performing the removal action work, *but:*

Requiring the client or owner to designate the location for offsite disposal of hazardous wastes and

Requiring the client or owner to sign any disposal manifest that may be required by the federal and state agencies for shipment or disposal of hazardous wastes

Under CERCLA, liability may be imposed on a contractor that selects the disposal site and arranges for wastes to be disposed of there. On the other hand, contractors and transporters may have a defense to CERCLA liability if they can demonstrate that the client selected the disposal facility and that the contractor followed all applicable laws and regulations (including the preparation and filing of manifests) when taking the wastes to a permitted facility.

The contractor should not sign the manifest as the generator of the waste. Avoid any appearance of becoming the owner or generator of wastes. If the contractor signs a manifest for the client it should be done only as the agent of the client or on behalf of the client. In this same regard, if the contractor removes barrels or similar items containing hazardous wastes, those barrels should be clearly marked with the owner or client's name and not the contractor's. This will be important if these barrels are later found at some facility that became a cleanup site in its own right. The party whose name is on the barrel is the most logical party to get the superfund notice letter from EPA.

10.8.6   Indemnification

A well-crafted indemnification agreement can be a great protection for the contractor. Not so long ago, it was common for contractors performing environmental remediation work to obtain indemnification from the client or owner for damages arising out of all acts or omissions except the contractor's sole negligence. Today, however, many clients and owners are demanding that the contractor indemnify them for all claims and damages regardless of whether the contractor was at fault.

Contractors should be aware of language that either directly or subtly requires this kind of indemnification. An example of such a clause is one that requires indemnification for all damages "arising from acts, omissions, errors, or negligence of the contractor." Another example is a clause that requires indemnification for all damages "arising out of the performance of the contract." Both of these clauses appear to create strict liability for the contractor and require the contractor to indemnify the client regardless of whether the contractor was negligent or otherwise at fault.

Agreeing to indemnify the client for claims and damages that the contractor would not be liable for at common law (e.g., based on negligence) may be classified as a contractual liability that is barred from insurance coverage pursuant to the terms of the applicable liability policy.

Where there may be a likelihood at a project site of encountering contaminants or hazardous waste that the contractor is required to work around or with, a special indemnification provision may be required. One possible form of this provision may contain a clause whereby the client agrees to indemnify the contractor for claims and damages arising out of any release of pollutants or hazardous substance that existed at the project site prior to the commencement of the contractor's work, provided that the release is "not finally determined to have resulted from the sole negligence of the contractor."

Variations of the above language can require the client to indemnify the contractor, including the payment of legal defense costs, while a claim or litigation is pending. The client will be relieved of the obligation for this payment where an ultimate decision is issued by a fact finder concluding that the contractor was solely negligent.

In drafting any indemnification clause the contractor should exercise caution to comply with state law provisions that may limit the nature and extent of indemnification agreements that are acceptable under public policy principles. The majority of states will not permit contractors to be indemnified for damages arising out of their sole negligence. There are a few notable exceptions to the rule, however, such as the state of Texas.

10.9 Conclusion

In order to reduce the risk of environmental liability, a contractor can:

- Carefully define the scope of work and the standard of care to be applied under the contract.

- Price the job to reflect the liability risk and staff the job with employees trained to deal with any relevant environmental matters.

- Obtain indemnification agreements from the client.

- Include a limitation of liability clause in the contract.

- Be certain not to sign manifests. Require the client to sign all manifests for waste materials found at the site.

- Do not sign warranties or guarantees that the contractor will comply with all guidance documents of federal and state agencies. Also do not warrant that the site will be clean as a result of the work. Limit such representations to the materials that the contractor actually addresses through his or her work at the site.

- Obtain insurance that fills the gap created by the pollution exemption contained in liability insurance policies.

Professional Liability Insurance

Salvatore J. Perrucci
Willis Corroon Construction

11.1   What Is Professional Liability?

As most general liability policies have an exclusion for professional acts or omissions, the need arises for a special coverage to protect architects, engineers, construction managers, and design-builders. A professional liability policy in the basic sense covers damages of liability arising out of a negligent act, error, or omission of the insured in the rendering of or failure to render professional services. Another way to think of it is as malpractice insurance for engineers.

What damages does it cover?   A professional liability policy covers sums that the insured becomes legally liable to pay as a result of any negligent act, error, or omission. Unlike many liability policies, it is not limited to bodily injury or property damage only. Pure financial loss is also covered (i.e., product was designed wrong or doesn't produce the volume of widgets it was designed to produce).

Who needs it?   Professional liability is needed by the following:

- Architects and engineers
- Design-build programs
- Construction managers

11.2   Design-Build Insurance, Surety, and Risk Management Challenges

As design-build entities are discovering, there are formidable challenges to developing a cohesive insurance, surety, and risk management program that will address the exposures from design-build work.

This issue is heightened by the increased use of turnkey project delivery, where the owner seeks one entity to provide a seamless franchise in outsourcing a host of services. Design-builders are being called upon to provide project financing, site selection purchase, operation and maintenance, and possibly an equity ownership interest, in addition to architecture, engineering, and construction services.

11.2.1   Liabilities and damages

In this expanded role, the design-builder can be subject to an array of liabilities and damages attributed to contractual liability and various types of tort liability. For example, the design-builder is subject to contractual liability such as indemnity and performance damages from contractual obligations with owners, subcontractors, and suppliers.

The theories of tort liability include:

- *Negligence.*   This occurs when builders fail to carry out their functions in a manner consistent with normal and accepted standards of practice.

- *Strict liability/joint and several liability.*   A design-builder who engages in hazardous activities can be held liable for all injuries that those activities cause, even if the design-builder was reasonable in providing the services. Under the theory of joint and several liability, if there is more than one defendant liable for the injuries, the injured party may collect damages from any one, or all, of the defendants.

- *Statutory liability.*   Federal and state liability statutes can impose strict liability without regard to degree of fault. This is especially true for design-build environmental remediation projects where the parties can be held liable to CERCLA (superfund) and other state environmental statutes.

11.2.2   Operational exposures

With the design-builder providing a broad scope of services, there is an inherent interrelation of operational exposures associated with the work, which can include:

- Faulty design
- Faulty construction workmanship
- Environmental damage
- Injury to workers and third parties
- Damage to property of owner, contractor, supplier and third parties
- Performance deficiencies

Through contractual obligations and the sources of tort liability discussed earlier, the design-build entity and/or its subcontractors or subconsultants can be held responsible for the injuries and damages listed above.

11.2.3 Traditional insurance and surety applications

The purchase of insurance and surety is a method to transfer the risk associated with operational exposures to an insurer or bonding company. Traditional insurance and surety applications to address these exposures are shown in Table 11.1.

11.3 Design-Build Insurance Challenges

Although a variety of insurance products are available, the coverages afforded by one policy do not necessarily dovetail with another policy to provide a seamless fit. It is important to review the standard policy exclusions and limitations to determine the true extent of coverage. Many standard exclusions are broad in their application, which can substantially restrict the coverage. The key insurance challenge is to properly dovetail the standard policy exclusions to address the interrelated exposures of design-build work. Table 11.2 outlines ways to treat several standard exclusions contained in professional liability, general liability, and pollution liability insurance policies.

For example, standard architect and engineer professional liability insurance policies contain broad design-build exclusions. A typical design-build exclusion reads as follows:

TABLE 11.1 Traditional Insurance/Surety Applications

Coverage	Exposure
Commercial general liability	Third-party bodily injury and property damage Faulty construction workmanship
Workers' compensation and employer's liability	Injury to workers
Professional liability	Faulty design
Contractor's pollution liability	Third-party environmental bodily injury and property damage and cleanup
Builder's risk	First-party damage to property of owner, contractor, or supplier during construction
Efficacy/contingency risks	Liquidated damages, debt obligations
Surety	Performance and payment obligations of contractors

TABLE 11.2   Dovetail Standard Policy Exclusions to Address Interrelated Exposures

Coverage	Exclusion	Treatment
A/E professional liability insurance	Broad design-build	Delete or restrict to faulty construction workmanship
	Equity interest	Delete or set percentage ownership threshold for company
	Pollution	Buyback or combined professional and pollution insurance
General liability	Professional E&O	Carve out named perils professional insurance
	Pollution	Buyback named perils pollution insurance
		Buyback or builder's risk
Pollution liability insurance		Combined professional and pollution insurance

Design-B

This p

... any   a project for which the assembly, construction, erection, ~~fabrication~~, ~~installation~~, or supplying of materials was provided in whole or in part by:

1. the Insured, or

2. a subcontractor of the Insured, or

3. any enterprise and/or any subsidiary of any enterprise that any Insured controls, manages, operates, or holds ownership in or by any enterprise that controls, manages, operates, or holds ownership in the Named Insured:

Most underwriters will state that the primary intent of the exclusion is to not cover faulty construction workmanship. However, the standard design-build exclusions are so broad in their application that they can exclude all professional liability claims arising out of design-build projects. The recommended treatment is to delete the exclusion in its entirety or restrict its application to faulty construction workmanship. An example of a modified exclusionary wording that underwriters use for this purpose is as follows:

Modified Design-Build Exclusion

This policy does not cover:

... any faulty workmanship, construction or work not in accordance with the design of the project or the construction documents (including but not limited to the drawings and specifications) if such work is performed by:

1. an Insured, or
2. a subcontractor of an Insured, or
3. any enterprise and/or any subsidiary of any enterprise that any Insured controls, manages, operates, or holds ownership in; or any enterprise that controls, manages, operates, or holds ownership in an Insured.

The majority of architect and engineer professional liability policies in the market today also contain an exclusion for claims arising out of pollution. As the demand for environmental design projects has grown, however, many insurers have become amenable to adding this coverage back to their policies. Coverage for the pollution exposure is typically provided through an endorsement to the standard professional liability policy that buys back the pollution coverage by eliminating or modifying the pollution exclusion.

The underwriting community has also introduced policy forms that combine contractor's pollution coverage with professional liability insurance. This approach is particularly well-suited for design-build environmental remediation projects. A key advantage to this approach is that a single policy including professional and pollution liability exposures should eliminate the problem of disputes between insurers over which policy covers a particular loss.

11.3.1 Dedicated project insurance

For larger design-build projects, an increasing trend is for an owner or contractor to use dedicated project insurance (commonly known as wrap-up insurance) programs. A wrap-up provides a vehicle to protect owners, design-builders and their subcontractors and subconsultants under a coordinated insurance program. Typical coverages provided under a wrap-up include general liability, workers' compensation, employer's liability, and excess liability. A builder's variation to this approach is a rolling wrap-up where several similar, medium to small projects can be combined under one blanket insurance program.

The key benefits to a design-builder with a wrap-up approach include:

- A coordinated insurance mechanism to address interrelated exposures

- A reduction in the total cost of risk by instituting coordinated safety management and claims management activities

- An assurance that subconsultants and subcontractors have adequate limits and coverage in place

11.3.2   Emerging insurance products

In recognition of the expanded use of design-build project delivery, a few select U.S. and European insurers can provide professional liability policies to address:

- Contingent E&O exposure of design-builders that subcontract the design services to an architect or engineer firm.

- Contingent E&O exposure of an owner for the design activities of the design-builder and/or the architect or engineer subcontractor.

These insurance products have recently been introduced in the market; coverage terms and costs vary by each of the insurers providing the insurance and by the nature of the project(s). Policies are available on either an annual or project-specific basis.

11.3.3   Performance guarantees and bonds

Since the design-builder is the single point of authority, many owners require some form of guarantee of performance. Parental guarantees and letters of credit are frequently used. Performance bonds issued by a surety are also an option.

11.4   Design-Build Surety Challenges

The number of sureties who will write performance bonds for design-build work is relatively limited. The primary concerns of the sureties include:

- Underwriting a project that is yet to be designed, particularly if there are specific performance guarantees associated with the work.

- Design performance or efficacy guarantees. Surety underwriters are not accustomed to addressing these exposures and have difficulty quantifying the performance guarantee.

- Given the fast-track nature of many design-build projects, the penal sum of the bond is not always determinable.

Larger general contractors with established surety relationships are in a better position to obtain performance bonds for their design-build projects, whereas engineering firms that act as design-builders have a much greater challenge since they typically do not have established surety relationships. In addition, the capital base for these firms looks strikingly different compared to a traditional construction contractor.

11.4.1 Environmental surety

The surety market is even more restrictive for environmental design-build projects. Given the changing conditions that inevitably arise during the remediation process, there are inherent uncertainties as to whether the design and remediation approach will meet the performance standards.

There are three key underwriters that are willing to entertain surety for environmental remediation projects:

- Commerce and industry (part of the American International Group)

- Reliance

- Acstar

There are a few other select players who are evaluating the entry into the environmental surety sector.

11.4.2 Approach to surety underwriters

When working with the underwriting community, there are several key issues to emphasize in order to enhance underwriters' comfort level with the design-build project:

- *Quality of risk transfer.* Insurance and contractual provisions should provide quality risk transfer. Since surety underwriters are concerned with the design efficacy exposure, it is important to demonstrate limits and scope of professional liability insurance maintained by the design professionals. Likewise, on environmental projects underwriters will require that design-builders and/or their subs have in place pollution liability insurance of some sort (contractor's pollution insurance, environmental E&O, etc.).

  Underwriters will also want to review the contractual obligations the design-builder has with the owner as well as with subcontractors and subconsultants. The key contract terms to highlight are:

  Scope of work

  Pricing and payment

  Changes

  Differing site conditions

  Time and schedule management

  Contractual risk allocation

  Performance guarantees/damages

- *Qualification and experience of design.* Builders and their subcontractors and subconsultants should be experienced in performing design-build projects or similar type of work.

- *Quality management programs.* These will mitigate the potential for a performance shortfall.

- *Financial standing.* The project participants should be in good standing financially.

11.5 Design-Build Risk Management Issues

Insurance and surety should be viewed as a backstop mechanism to manage risk. The design-builder should take an aggressive front-end approach through the use of sound risk management techniques to avoid or mitigate liability and damages. Listed below are some suggestions.

- Select the right client and design-build project.

- Analyze subcontractor selection and management.

- Establish a partnering relationship with client to forge mutual goals and objectives.

- Evaluate contract limitations of liability.

- Develop project management capabilities (e.g., change order management, scheduling and coordination of construction trades) to mitigate the potential for cost overruns or delays.

- Coordinate health and safety and quality assurance programs.

- Use dispute resolution techniques to address problems quickly and fairly.

- Implement aggressive claims management.

Instituting sound risk management practices has a direct impact on how insurance and surety underwriters assess risk and develop premium and coverage terms. Demonstrating positive risk management protocols will enhance the design-builder's ability to obtain insurance and surety for his or her work.

11.6 Conclusion

The expanded use of design-build project delivery is driving the demand for insurance and surety products to address the unique exposures associated with design-build work. Currently the insurance and surety markets are in an evolutionary phase which heightens the importance for builders of instituting aggressive risk management practices as a front-end attack to reduce exposure to liability and damages. With the increased demand, we will see more insurance and surety companies offering new products and new capacities. This is good news for design-builders as it will enable them to develop cohesive insurance, surety, and risk management programs.

Contractor-Controlled Insurance Programs (CCIPs)

William F. Ward, III
Willis Corroon Corporation

12.1 What Is a Contractor-Controlled Insurance Program?

The cost of insurance and increasing costs of construction have led many contractors to seek out innovative means of handling the associated exposures. *Contractor-controlled insurance programs* (CCIPs) are quickly becoming a favored method of treating construction risk.

These programs pool or consolidate the general liability, workers' compensation, and builder's risk coverages for all subcontractors into one program, insured by one carrier (per line) and managed by the contractor. There are a number of risk management advantages associated with CCIPs:

- *Reduced insurance costs.* Because of the economies of scale, contractors will enjoy more negotiating leverage with underwriters. Carriers can reduce the administration and service expenses with CCIPs and, through competition, pass these savings on to the contractor.

- *Expanded coverage and higher limits.* The same leverage producing premium savings is used to broaden coverage. Favorable loss experience on projects protected by CCIPs has underwriters eager to write these programs. This gives the contractor significant leverage when negotiating coverage. For this reason, CCIPs will provide higher limits and broader coverage to the contractor and the subcontractors.

- *Centralized safety program.* Losses drive up project construction costs by increasing claim and litigation expenses, causing construction delays and reducing efficiency. CCIPs put the contractor in a better position to influence project safety activities to reduce costs. By coordinating the safety program, the contractor eliminates multiple insurance safety representatives with different levels of experience and dedication from the project. A CCIP will improve the contractor's position to negotiate better loss control services from the carrier.

- *Equal treatment.* With a CCIP, even the smallest subcontractors receive the attention usually reserved for the largest insureds.

- *Improved cash flow.* The CCIP premium volume can result in favorable payment terms. With a high deductible or paid loss retrospective-rating plan, the contractor may be able to hold significant amounts of insurance dollars until claims are actually paid.

- *Enhanced claims administration.* Contractors of CCIPs will enjoy a higher standard of claim service. A CCIP centralizes the claims administration process to ensure that claims are investigated properly and all claimants treated fairly.

- *Unified contractors.* Projects insured under conventional programs tend to have a higher incidence of cross litigation. A significant amount of this litigation is initiated by insurance companies on behalf of subcontractor insureds. CCIPs involve one underwriter; this reduces the carrier's incentive to pursue cross complaints.

- *Improved disadvantage enterprise participation.* Many owners or sponsors of large construction projects, especially in the government sector, encourage minority participation. Many of the smaller enterprises lack the required project insurance coverage to do likewise. A CCIP diminishes this concern.

The use of a CCIP presents some challenges for the contractor. For instance, some subcontractors may experience a diminished incentive to work safely because the bulk of the risk is transferred to the contractor's program. This concern can be eliminated or reduced with an aggressive safety program. Willis Corroon can design an incentive program that promotes the contractor's risk management goals and objectives.

There is a concern that CCIPs require a greater degree of involvement on behalf of the contractor. However, a properly managed program will produce enough savings to justify the cost of the additional administration. Many benefits of CCIPs are not quantifiable. Therefore, we encourage you to consider all the advantages of a CCIP before deciding what method of risk management is best for the project(s). We have summarized the advantages and disadvantages of a CCIP in Table 12.1.

TABLE 12.1   Advantages and Disadvantages of a CCIP

Advantages	Perceived contractor disadvantages	Perceived subcontractor disadvantages
Uniformity of coverage insurance	Assumption of safety	Loss of competitive edge
Economies of scale	Separate premium payouts	Multiple audits
Savings through loss-sensitive programs	Duration of rating	Incentive for safety diminishes
Elimination of gaps responsibilities and overlays	Poor experience	
Community relations	Administrative burden	
Minority contractor participation		
Elimination of coverage disputes		
Reduced subrogation		
Improved safety program		
Effective claims administration		
Centralized insurance		
Unification of interest		
Control of coverage		

12.2   How a CCIP Works

The contractor procures the coverage on behalf of the subcontractors working on the project insured with a CCIP. In return for this coverage, the subcontractors agree to remove all associated insurance costs, including profit, from the bids.

The CCIP provides workers' compensation, general liability, umbrella liability, and builder's risk. However, the contractor controls the program and may expand or restrict the program as desired. For example, the program may be expanded to include:

- Pollution
- Railroad protective
- Professional liability
- Force majeure
- Owner's protected liability

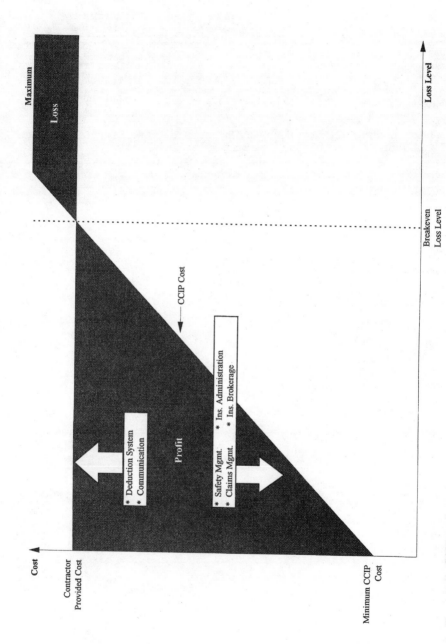

Figure 12.1 CCIP profit model.

190

Automobile liability, contractor's equipment coverage, and bonds remain the responsibility of the individual subcontractors.

The financial incentives of CCIPs are contingent upon the type of project, size, location, and legal environment. For a diagram of the CCIP profit model, see Fig. 12.1. Most sponsors of CCIPs enjoy direct savings of 20 to 40 percent over a traditionally insured program.

13

Insurance Administration and Accounting

William J. Palmer
Ernst & Young LLP

Insurance administration and accounting can be difficult prospects for many contractors, especially small contractors. Insurance is a complex subject as anyone who has attempted to read an automobile insurance policy will know. In larger construction companies that can afford to hire an individual who knows insurance, this does not create much of a problem. In small and startup situations, it is important that management understand that the typical construction "stiff" does not like to read small print. Management should first select an agency with a specialty in construction and hire a bookkeeper or outside accountant who, in addition to accounting skills, has a fairly sound knowledge of the administrative and accounting aspects of insurance.

13.1 Insurance Administration

Because the insurance broker is customarily the agent of the insured and not of the insurance company, it is important that this individual be both responsible and competent. It is also important that the broker be aware of the needs of construction companies both small and large as well as differing in type (general, electrical, mechanical, etc.).

Once the insurance coverage is placed, it is usually up to the accounting department to see that the necessary reports are made and that any claims are properly documented. As previously stated, in very small or startup construction companies, accounting services are usually performed by an outside accountant or CPA who will usually specialize in handling small construction companies.

Premiums on workers' compensation insurance are based on payrolls. However, despite the language of some policies and some reporting forms, the base on which premiums are computed is usually not the total dollar amount of the payroll, but rather the amount arrived at by multiplying the total hours (whether straight time or overtime) by the straight-time rates. For that reason, many contractors now compute gross pay by computing total hours at straight time. The premium for the overtime is then added to the straight-time amount. This makes reporting the monthly payroll amount easier for computing the premium.

Other types of insurance for which premiums are usually based on payrolls include public liability and property damage (other than automobile), employer's liability and occupational disease, and group health and accident insurance. A few states like California have a compulsory state health and accident coverage tied to their unemployment insurance laws. In these states, health and accident coverage is automatic unless written, through special permission, by private insurance companies. As a practical matter, little if any private insurance is written in this field because of onerous legal requirements. When health and accident coverage is privately written, the report is usually a copy of the state unemployment insurance tax form. As for the other reports based on payroll for insurance purposes, each company has its own reporting requirements, although the differences tend to be minor. In any event, complete reporting instructions should be obtained from the contractor's insurance broker.

For accurate and timely reporting of payroll data for insurance purposes, the accrual of insurance should be made on each payroll and the preparation of the insurance reports should be an item on the bookkeeper's checklist.

Public liability and property damage insurance on automotive equipment is usually covered separately for each piece of equipment reported, under a blanket policy. Each piece must be reported as soon as use starts; special provisions are made for vehicles not owned by the company but used on company business (nonownership coverage). The so-called equipment floater is a policy that protects contractors from loss or damage to their equipment from a wide variety of causes including, generally, fire and storm. There are, however, certain risks that are not covered by the general floater policy. To facilitate the preparation of equipment insurance reports, it is important that equipment records show the value of each piece of equipment for insurance purposes. At the beginning of the policy year, a list of equipment to be covered and the value of each piece is furnished to the insurance company. Thereafter, a report of changes in items covered and their values is made. With most equipment, this procedure is followed monthly. With automotive equipment, it is sometimes necessary to report each piece of equipment as soon as it goes into service, particularly if public liability and property damage insurance is placed on the automotive equipment at the time the floater goes into effect.

Builder's risk insurance is usually placed on individual jobs; the only reports required are the periodic (usually monthly) statements of the value of material stored at the jobsite and the cost of completed work to date. Any substantial amount of material stored in a warehouse or elsewhere should have the usual fire and comprehensive coverage. If the amounts vary, a monthly report is usually required on these types of insurance.

Fidelity bonds may be set on the basis of individual employees, or they may cover any employee who occupies a given position. To cover individuals, a copy of the hiring slip on all bonded employees should go to the bonding company. To cover jobs, a periodic report showing the names of the employees covered and their positions is customarily required (usually quarterly or semiannually) by the bonding company.

All-risk insurance is sometimes used in lieu of some of the other types of coverage. This type of coverage has many advantages but should be used with full knowledge of its limitations.

Other types of insurance—such as comprehensive, fire, burglar, robbery, business record, and business continuation—are on an annual basis; only their expiration dates need to be checked.

It is a good idea to charge one person with full responsibility for all insurance matters. In addition to being familiar with types of insurance, premiums, and so on, that person should have the ability to compute the payroll burden percentage necessary to charge out these costs to jobs and other cost or profit centers. This person should maintain a checklist of all coverages, the due dates of reports and returns, the existence of coinsurance clauses in reporting-type policies, and similar matters. It is this person's responsibility to follow up on all reports, claims, cancellations and renewals, and all retrospective and cancellation refunds. To help management keep abreast of the insurance situation, a number of large- and medium-sized firms (and some small firms as well) have an insurance summary included in their monthly financial statements showing all policies and their current status. Figures 13.1 and 13.2 provide typical insurance register pages for facilitating administration of insurance. Figure 13.3 presents a typical monthly report for workers' compensation.

13.1.1 Claims

As soon as an employee is injured, entirely apart from the reports required by the *Occupational Safety and Health Administration* (OSHA), a complete and accurate accident report must be made to the insurance company. All companies writing workers' compensation

	DATE 19__		1	2	3	4	5	6	7	8	
			PROPERTY INSURED	COMPANY	TYPE OF INSURANCE	POLICY NUMBER	PERIOD COVERED	PREMIUM	PAYMENTS	BALANCE	
1											1
2											2
3											3
4											4
5											5
6											6
7											7
8											8
9											9
10											10
11											11
12											12
13											13
14											14
15											15
16											16
17											17
18											18
19											19
20											20
21											21
22											22
23											23
24											24
25											25
26											26
27											27
28											28
29											29
30											30
31											31
32											32
33											33

Figure 13.1 Example of monthly maintained insurance register.

PREMIUM REPORTING INSTRUCTIONS

CONFIDENTIAL-FOR XYZ Co. USE ONLY
CLIENT: TRANSIT SERVICES
JOB 28554
LOCATION: Anytown, USA

(DOMESTIC) SERVICES: CM
ENTITY: XYZ, Co.
DATE: September 1, 1992

COVERAGE	CARRIER	POLICY & PERIOD	PREMIUM RATE	BASIS OF PREMIUM	FORM NO.	DISTRIBUTION
1. WORKERS' COMPENSATION & EMPLOYEES LIABILITY	Industrial Indemnity	CB 912-7/300 4/1/92-4/1/93	PER ATTACHED	As indicated under "Basis of Premium" in attached WC Schedule	XYZ Co. Form I-4-1 & I-4-2	MONTHLY: (Due 10th of each month) - Original and 2 copies to Broker - 1 copy to S.F. Ins. Dept.
2. PUBLIC LIABILITY (including Auto Property Damage)	Industrial Indemnity	IL 909-3216 4/1/90-4/1/93	$1.95	Rate per $100 of Gross Payroll (See Note #1)	XYZ Co. Form I-4-1	MONTHLY: (Due 10th of each month) - Original and 2 copies to Broker - 1 copy to S.F. Ins. Dept.
3. THIRD-PARTY PROPERTY DAMAGE	Lloyd's London	LUS 1136 4/1/91-4/1/94	$1.25	Rate per $100 of Gross Payroll (See Note #1)	XYZ Co. Form I-4-1	MONTHLY: (Due 10th of each month) - Original and 2 copies to Broker - 1 copy to S.F. Ins. Dept.
4. EXCESS & ADDITIONAL EXCESS PUBLIC LIABILITY AND PROPERTY DAMAGE	Lloyd's London	LUS 1083/4/5 4/1/90-4/1/93	$1.32	Rate per $100 of Gross Payroll (See Note #1)	XYZ Co. Form I-4-1	MONTHLY: (Due 10th of each month) - Original and 2 copies to Broker - 1 copy to S.F. Ins. Dept.
5. CON TRACTORS "ALL-RISK" EQUIPMENT (a) Hired Automotive & Construction Equipment, Tools & Office Equipment & Furniture (b) Job-owned or Client-furnished Auto & Construction Equipment, Tools, & Office Equipment & Furniture	Switzerland General Insurance Co.	LUS 1081 4/1/90-4/1/93	(a) $2.25 (b) $0.10	Rate per $100 of rental cost on bare rental basis / Rate over $100 of original cost	XYZ Co. Form I-4-1	MONTHLY: (Due 10th of each month) - Original and 2 copies to Broker - 1 copy to S.F. Ins. Dept.
6. BUILDER'S RISK (Loss or damage to the Work)			N/A		XYZ Co. Form I-4-1	MONTHLY: (Due 10th of each month) - Original and 2 copies to Broker - 1 copy to S.F. Ins. Dept.

Notes:

(1) Under "Basis of Premium," except as otherwise indicated, use GROSS PAYROLL as defined in Controller's Procedure 99.02. With respect to Workers' Compensation, please see definition in attached Schedule of WC Rates

(2) Coverage 1 - Workers' Compensation premium should be adjusted by Experience Rating and other factors shown on attached Rate Schedule to arrive at net premium due.

(3) Coverage 5 - No report necessary for BLSI-owned tools and equipment rented to Job. For tools and equipment rented from third parties, use "Hired" rate (a). For tools and equipment owned by Job or furnished by Client, use "Job-owned" rate (b), unless agreement with Client relieves XYZ Co. of responsibility to furnish this insurance.

NAME & ADDRESS OF XYZ Co. BROKER:
J. Doe, Inc., Two Market St., Suite 10.
San Francisco, CA 94115

Figure 13.2 Example of desktop computer–maintained insurance register.

MONTHLY WORKERS' COMPENSATION REPORT
STATE DETAIL SUMMARY

STATE: 015 TEXAS
POG/OFFICE: 1A1 S F DOMESTIC
COMPANY: XYZ, Co.
JOB NUMBER: 18966

	CRFT CODE	WC CODE	ST	**PAY** ST/OT	PT/OT	GROSS PAY	WORK COMP SUBJECT TO PREM	RATE	EARNED PREMIUM	WEEKS WORKED
WC.CODE TOTAL	UO	7,539.00	5,316.92	1,960.62	0.00	7,277.54	7,277.54	3.81	277.27	4.00
STATE/POG/CO/JOB NUM TOTAL			5,316.92	1,960.62	0.00	7,277.54	7,277.54		277.27	4.00
JOB NUMBER: 18966										
EARNED PREMIUM		277.27								
PLUS EMPLOYERS LIABILITY	0.03	8.31								
SUBTOTAL		285.58								
CALC. EXPERIENCE MOD.	0.71	202.76								
LESS PREMIUM DISCOUNT	0.14	28.99								
ADD SPEC. ASSESS. ON EXP. MOD.	0.03	5.27								
TOTAL PAYMENT AMOUNT		179.04								

Figure 13.3 Example of desktop computer–prepared workers' compensation monthly report.

insurance require such reports and check to see that they are made. Usually the questions asked by the accident report will provide a checklist of all the necessary information required by the insurance company.

It should be remembered, however, that injuries due to serious and willful negligence on the part of the employer are not normally covered by workers' compensation insurance. It must also be remembered that a great many innocent details can add up to a circumstantial case of serious and willful negligence. For that reason, any injury that may be serious is worthy of a special report of conditions including pictures of the site of the accident and surrounding area; names, addresses, and telephone numbers of witnesses; memoranda regarding safety measures in effect at the time; and anything else that might indicate the high rate of losses that can result in lost jobs.

If a person injured on the job is the employee of a subcontractor or is merely a business visitor, the prime contractor must obtain complete details in provable form. Otherwise, the contractor may be faced with a lawsuit months after the event and have nothing with which to determine liability or on which to base a defense against an accusation of negligence. Therefore, the information and evidence assembled should always be the same regardless of who is insured at the jobsite. If the injured person is not an employee, the accident should be reported as completely as possible to the company carrying the contractor's public liability insurance.

Automobile accidents must be reported on special forms provided by the insurer; in most states either the state or local police require additional reports to be made on their forms. Many states place liability on the owner of the vehicle as well as the driver if an accident is caused by negligence. In all states the general rules of law place responsibility on employers for acts of their employees in the course of employment. Because of this fact, most contractors carry insurance against liability for accidents their employees may have while driving personal cars on company business. If nonownership coverage is carried, reports of employees' on-duty accidents in their own vehicles should be reported as completely and carefully as those in company vehicles.

Claims against common carriers or insurance companies for shortage and damage to goods in transit are usually first made against the carrier. In companies that maintain a warehouseperson, it is usually his or her job to see that such claims are made. In small companies, the losses on materials and supplies in transit are usually reported to the office by a supervisor, and the claims against the carrier are made by the bookkeeper.

Claims under marine insurance policies are probably the least understood by contractors. Any contractor undertaking waterfront work should employ an insurance broker who is well-versed in marine insurance matters. Many of the rules of negligence in maritime law are the same as those upon land, but the tests used in applying the rules often differ substantially. It can be stated here only that such differences do exist and that they call for the services of a specialist.

Storm and fire damage is reported on an affidavit customarily referred to as a *proof of loss*. As soon as possible after the storm or fire, the damage is surveyed and estimated. Sometimes settlement is made from the estimate but it is not uncommon for the insurance company to insist that the damage be repaired and that a detailed schedule of costs be submitted. If this procedure is followed, it is necessary to set up a work order to cover the repairs and charge all the repair costs to the work order. In such work, care must be taken to see that overhead and equipment rates are adequate and that the impact of cost on jobs in progress, if any, is given proper recognition.

If the insurance company pays on the basis of the estimate of damage, the contractor's unit costs often form the basis for computing the total amount to be paid. Unless the unit costs contain all the applicable costs, including appropriate overhead allocations, the contractor stands to lose by their application. There is a tendency on the part of many contractors' accountants to take the attitude that as long as the insurance company is liable for a loss, the matter is no concern of theirs. Actually, when the insurer admits liability, the accountant's work may be just beginning. The next step is to prove the amount of the loss.

Good cost records and complete records of equipment and inventories are essential to establish what has been lost and what its cost is. It is true that most insurance claims are based on present value instead of cost. It is also true that, on the type of loss usually covered by builder's risk policies, the number of units lost and their replacement costs will be estimated by engineers. When differences arise, as they frequently do, between the estimates of the insurance company's engineers and a contractor's engineers, good accounting records provide the means for reconciling them. Lacking such records, the contractor usually has to choose between an unreimbursed loss and a lawsuit that might result in further loss.

Another problem that sometimes arises on damage claims is the allocation of lump-sum insurance settlements between the prime contractor and subcontractors. There is often a feeling (sometimes justified) on the part of the subcontractors that the prime contractor is trying to profit at their expense in allocating the loss. If an overcollection has been made, it is quite often better to refund it to the insurance company than to divide it among the prime contractor and subcontractors. There are two good reasons for this policy. First, contractors who establish a reputation with their insurers for strictly scrupulous dealing have a better chance of getting paid on doubtful claims. Second, excessive payments secured by overstated or fraudulent claims can be recovered by the insurance company. If under such circumstances some part of an excessive payment has been passed on to a subcontractor, the prime contractor may be forced to repay the insurance company but unable to recover from the subcontractor.

The construction accountant's interest in insurance claims cannot end when the amount is determined. This person must record the claim as an asset, list it on the income tax returns, and report it in the financial statements. Normally a claim is treated as a current account receivable and since it is a nonoperating item is included in "other accounts receivable." However, it may not always be so. For example, if overlapping coverage causes a dispute among insurers over who must pay, collection may be delayed for many months. Or the insurance company may deny liability or try to reduce the amount of the claim because of coinsurance provisions. Collection of a claim under such circumstances may require litigation and the claim could be properly treated as a current asset, either on financial statements or in working capital forecasts. If the insurer denies liability, the most conservative balance sheet treatment is to show the claim as an asset reduced by a reserve for loss that is equal in amount. Normally such losses should be claimed as income tax deductions in the year in which they occur; any recovery in a subsequent year should be treated as income in that year.

13.2 Subcontractors' Insurance Certificates

Most subcontracts require that the subcontractors carry certain basic coverages, such as public liability, property damage, and workers' compensation. To assure the prime contractor that such insurance is, in fact, in force, the subcontractor secures from his or her insurers either copies of the policies or certificates that the policies are in force, or both. The prime contractor should make sure that the certificates of insurance contain the provision that the prime contractor will be given notice before the policy is canceled or allowed to terminate for any reason.

In establishing the requirements for subcontractors' insurance, possible complications from overlapping coverage must be considered. If any possibility exists of dispute over which insurer has the primary liability, that question should be recognized and settled before there is any claim. Usually compensation insurance would be required by the law, and public liability and property damage would apply primarily to accidents caused by the subcontractor's employees. If builder's risk is desired, it might well be carried by the prime contractor with resulting adjustment of the subcontract prices.

Most subcontracts contain a so-called hold-harmless clause under which subcontractors assume responsibility for any loss or damage caused by their operations. If the size of the operation warrants, this clause can be reinforced by insurance certificates, and in larger operations a control file may be maintained to see that all necessary subcontractor coverage is maintained. However, if a prime contractor wants to be fully protected, the subcontract form should stipulate all required insurance coverages and provide that, if they are not maintained by the subcontractor, the prime contractor has the privilege of purchasing the required coverage at the subcontractor's expense.

If the subcontract contains such a clause, it should be simple to decide what to do when a subcontractor's insurance is canceled. Often, however, it is not simple, because a subcontractor whose insurance has lapsed has probably exhausted his or her credit with the prime contractor, and may have built up liens far beyond the subcontractor's capacity to pay and far in excess of the retained percentage (retention). Insurance premiums are not the sort of expense normally covered by payment bonds, even if subcontractors are bonded—and frequently they are not. The prime contractor can usually bring a defaulting subcontractor's risks under the coverage of the contractor's own insurance by taking over the work. Then if the subcontractor is relet, the risk can again be shifted to the insurers of any new subcontractor who takes over from the old one.

Some questions may arise about the protection of the subcontractor against the risks which the prime should cover. Usually the subcontractor's own casualty coverage will hedge the most important risks. However, if the question is one that might, under the circumstances, become important, it should be covered in the provisions of the subcontract document and an attempt should be made to include it in the risks covered by the payment bond.

13.3 Accounting for Construction Insurance Costs

The primary objective of accounting in any business is to help that business make the maximum profit after taxes. Unless accounting makes its full contribution to that objective, its costs cannot be justified. Accounting for insurance costs must meet the same criteria; that is, the cost of accounting for the insurance must be offset by the benefits of that accounting. To properly understand accounting for insurance costs, it is important to understand broad cost category definitions. They are:

1. *Direct costs*—Also referred to as variable or incremental

2. *Indirect costs*—Also referred to as semivariable or semi-incremental

3. *Overhead*—Also referred to as fixed or sunk costs

Figure 13.4 shows the relationship of these costs to a contractor's financial attachments.

13.3.1 Direct costs

A direct cost is one that can be specifically identified with a construction job or with a unit of production within a job. This is an important

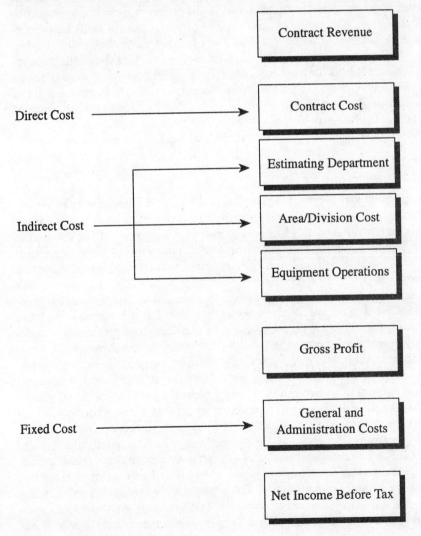

Figure 13.4 Relationship of costs.

distinction. From the home office point of view, any cost that can be identified with a construction job is a direct cost. From the construction superintendent's point of view, however, costs are measured differently; if a cost cannot be identified with units within the job, then the superintendent would not consider such a cost to be direct, but would rather consider it to be due to general conditions or jobsite overhead.

Direct costs will also vary in direct proportion to the amount of work performed. Therefore, if no work is being performed, there will be no direct costs. As we shall see later in this chapter, however, when it comes to insurance, that statement is only correct from the home office point of view. Examples of direct construction costs are craft labor, permanent materials, subcontracts, consumable materials, and certain types of insurance costs. Many contractors will argue that construction equipment is also a direct cost. Strictly speaking, however, construction equipment does not meet the definition of direct cost since the bulk of the equipment costs (depreciation, taxes, etc.) continues whether or not a contractor has any jobs to work on. That same statement is true with respect to certain types of insurance in which the premiums cover time.

13.3.2 Indirect costs

An indirect cost is a cost which can be identified with jobs in progress, but not with any specific job or individual work item within a job. Here again, whether a cost is indirect or not depends on whether the cost is being looked at from the home office or jobsite point of view. For example, a performance bond premium charged to a job will be considered a direct cost by the home office. However, the superintendent would look upon this as jobsite overhead since it does not relate to units of production at the jobsite. With these facts in mind, indirect costs can have an important bearing on contract costs and therefore pricing. It is especially significant where the total bid is based on numerous individual bid items within one project with differing unit prices for each bid item.

Examples of indirect costs are salaries of jobsite management (superintendents may manage more than one job at the same time), utilities and telephones, construction equipment, construction equipment insurance, umbrella excess liability insurance, etc. Indirect costs will generally increase in increments as the construction company grows. For example, suppose a construction company has one superintendent who can oversee four jobs in progress at any point in time and is paid $4000.00 per month. If the construction company has only one job in a particular period of time, the indirect cost for the superintendent is still $4000.00 a month. In order to run more than four jobs but fewer than eight jobs, another superintendent must be hired, so that the indirect cost for superintendents increases to $8000.00. If in our hypothetical example the contractor continued to have jobs of a similar size and nature, he or she would have to hire an additional superintendent when job number 9, 13, etc., is reached. Some forms of insurance follow this same pattern. There are also some forms of insurance which are a blend of fixed, indirect, and direct.

13.3.3 Overhead costs

Overhead costs are simply those which cannot be identified with construction jobs either directly or indirectly. Examples of overhead costs are salaries of corporate officers, labor relations, accounting department, insurance department, fidelity bonds, and other forms of insurance such as fire insurance on the company's headquarters, etc. Salaries in these departments are usually attributed to "home office workers" by construction people.

Typically, overhead costs will remain constant for a fairly large range of activity within the construction company's environment. Obviously in the long run all costs become variable or semivariable. However, in the near term, fixed costs tend to increase only with inflation or new types of home office services.

To summarize, all insurance costs should be identified as either direct, indirect, or overhead. Unless these distinctions are made in the accounting system, a contractor may not recover all of the job costs and be competitive in bid prices. The following sections of this chapter discuss various types of insurance and bonding costs and the approach to recovering these costs in the competitive bid process.

Many types of insurance and bonding premiums can be specifically identified with the cost category to which they relate (direct, indirect, and fixed). Examples of these are performance, bid, or payment bond premiums obtained for a specific contract, liability insurance required by contract terms, officers' life insurance, home office fire and theft insurance, and fidelity bonds. These types of insurance clearly relate to providing coverage for specific and identifiable items. For example, the premium for the bid bond would be charged to the contract cost if the contractor is successful in winning the bid; if not successful the premium would be charged to the estimating department—an indirect cost. Similarly, the required performance bond premium would be charged to contract costs for that job. Officers' life insurance to cover the president and other related executives would be charged to general administrative expense since the salaries of those individuals would be accounted for there. However, many types of insurance are payroll-related in nature and cannot be identified with a specific project or cost center. Therefore, these costs must be allocated and are usually allocated as a percent of bare labor. For example, the employer's share of FICA (social security) is clearly a payroll-related cost but can only end up in the proper accounts of a contractor's statement if it follows the bare labor associated with the employee. Other examples are workers' compensation, vacation, profit sharing, and other similar costs. A fairly comprehensive listing of these costs appears in Figure 13.5.

Generally, these costs, as incurred, are charged to a clearing account in the contractor's overhead chart of accounts (general administrative

WORKERS' COMPENSATION INSURANCE
SCHEDULE OF CLASSIFICATIONS AND RATES

ANY STATE

JOB OR OFFICE:
EFFECTIVE DATE:
INSURANCE CARRIER:
POLICY NUMBER:

BIDDING - New Jersey
April 1, 1994 - April 1, 1995
Company
CB 913-6600

Code No.	Classification	Basic Rates
8810	Clerical office employees, including office engineers, draftsmen and other employees working in the office 100% of their time.	.36
8742	Salesmen, collectors or messengers, including inspectors and expediters, who are required to leave the office part of the time in the course of their work	.64
8601	Engineers, architects and surveyors who are required to leave the office part of the time in the course of their work	1.17

PREMIUMS ARE TO BE COMPUTED AS FOLLOWS, IN THE ORDER SHOWN:

	Rates apply per $100 of gross payroll excluding premium overtime and travel and subsistence allowances
Experience Modification:	Apply a factor (multiplier) of **.431** (equivalent of a **56.9%** credit) to the total basic premium.
Premium Discount:	9.5% credit applied to premium total after application of experience modification to arrive at net premium due.
Second Injury Fund:	Apply a separate charge of 9.51% to the premium developed after the application of the Experience Modification and show the product as a separate item identified as "Second Injury Fund". Do not apply the Premium Discount to that amount; it is not considered premium but is a form of tax.
Uninsured Employers Fund:	Apply a separate charge of 0.00% to the premium developed after the application of the Experience Modification and show the product as a separate item identified as "Uninsured Employers Injury Fund". Do not apply the Premium Discount to that amount; it is not considered premium but is a form of tax.

Figure 13.5   Example of policy manual instructions for workers' compensation insurance.

section). At the end of each payroll accounting period, the payroll-related costs charged to this account are relieved at a standard percentage commonly referred to as *payroll burden rate*. The formula for deriving this percentage to apply burden to labor so that it ends up charged in the proper accounts is as follows:

$$\frac{\text{Payroll-related costs}}{\text{total bare labor}} = \text{burden percentage}$$

A contractor's bare labor is defined as the total gross pay of all employees at straight-time rates for the period to be accounted for. Therefore, the typical contractor, in determining the burden rate for the coming year, would take the prior year's gross pay (net of overtime premium pay) from the payroll records and increase or decrease it based on expected increases in employees' salaries and changes in the contractor's volume for the coming year.

To illustrate further the accounting for payroll-related taxes, assume that the contractor has expected payroll-related expenses for the coming year as set forth in Figure 13.6 and expects that the bare labor for the year will be $1 million. The burden percentage that must be applied to bare labor to recover the direct and indirect contractor labor is 43.7 percent. Applying the formula set forth above, this is determined as follows:

$$\frac{\$\ 437,000}{\$1,000,000} = 43.7\%$$

As each payroll is then processed, its bare labor is multiplied by 43.7 percent, added to each employee's bare labor amount, and charged to the appropriate job, indirect cost category, or overhead category.

Many larger contractors prefer to be more precise in their calculations and will break these percentages down by type of payroll-related cost. For example, some employees may not qualify for social and retirement benefits (profit-sharing plan, retirement plan, group life, etc.). Therefore, looking at Figure 13.7 the burden rate of 43.7 percent would be reduced by 19.6 percent and as a result, the burden rate applied to these employees would be 24.1 percent. This can be an important distinction to make—if these employees are part of job cost, not adjusting the burden rate to 24.1 percent can overstate contract costs significantly for bidding purposes. In today's environment, a sharp pencil can make the difference between a win or a loss in a competitive bid situation. Figure 13.7 sets forth typical burden percentages for the various categories of payroll-related costs that a contractor incurs.

Payroll Taxes   **$86,000**
FICA (Social Security)
Federal Unemployment Insurance
State Unemployment or Disability

Payroll Insurance   **$37,000**
Workers' Compensation and Employers Liability Insurance
Comprehensive General Liability Insurance
Additional Liability Insurance

Paid Lost Time   **$118,000**
Vacation
Holiday
Other (sick leave, military, jury duty, family death or serious
illness, etc.)

Social and Retirement Benefits   **$196,000**
Profit-Sharing Plan
Retirement Plan
Group Life, Dental, Medical, Travel Insurance, etc.
                                                                     _____

TOTAL   **$437,000**

Figure 13.6   Summary of payroll-related expenses for a California contractor.

Payroll Taxes
$$\frac{\$86,000}{\$1,000,000} = 8.6\%$$

Payroll Insurance
$$\frac{\$37,000}{\$1,000,000} = 3.7\%$$

Paid Last Time
$$\frac{\$118,000}{\$1,000,000} = 11.8\%$$

Social Security and Retirement Benefits
$$\frac{\$196,000}{\$1,000,000} = 19.6\%$$

TOTAL
$$\frac{\$437,000}{\$1,000,000} = \mathbf{43.7\%}$$

Figure 13.7   Calculation of burden percentage for allocation of payroll-related expense to cost/profit centers.

To illustrate the accounting flow in the contractor's records for payroll-related costs, Figure 13.8 provides a diagram illustrating the payroll-related expenses charged to a clearing account. It then shows the flow to the various categories of cost in a contractor's financial statements as charged out at 43.7 percent of bare labor.

Figure 13.9 shows the dramatic effects on a contractor's financial statements by the proper allocation of payroll-related insurance and other costs.

It is especially important to properly account for payroll-related insurance and other costs in the cost-plus fee environment. Contractors performing negotiated work, design-build or fast-track jobs will usually have contract terms providing for payment on a cost-plus fee basis. These contracts are subject to audit and, because payroll-related insurance and other costs are such significant items, many auditors will focus on them. They will be looking for overrecovery of costs due to the use of standard percentages.

Equipment insurance premium costs, insurance for warehouses storing job materials, and other premiums associated with contract costs should be allocated in a similar fashion. For example, equipment insurance premiums would be the numerator and all other equipment costs for the accounting period would be the denominator. The resulting percentage would then usually be applied to the equipment cost to determine the overall equipment rate to be charged out to contracts. The same is true with materials. The related insurance premium would be

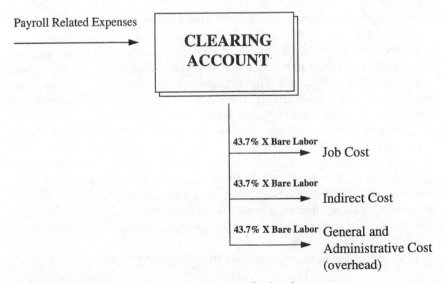

Figure 13.8   Flow diagram of accounting for payroll-related expenses.

Income Statement

	Unburdened	43.7% Applied Burden	Burdened
Contract Revenue	$7,500,000		$7,500,000
Contract Cost			
Direct:			
Labor	700,000	305,900	1,005,900
All other (materials, equipment, subcontracts, miscellaneous)	5,353,400		5,353,400
	6,053,400	305,900	6,359,300
Indirect:			
Division Salaries	100,000	43,700	143,700
All other	97,000		97,000
	197,000	43,700	240,700
Total Contract Costs	6,250,400	349,600	6,600,000
Gross Profit	1,249,600	(349,600)	900,000
General and Admin.			
Officers and Admin. Salaries	200,000	87,400	287,400
Payroll Related Costs	437,000	(437,000)	
All other	394,600		394,600
	1,031,600	(349,600)	682,000
Net Income before tax	$218,000	$0	$218,000

Figure 13.9 An illustration of the effects of allocating payroll-related insurance and other costs using a burden rate of 43.7 percent.

added to all other warehousing costs and a factor then applied to materials charged out to jobs to recover insurance and handling costs.

Many contractors use a standard rate for payroll burden, equipment insurance, materials insurance and handling, and other contract costs which must be allocated. These rates can be obtained from various trade association manuals. In the author's opinion, it is fool-hardy to use such rates because they are based on overall averages of various contractors. These rates can vary significantly due to size of a contractor's payroll, age of the equipment (if fully depreciated, etc.), and other factors. Also, many contractors fail to update their burden rates on a periodic basis. The contractor's accountant should track monthly bare labor versus the planned bare labor. It is a simple matter to recalculate payroll-related burden on a monthly basis so that the contract estimators have the latest and most accurate percentages for purposes of determining bid pricing.

In summary, it is important that the contractor:

- Has a knowledgeable insurance and surety agent

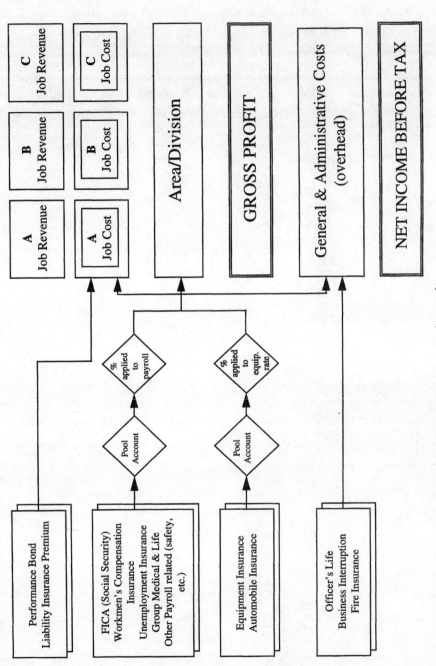

Figure 13.10 Flow of insurance and bond premiums in a contractor's records.

211

■ Has an individual in the company fixed with the responsibility for keeping the contractor's insurance current and adequate

■ Has a bookkeeper or outside accountant knowledgeable in the proper accounting for insurance and bond premiums

Last, the contractor should meet periodically with his or her insurance agent and surety to take a big picture review of all the insurance needs so that insurance and bonding is adequate and premium costs are kept to a minimum.

Figure 13.10 provides an overview of insurance premium cost flow in a contractor's accounting records.

14

Property/Casualty Insurance and Risk Management

Douglas H. Hartman, ARM
Ernst & Young LLP

14.1 Introduction

Risk management involves more than the purchase of insurance. There are other risk financing and risk control techniques available to builders and contractors. In this chapter the author explains the elements of risk management and types of property and liability insurance perils and hazards facing contractors and builder/owners. The applicable insurance policy forms are also described. The services that can be secured from insurance companies and brokers are explained, and a ten-step broker bid procedure is outlined to guide the business owner in competitively placing property casualty insurance. This chapter focuses on the needs of business owners who do not necessarily have full-time risk managers.

What is risk management? The goal of risk management is to minimize the adverse effects of accidental losses. While the purchase of insurance is certainly the most important tool or risk financing technique, well-managed builders and construction firms also pay special attention to identifying and avoiding hazardous conditions and unsafe methods of construction. See Table 14.1.

Table 14.2 lists the major causes of accidental losses, the insurance policy forms, and the parties these policies cover.

Today's insurance policy forms differ significantly from those in use as recent as 10 years ago. The reasons why coverage forms changed so rapidly had to do with the huge volume of litigation between insurers and their insureds over such issues as pollution, asbestos, occupational illness, harassment, and civil rights actions. Here is the legacy of the tumultuous 1980s:

TABLE 14.1   **The Risk Manager's Tools**

Risk financing	Risk control
Purchasing insurance, i.e., choosing an insurance agent, deciding which insurance quotation to accept, and determining how much insurance to buy	Conducting jobsite inspections for both liability and property insurance purposes; reviewing construction plans and blueprints before the start of construction
Estimating the cost of insurance before bidding on a job	Avoiding risk by eliminating hazardous practices or even declining to bid on certain business
Using deductibles, or becoming a qualified self-insurer of workers' comp., or purchasing a retrospective-rated policy	Establishing worker safety policies and procedures
Making another party purchase insurance protecting your interests	Investigating accidents and reviewing "near miss" incidents to determine the human error, material failure, and management lapses that caused accidents and incidents to occur

- There now are annual aggregate limitations—as well as per-occurrence limitations—on *all* liability insurance policies, CGL and excess. For example, most contractors purchase (and are contractually required to purchase) a minimum of $1 million per-occurrence insurance. Larger contractors purchase primary policies that may have annual aggregate limits greater than the per-occurrence limits; hence, their limits would be expressed by the abbreviation $1MM/$2MM, for $1 million per occurrence subject to an annual aggregate limit of $2 million. As a result of these changes, contractors today are purchasing more amounts of annual aggregate excess insurance than in the past.

- Legal defense expenses on the *primary* liability insurance policy are *in addition* to the limits listed on the insurance policy. In contrast, some excess liability insurance policies include legal defense expenses.

- The Insurance Service Office's (ISO) standard policy forms now include XCU and contractual liability, but contractors must obtain coverage enhancement for liability of others assumed in business contracts other than tort liability (liability that exists in absence of a contract).

- All ISO policies contain an absolute pollution liability exclusion; however, it is possible to pay additional premium to buy back pollution liability where the release of vapors, gases, or liquids is sudden and accidental.

TABLE 14.2   Hazards, Insurance Policies, and Interests Insured

Hazard or peril	Insurance policy	Parties covered
Bodily injury to the public, invitees, and trespassers; also, damage to the property of others	Commercial general liability (CGL), usually written in a first or primary limit of $1 million per occurrence with a $1 million annual aggregate limit	A basic requirement of all parties under AIA agreements: the named insured is the party purchasing the policy, additional insureds are also covered if listed by endorsement
Liability of others that you have assumed under the indemnification agreements of a contract	The CGL policy of insured contracts must be amended so as not to limit indemnification to tort liability only	Indemnitors under AIA and other written contracts; persons, governmental bodies, and other parties named as additional insureds, as well as others indemnified by contract
Explosion, collapse, and underground cave-in (XCU)	Included in the CGL policy	Third parties; does not cover damage to your own property
Personal injury: false arrest, humiliation, and other claims brought by outside parties	Usually included in the CGL	Third parties, not including your employees
Suits brought by workers' compensation insurers alleging your gross negligence	Employee's liability section of the workers' compensation policy	Employees and their spouses
Bodily injury to the public, invitees, and trespassers; damage to the property of others; gross negligence as employer	Umbrella and/or excess liability, in limits which reflect the degree of hazard associated with the specific operations the contractor has been hired to perform—total limits may be layered	This coverage is in excess of the CGL policy and, typically, employer's liability and automobile liability; coverage excess of these primary policies is usually also a basic contract requirement
Sexual harassment, racial discrimination, and claims arising out of the Americans with Disabilities Act (ADA)	Employment practices liability: a new separate form of insurance that cannot be added to the CGL policy, nor will a workers' compensation policy respond to these claims in the absence of bodily injury	Protects the interests of employers
Bodily injury, death, and damage to the property of others arising out of the insured's operation of the insured vehicles	Automobile liability	Contractor's registered trucks and automobiles and not bulldozer, cranes, and other equipment

continued

TABLE 14.2   Hazards, Insurance Policies, and Interests Insured *(Continued)*

Hazard or peril	Insurance policy	Parties covered
Liability associated with the work performed by the insured off-premises, after the work has been completed	Completed operations	Contractor and subcontractor
Sudden and accidental, and gradual seepage pollution	Environmental impairment liability (EIL)	Covers bodily injury and third-party property damage to the owner of the construction site as as damages to the surrounding environment
Violations of the Employee Retirement Income Security Act of 1974 (ERISA) regarding the administration of employee benefit plans	Fiduciary liability	Protects the fiduciary liability of the employer, the plan fiduciaries, and the plans themselves
Financial loss caused by design errors	Architect's errors & omissions liability	Covers liability associated with the design or plans of the insured architect
Construction of buildings in progress	Builder's risk	Provides coverage for loss or damage to the insured property and other specialized coverages as required by the construction project
Machinery and equipment not otherwise insured; usually large types of equipment	Inland marine	First-party coverage for loss or damage to the property insured
Bodily injury and damage to the property of others arising out of the operations that the contractor performs for the owner, or out of the owner's general supervision of the project	Owners and contractors protective liability	Protects the interest of the owner, not the contractor
Bodily injury and damage to the property of others arising out of the operations at the location where the contractor performs	Railroad protective	Protects the interest of the railroad, not the contractor

- A new form of coverage has evolved: employment practices liability insurance. This product usually provides coverage for claims such as wrongful termination and discrimination and other employment-related practices, policies, acts, or omissions.

14.2 Importance of Insurance Agents/Brokers and Insurance Companies

Insurance agents and brokers. Insurance agents and brokers essentially perform the same services. Technically, agents are employees of the insurance company(ies) they represent. Brokers technically work for the insurance buyer and are capable of arranging insurance with just about any insurance company. All agents/brokers must be licensed by the state in which the business is placed. Additionally, there are surplus lines agents who write specialty lines of coverage. Local agents/brokers go to these surplus lines agents when necessary.

Table 14.3 lists the largest insurance brokers according to *Business Insurance*. Under most state insurance laws, the agent or broker has a legal obligation to be diligent in the placement of coverage. In other words, if an agent/broker should fail to offer a contractor a necessary insurance policy or if an agent/broker should arrange for the wrong insurance policy form, the agent/broker can be held liable for a professional error or omission if an uninsured claim occurs. Builders and contractors can therefore be reasonably comfortable with the idea that their agent/broker will make an effort to assure that there are no errors in the placement of insurance coverage.

Additionally, insurance agents and brokers offer a wide variety of risk management support services such as:

- Issuance of certificates of insurance, endorsements, and other evidences of coverage

- Premium cost estimation and billing

TABLE 14.3   Top Ten Insurance Brokers According to *Business Insurance* (June 18, 1994)

Broker	1993 Revenues
Marsh & McLennan Companies, Inc.	$3,175,300
Alexander & Alexander Services, Inc.	1,341,600
Sedgwick Group, P.L.C.	1,216,169
Rollins Hudig Hall Group, Inc.	1,215,000
Willis Corroon Group, P.L.C.	1,057,000
Johnson & Higgins	962,000
Bain Hogg Group	397,097
Acordia, Inc.	364,779
JIB Group, P.L.C.	346,962

- Representation in matters pertaining to claims
- Advice in the selection of markets and coverage needs
- Preparation of coverage specifications and other underwriting data
- Loss prevention and loss control

Insurance companies. There are two kinds of insurance companies. Some, like Liberty Mutual Insurance Company, write business directly without the intermediation of an insurance broker. These are the direct writers. Their employees are the licensed agents and will not arrange for insurance with another insurance company. The second kind of insurance company requires placement to be made by either a licensed agent or broker; the insured cannot place or negotiate coverage directly with the insurer. Some brokers, however, may not have enough volume of business to place business with all of the companies listed in table 14.4.

Table 14.4 lists some of the largest property and casualty insurers in alphabetical order.

Surety companies. These are used to provide monetary compensation in the event that certain acts or obligations are not fulfilled by the bonded party. Whereas insurance policies are contracts of indemnification between two parties, surety bonds are guarantees that involve three parties. The principal is the party who is obligated to perform whatever acts have been specified, namely, those acts that required the securing of the bond. The obligee is the party that will be compensated for the loss should the principal to the bond fail to live up to its obligations. The surety is the party that provides and guarantees the bond and commits itself to pay for the debt or obligation of the principal if the principal fails to perform.

Certain bonds are used specifically to secure the obligations of general contractors; these are referred to as contract bonds. Two of the most commonly used contract bonds are bid bonds and performance bonds. When the contractor is being considered for work on a particular construction project, the contractor may be required to supply a bid bond. This type of surety bond guarantees that the contractor will actually enter into a contract should the contractor be selected from among others to do the work. The performance bond is required to begin the work and guarantees that the work will be completed as it was specified in the contract.

Other bonds which may be required are payment bonds and maintenance bonds. The contractor may be required to provide a bond which guarantees that all materials and labor used for the project will be paid

TABLE 14.4   Top 50 Property/Casualty
Insurance Companies

Aetna Insurance Company
Alfa Insurance Group
Allendale Mutual Group*
American International Group
American Reinsurance Company
Chubb Insurance Group
Cigna
Cincinnati Insurance Group
Dorinco Reinsurance Company
Empire Fire & Marine Group
Erie Insurance Group
GRE Insurance Group
Guaranty National Company
Gulf Insurance
Hartford Insurance Company
Home Insurance Company
Integon Indemnity Group
Interinsurance Exchange
Liberty Mutual
Mercury Casualty Group
National Indemnity Company
New York Central Mutual Insurance Company
New York Marine & General Group
Northland Insurance Group
Old Republic
St. Paul Companies, Inc.
United Fire & Casualty Group
United National Insurance Group
Universal Underwriters Insurance Group
USF & G
W. R. Berkley Corp.
Zenith National Insurance Group

* Direct writer

on time; this is a payment bond. Some insurers include both the payment bond wording and performance bond wording within the same bond. The maintenance bond finally guarantees the work upon completion, sometimes for up to one year, and provides the owner with security if the work is faulty or defective. In addition, separate subcontract performance and payment bonds also exist for subcontractors to the general contractor.

The contractor, on the other hand, may also obtain a release and retainage bond to guarantee that the hold-back is paid by the owner up front. This may be especially useful where the contract specifications include very costly materials, or the project is so extensive as to require excessive expenses.

The market for surety bonds, compared to the rest of the marketplace, is limited.

14.3   How to Conduct a Competitive Insurance Placement (broker bid)

According to the most recent survey of the members of the *Risk & Insurance Management Society* (RIMS), contractors and developers spend more for insurance than most industry groups. It is important for the business owner to periodically test the market to assure that premiums are competitively priced.

Property/casualty insurance can be placed through a bidding process similar in many respects to the construction contracts which are put out to bid; however, it is not a good idea to put the insurance out to bid every year. This is because constant circulation of a commercial account within the industry is not reviewed favorably.

The starting point is an examination of what one wants to accomplish. For example, if premium rates have gone up, the objective of a broker bid could be cost savings. Or, if the insurance agent or insurance company has not been providing a satisfactory level of services, then an objective of the broker bid should be to seek an agent or company that can provide better services.

Listed below is a 10-step process which can serve as a planning guide.

1. Create an insurance policy list, naming the insurance company, the limits of liability, the premium, and the expiration date. Organize the first draft list by date of expiration.

2. Set up a schedule. Insurance companies often require 60 days lead time so that they can get out to inspect the properties and construction sites. Allow another 30 days for you to evaluate the competitive quotes.

3. Synchronize the expiration dates of the insurance policies so that your policies all have the same expiration date. If necessary, ask your insurance broker to arrange for insurance policy extensions or consider canceling certain policies so they expire with all the other policies you include in the broker bid.

4. Compile the exposure data and information about your company that will be useful to the underwriters who will prepare insurance quotations. It will be important to give each competing insurance broker the same basic information. The agents/brokers will later tell you exactly what they will need. This will include workers' compensation payrolls, the number of vehicles to be insured, a list of current projects and their contract amounts, insurance company claim reports going back three or more years, and audited financial statements.

5. Make a short list of qualified insurance brokers. Unless the present insurance broker has not performed to your satisfaction, the incumbents should be at the top of the short list. Speak to your accounting firm, your lawyers, and others in the construction business about who is considered a good local broker. Include a direct writer or two on your short list.

6. Call up the short list participants to identify the person to whom to send an invitation letter. Then prepare a letter conveying basic payroll, sales, and financial information. Additionally, send each broker your list of insurance policies, but with the premiums omitted. Tell them what your objectives are and invite them to come to your office for a one-hour meeting. Ask them to bring a list of their first, second, third, etc., choices of insurance companies for each type of policy.

7. Based on the interview, narrow down the short list. Then assign each broker to an insurance company of his or her choosing. If two brokers have each chosen the same insurance company, decide which company to assign based on their order of preference. The brokers may need a letter from you in case their authority is questioned by the insurance companies, so make the appointment and assignments in writing. In the letter, tell them when the quotes are due.

8. Organize all of the underwriting information that was compiled in step four. Also, attach a copy of your most recent financial statement and promotional information about your company. The broker/agent will tell you what else they will need. Make certain all participants get the same information.

9. Thirty days prior to expiration, schedule a meeting with the participants to receive their quotations. In addition to the producer or salesperson, each brokerage firm should bring the account manager who will actually service your account. If you are interested in support services, the specialists in loss control, risk analysis, and claims should also be invited to attend this meeting.

10. Evaluate the competing quotations and select the best offering or combination of quotes. Get signed binders from the winning insurance agent/broker.

This 10-step process can be used by builders and contractors in most states; however, from time to time local market conditions may not favor the broker bid process—and with some types or lines of insurance this process may not work especially well.

Independent risk management consultants.   Business owners should feel confident in using their financial skills and general business judgment in evaluating competing insurance proposals. On the other hand, there are a number of situations in which it may be worthwhile to engage the services of an independent risk management consultant. Consultants can analyze your coverage, assess your need for a broker bid, and actually manage the 10-step process for you. Table 14.5 shows the largest independent consulting firms.

Other resources.   There are a good deal of reference materials available to the business owner. Your agent/broker should be able to provide you with this information, or you can subscribe to a publication. Below are two publications and how to get them.

Business Insurance is a weekly newspaper. Call (800)678–9595 to subscribe. The yearly subscription rate is $80.00.

Construction Risk Management is a reference manual published by the International Risk Management Institute, Inc. Call (800) 827–4242 to subscribe. For $221.00 you get 9 two-volume sets, periodic updates, and a quarterly newsletter. Thereafter, the annual renewal subscription is $126.00.

14.4   Property/Casualty Outlook for Builders and Contractors

The property/casualty insurance marketplace has been subject to cyclical changes in premium rates. When premiums start to firm up, this is known as a hard market in the insurance business. In the mid-1970s and mid-1980s the hardening of the casualty market was quite rapid and dramatic. As a consequence, many business owners found them-

TABLE 14.5   Independent Risk Management Consulting Firms

S. B. Ackerman Associates, Inc.
J. H. Albert International
Aronia Corp.
Coopers & Lybrand LLP
Deloitte & Touche LLP
Ernst & Young LLP
KPMG Peat Marwick LLP
Millemin & Robertson/Betterley
Towers Perrin/Tillinghast
The Wyatt Company

SOURCE: *Business Insurance,* March 7, 1994.

selves paying twice or three times as much for liability insurance. The market for workers' compensation insurance has also had its peaks and valleys, made complicated by a patchwork of differing state rates and rules and availability problems aggravated by bad politics.

At this writing, the property/casualty insurance cycle is in the ninth year of steadily declining liability insurance rates, i.e., a soft market. Property rates began firming in 1992; hence, some industry analysts have declared that we have entered a hard market but only for certain kinds of property insurance. What this means for contractors is that conditions could remain favorable in the workers' compensation and liability lines for the indefinite future with little disruption in coverage and moderate firming of rates. During this period of stability, business owners should review their coverage to assure that it is as broad as can be. Insurance price concessions should be less for those who conduct broker bids. This may be a good time to review your risk management policies and procedures to determine if you are receiving an acceptable level of support from your agents/brokers and insurance companies and that your risk control practices are reducing claims activity.

15

Claims Management

B. Calvin Deaner
Willis Corroon Corporation

15.1 Purpose

A sound claims management program is essential and is probably secondary only to an effective loss control program. Aside from the obvious direct costs of claims, a proper claims administration program can and does affect the price of both primary and excess insurance programs. The cost dedicated to an effective and efficient claims administration program will reduce insurance cost substantially over a period of years.

The following pages have been prepared to offer proven procedures that can and will assist in the reduction of your insurance costs directly to the bottom line of profit and the protection of corporate assets.

15.2 Procedures

Establishment of procedures endorsed by top management for prompt and efficient claim reporting is imperative. One of the components of cost control is the control of claims without legal representation. Delays can only lead to increased costs.

A well-prepared claim procedures manual is helpful in the implementation of proper reporting. These procedures should be specific and concise. A primary focus of the project personnel is to complete the construction contract, so the procedures need to be in such a format that they are easily understood and not overly detailed. We have included a guideline in Sec. 15.9 at the end of this chapter.

Many times there are needs to preserve the evidence and to secure information from witnesses and employees immediately following an accident or incident. Construction sites change rapidly; securing and safeguarding evidence can make a difference in the outcome of a case and greatly impact the costs. As a guide, Figure 15.1 (p. 242) contains a claims investigation section.

Establishment of claim coordinators with the insurance carrier will provide an expeditious resolution to any claim dispute. The broker in the marketplace for contractors allows the ability to establish these contact people who have the authority to move swiftly toward the resolution of any problem and the ability to act upon the issues for expediting resolutions. A claim coordinator will be the contact person for the establishment of thresholds of reserving and settlement discussions. Designated persons and the insurance company representatives can many times exchange ideas and information in discussion that can lead to cost reductions of insurance issues. For instance, there may be an employee who has information regarding a fellow employee that can be learned only through relationship with the workers' compensation claimant—information that the insurance company employee would not have the ability to access or determine.

Quality assurance audits of the case file on a periodic basis, no less than semiannually, is essential. These reviews will provide a detailed analysis of the carrier's performance in the processing and handling of claim files. The following areas will be considered:

- *Evaluate reserving practices.* Reserving is certainly not an exact science; however, the experience of the broker personnel should provide a fairly accurate cost projection for each case. These reviews will identify and target the cases that have the high cost potential.

- *Evaluate the promptness of the carrier in processing and handling of cases.* As stated earlier, promptness is the essential control factor of any case and, therefore, can lead to a reduction in costs.

- *Evaluate the thoroughness of the investigation.* Are the carriers making proper investigations to determine responsibility, recovery possibilities, and the need to involve other parties in the process?

- *Evaluate negotiations and settlement tactics.* Are the carriers moving forward toward settlement of the cases that need to be settled? Are they taking a firm stand on those cases that need to be resisted and defended?

The selection of the proper attorney for the case at hand is important for favorable results. The broker should coordinate with the carrier for client input in the selection of attorney on cases with specific interest, i.e., successful results of prior similar issues or causes of action.

There will be cases where coverage issues raised by the insurance carrier, based upon allegations of a lawsuit or factual issues of an accident or incident, will give rise to a reservation of rights and perhaps a declination of coverage. The broker should act on the behalf of contractors to aggressively protect their interests as per the insurance policy terms and conditions.

15.3 Workers' Compensation

Few issues in claims administration are more critical than seeking ways to control workers' compensation costs. Over the past eight years medical costs alone have increased from $6 to $45 billion and continue to rise. All jurisdictions have increased benefit levels, and the rates are exceeding those of inflationary factors. Decisions of the hearing board, commissioners, and courts of law continue to maintain liberal positions in the finding of compensability in medical injuries such as carpal tunnel syndrome, emotional trauma, and ergonomics.

The broker should continuously seek ways to control workers' compensation costs. These costs determine the experience modifier. If experience modification reaches an unacceptable level, the ability to compete on jobs is lost. In fact, some owners whose jobs include state and federal projects will not permit bidding by contractors if their experience modifier exceeds a certain level. We offer the following recommendations for cost reduction from the claims administration section of your risk management program.

- Educate your people as to what benefits are available to them if they are injured while working. Most employees are well informed on coverages such as medical benefits, life insurance, etc., but seldom does the employer explain what happens when a job-related injury or sickness occurs.

- Control the processes involved in workers' compensation claims: establish prompt reporting procedures; bring about early contract with contact; show empathy to employee; explain when the employee can expect that first indemnity check. *Control is the key to cost containment.*

- Establish procedures to alert the carrier of fraudulent claims. Use words on first report such as "alleged." Don't make statements of fact if there are any questions about accidental injury. Call the carrier for additional conversation if you are concerned about the circumstances of any particular loss.

- Institute a return-to-work program.

- Encourage proactive return to work. Know duties of the employee; relate to doctor for assessment and collaborate with employer and doctor for return to productive capacity or an alternative or modified position. Establish a sound drug testing program.

15.4 Casualty

Concentrated efforts by the legal profession to attack the exclusive remedy of workers' compensation defenses have eroded this valuable defense of employees' actions that arise from injuries on the job. This defense is the theory that all employers have the legal obligation to provide their employees with a safe place to work. Failure to do so creates a legal liability that has opened an area of litigation that did not exist 10 to 15 years ago. Many jurisdictions have ruled that the obligation of safety cannot be delegated, nor can safety be passed through to the other parties by contract. These issues must be addressed on a state-by-state basis.

In an effort to shift costs such as the employee action over cases and other tort actions that may arise from construction projects, we offer the following recommendations to reduce financial risks.

- Transfer risk through clauses dealing with indemnity and insurance found in all construction contracts. These clauses attempt to protect the owner, general contractor, or subcontractor(s) from claims arising in the performance of contract terms and conditions made by the general contractor to the owner and by the subcontractor to the general contractor. There are also such agreements between subcontractor levels.

- Each jurisdiction has statutes and case decisions that affect the enforcement of these provisions. The decisions change quite frequently and need to be updated as case decisions are handed down for your jurisdiction. Consultation with your attorney is prudent and recommended to best conform with your jurisdiction requirements. A few states still hold that such agreements to contract away your legal responsibility are against public policy and are null and void.

    Notwithstanding conforming with state statutes and case decisions, the indemnity clauses are not absolute protection as you are responsible for your own negligence. Many times questions of sole negligence cause the greatest problem in seeking this protection. More and more of the insurance carriers are resisting the defense indemnity based upon the hold-harmless agreements. These carriers raise the issue of sole negligence and, therefore, refuse to provide the protection that is sought by the hold-harmless agreement language.

    For these reasons the additional insured protection is needed without exception. It is most desirable to insure additionals on a primary basis, stating that your insurance is noncontributory. If the insurance is not on a primary, noncontributory basis, the insurance carrier for the indemnity will raise the issue of concurrence. Concurrent coverages can be addressed either by limits or on an equal shared basis, depending on the interpretation of the jurisdiction involved.

15.5 Certificate of Insurance

Hold-harmless agreement clauses are only as good as the ability of the indemnitor to assume the financial responsibilities as per contract agreements; therefore, owners require general contractors to provide certificates of insurance, general contractors require subcontractors and suppliers, subcontractors require sub-subcontractors, and so on, as evidence of insurance. The certificate is only evidence of insurance and does not change policy terms and conditions. Ideally, in those situations where the certificate indicates that a party is named as additional insured, endorsement as opposed to certificate should be secured. Most carriers will honor the certificate as being enforceable; however, the endorsement is desired.

Following is a checklist for doing certificates of insurance.

1. Determine the rating of the company providing the coverage.

2. Is the general liability coverage on an occurrence basis rather than a claims-made basis?

3. Check policy dates: is coverage in effect at the time the contract is issued? How do the terms relate to the length of the contract?

4. Review limits: are they ample and do they conform with contract requirements?

5. Automobile insurance: if this exposure exists for contract performance, then coverage required and limit should be examined closely. Name-insured status should be required if there is use of an automobile that creates an agency relationship.

6. Excess coverages: check on dates, i.e., do they conform with contract terms and conditions, what is the length of contract, and are limits adequate.

7. Workers' compensation: check dates of coverage. Do they conform with the project contract dates? If the subcontractor is not insured, then responsibility of providing coverage will fall upon the general contractor.

8. Under "miscellaneous items" of the certificate, check the wording for additional insured on a primary noncontributory basis for all coverages except workers' compensation. If coverage cannot be secured on a primary noncontributory basis, then additional insured coverage should be required without exception. The owner, general contractor, and subcontractor are to be added as additional insureds on a pri-

mary noncontributory basis. Waivers of subrogation for workers' compensation and builder's risk should be included. Confirmation that workers' compensation and builder's risk policies have been endorsed for the waiver is desired. If the workers' compensation and builder's risk policy have not been endorsed, the enforceability of this waiver is not binding upon the insurance companies, even though the contract provisions may state otherwise.

15.6 Control Programs

Certificate holders make the mistake of requiring certificates of insurance when the contract is executed, but do not obtain new ones when the policy expires. This creates additional risk of possible nonrenewal of material changes in the coverage, terms, and conditions. For this reason, it is important to implement a suspense system to acquire new certificates during the course of construction.

15.7 Builder's Risk

Builder's risk insurance is needed during the construction phase of a project. This insurance is written on the basis that the values increase as the building is constructed. It is important to check policy terms and conditions to make sure they comply with contract requirements. One of the most important requirements is that if builder's risk is the only provided coverage, then the general contractor and all tiers of subcontractors must be named as additional insureds. This gives the parties the same rights as the owner with the exception of cancellation and alteration of coverage. This is extremely important in jurisdictions where there are cases holding which require that the general contractor and subcontractor are only protected for the scope of their work. Also, it is equally important to determine the deductibles that apply to certain perils and what arrangements are made with the owner for the satisfying of these deductibles if it is an owner-provided coverage.

15.8 Litigation Management

Some ways to manage litigation include the following.

- Employ experienced construction attorneys
- Require an assessment from the attorney at the time of assignment addressing recommendations and strategy for handling the case
- Control all activities and require approval of all discovery, depositions, etc.
- Require monthly statements of costs and activities

- Agree on fees and costs at the time of assignment
- Standardize reports for depositions, pretrial reports, and medical reports and examinations in summary form
- Consider arbitration or mediation which is less expensive and much quicker to reach resolution

15.9 General Claims Procedures

This information is to guide you in the reporting of insurance claims for the various types of insurance coverages applicable to your operations. The prompt and accurate reporting of claims is critical to comply with state regulations and our insurance contract provisions as well as to provide the insurance carrier with maximum control claims, thereby reducing the overall cost.

Responsibility: The *project superintendent/foreperson* or *department head* is responsible for the prompt and accurate reporting of claims.

Company headquarters contacts: Assistance and information concerning operations and insurance programs can be obtained by calling either of the following:
XYZ INSURANCE AGENCY
100 MAIN STREET
ANYTOWN, NY 88888
(555)555-5555

Insurance coverage: Our company maintains a very comprehensive insurance program.

- *Workers' compensation.* For injuries or occupational diseases sustained by our employees while at work
- *General liability.* For claims made against our company for damage to the property of others or injury to nonemployees
- *Automobile.* For any damages or claims related to company cars, trucks, or rental cars used for business
- *Equipment.* For loss or damages to contractor's equipment (owned and rented); for example, backhoe, crane, etc.
- *Property insurance.* For loss or damage to property, either real or personal, owned by the company (i.e., building and contents)
- *Builder's risk.* For loss or damage to property while in the course of construction
- *Other insurance.* Crime (employee dishonesty and forgery) and electronic data processing (including media and data)

I. Workers' Compensation

The company is required by law to provide prompt payment of workers' compensation and occupational disease benefits to any employee who is injured in the course of employment. These benefits include loss of wages and medical costs as outlined by state regulations.

A. Claims reporting forms: *First report of injury*—This form must be completed and copies mailed promptly as soon as complete injury information is available. (Attach copy for state form.) Be sure to:

1. Provide complete information.
2. Describe the accident in detail. Tell *what* happened and *how* it happened.
3. Contact doctor if necessary to determine extent of injury.

B. Report serious injuries by telephone to risk management *and* corporate offices. Serious injuries should be reported by telephone to the office as soon as possible so that the other claim can be given immediate attention and then followed up by the written report. Examples of serious injuries are as follows.

1. Death
2. Injury to head, eye, or ear if medical complications are apparent
3. Burns requiring hospitalization
4. Amputation of any part of the body
5. Lost time expected to exceed two weeks
6. Multiple injuries
7. Complicated fractures
8. Electrical shock
9. Spinal cord injury

C. Questionable injuries: If you have any reason to suspect a fraudulent claim, this should be indicated in a *separate* report with a full explanation of your reasons for questioning the claim.

II. Automobile Insurance Claims

Our automobile insurance provides coverage for claims and damage in connection with company vehicles (owned or rented) as defined in the following statement.

Liability: Covers our legal obligation to pay for bodily injury or property damage to others as a result of the use of a vehicle by or on behalf of the company.

A. Claims reporting form: Use the ACORD form which gives instructions for completion. Be sure to provide complete information. The report should be sent to broker as soon as possible.

B. Report serious accidents by telephone to office. Serious accidents should be reported by telephone to the office as soon as possible so that the claim can be given immediate attention and then followed up by the written report to broker. Examples of serious accidents are as follows.
  1. Serious injury to any person. Generally, this would be any injury requiring hospitalization.
     Note: If injury is to an employee performing company business, a workers' compensation report must also be filed.
  2. Major damage to vehicles or other property.
III. General Liability Insurance
The general liability insurance provides coverage for the company's legal obligation to pay for bodily injury or property damage sustained by others as a result of our operations. It does *not* cover:

- Injury to our employees (see workers' compensation)
- Damage to the facility being constructed (see builder's risk)
- Damage to equipment owned or rented by us (see property insurance)

A. Claims reporting form: Using the ACORD form which gives instructions for completion, be sure to provide complete information. The report should be sent to the insurance carrier and your broker as soon as possible.
   Note: This type of report should be filed whenever there is a possibility that a claim could be made against us because of injury or property damage, even though you feel our company might not be directly liable.
B. Report serious accidents by telephone to insurance carrier and your broker. Serious accidents should be reported by telephone as soon as possible so that the claim can be given immediate attention and then followed up by the written report. Examples of serious accidents are as follows:
  1. Serious injury to any person. Generally, this would be any injury requiring hospitalization.
  2. Major damage to property of others.
  3. Accidents that involve interruption of services or operations of others such as electrical, telephone or other utilities, and roadways.
C. Documentation: As soon as the circumstances allow, be sure to take the time to make written notes of all details relevant to the claim.

1. Be sure to include names of people who might have knowledge of circumstances.
2. Take photographs, if appropriate.
3. Keep and store all physical evidence related to the claim.

IV. Equipment Insurance

The company provides insurance coverage for loss or damages to equipment, tools, and office furnishings on the job as follows:

Company-owned equipment: We have blank coverage for all company-owned equipment, tools, and office furnishings.

Rented equipment: Coverage is available for equipment rented from others. Generally, this would be for large pieces of equipment where the rental agreement makes you responsible for the insurance coverage.

A. Reporting claims: Any loss or damage to insured equipment must be reported immediately by telephone to the issuance carrier and your broker. Use the ACORD form—property loss notice. When reporting losses, be prepared to give the following information.
   1. Date and time of loss or occurrence
   2. Cause of loss—describe what happened
   3. Description of what was damaged or stolen
   4. Estimate of amount of loss
      Note: If loss or damage appears to be the result of a criminal act, notify the local law enforcement officials immediately.

B. Documentation: As soon as circumstances allow, be sure to take the time to make written notes of all details relevant to the claim.
   1. Be sure to include names of people who might have knowledge of the circumstances.
   2. Take photographs, if appropriate.
   3. Keep and store all physical evidence related to the claim.

V. Property Insurance

The company provides insurance coverage for loss or damage to real and personal property.

A. Reporting claims: Any loss or damage to insured property should be reported immediately by telephone to the insurance carrier and your broker. Use the ACORD form—property loss notice. When reporting losses, be prepared to give the following information.
   1. Date and time of loss or occurrence
   2. Cause of loss—describe what happened
   3. Description of what was damaged or stolen

4. Estimate of amount of loss
   Note: If loss or damage appears to be the result of a criminal act, notify the local law enforcement officials immediately.

B. Documentation: As soon as circumstances allow, be sure to take the time to make written notes of all details relevant to the claim.
   1. Be sure to include names of people who might have knowledge of the circumstances.
   2. Take photographs, if appropriate.
   3. Keep and store all physical evidence related to the claim.

VI. Builder's Risk
The company provides insurance coverage for loss or damage to materials and supplies while in the course of construction and installation.

A. Reporting claims: Any loss or damage to insured property should be reported immediately by telephone to the insurance carrier and your broker. Use the ACORD form—property loss notice. When reporting losses, be prepared to give the following information.
   1. Date and time of loss or occurrence
   2. Cause of loss—describe what happened
   3. Description of what was damaged or stolen
   4. Estimate of amount of loss
      Note: If loss or damage appears to be the result of a criminal act, notify the local law enforcement officials immediately.

B. Documentation: As soon as circumstances allow, be sure to take the time to make written notes of all details relevant to the claim.
   1. Be sure to include names of people who might have knowledge of the circumstances.
   2. Take photographs, if appropriate.
   3. Keep and store all physical evidence related to the claim.

VII. Other Insurance
The company maintains various other coverages such as crime and electronic data processing insurance.

A. Reporting claims: Any loss or damage to insured property should be reported immediately by telephone to the insurance carrier and your broker. Use the ACORD form—property loss notice. When reporting losses, be prepared to give the following information.
   1. Date and time of loss or occurrence
   2. Cause of loss—describe what happened

3. Description of what was damaged or stolen
4. Estimate of amount of loss

   Note: If loss or damage appears to be the result of a criminal act, notify the local law enforcement officials immediately.

B. Documentation: As soon as circumstances allow, be sure to take the time to make written notes of all details relevant to the claim.

   1. Be sure to include names of people who might have knowledge of the circumstances.
   2. Take photographs, if appropriate.
   3. Keep and store all physical evidence related to the claim.

VIII. Lawsuits

Lawsuits are generally served upon the company through a U.S. marshall or deputy sheriff or through the state in which you have operations via the U.S. Postal Service. The statutory time allowed for filing an answer to a lawsuit commences to run upon service to the defendant. It is, therefore, necessary that the actions described below be done *immediately.*

A. Telephone your risk management and corporate counsel and advise them of the details of the lawsuit, e.g., plaintiff's name, basis of complaint (automobile accident, fall down injury), dollar amount involved (shown on the last page), and if the claim had been reported to the insurance carrier previously. Follow the counsel's instructions as to how to proceed.

B. Immediately deliver the original to the risk management department, with copies to be sent to corporate counsel, the insurance company, and your broker.

C. Do not discuss the lawsuit with any persons other than your employer or representatives of your insurance carrier.

Failure to carry out these instructions may result in a judgment being awarded by default and voiding of insurance coverages to pay such judgment.

IX. Accident Investigation

An accident investigation is an analysis evaluation and report of the accident based on information gathered by the investigator. The quality and usefulness of the information is directly related to the degree of thoroughness of the investigation.

A serious accident would be investigated immediately after its occurrence. Information is more accurate when there is less time between the accident and the investigation. Facts are more accurate because people have not had time to be influenced by the opinions of others, memories are clearer, and more details are remembered. Also, prompt arrival at the scene allows the investigator to observe evidence before it has been removed or altered.

The number of persons required to make an investigation will depend on the nature of the accident, its magnitude, and its technical complexity. Project personnel to include on the investigation team are the following:

- Superintendent
- Risk management
- Employer's supervisor

It is essential to preplan investigations so that the appropriate persons know how and to whom to report accidents, that people responsible to investigate know what to do, and that equipment needed for the investigation is available when and where needed. Consider the following areas to make sure it is complete.

A. Locate and identify evidence relevant to the accident. Evidence will be available from:
   1. People involved—injured, witnesses, and other workers
   2. Materials involved
   3. Environmental factors involved—weather, light, heat, noise, etc.
B. Examine the evidence to determine its impact on the accident sequence.
C. Reconstruct the sequence of events based on the evidence.
D. Develop conclusions from the evidence.

X. Serious Accident Procedure

In case of a serious accident as defined previously in Sec. I.B., the following must be done:

A. *The injured is to receive proper care.*   The first concern is to see that the injured receives immediate medical care.
B. *Protect other people and property, and secure the accident site.* Rope off or barricade the area to keep bystanders from being injured or destroying evidence.
C. *Proper reporting is essential.*   As a reminder, call risk management and the insurance manager immediately to report the loss. Follow with a written report, including a copy to the insurance carrier and your broker.
D. *Make a walk-through of the accident site.*   Conditions at the scene will change rapidly. Record as much information as possible before any cleanup or changes take place. This information should be documented in such a way that the evidence can be presented and allowed at a later date.
E. *Identify as many people as possible who might have information concerning the accident.*   Record all persons who may

have witnessed the accident. If witnesses are not your employees, obtain their addresses, telephone numbers, and their employer's name.

F. *Photograph all evidence.* Photographs of the general area, major elements of the accident site, and articles of evidence should be taken as soon as possible after the accident. These should be made before any items are moved if at all possible.

G. *Make a diagram of the accident site.* A sketch should be made of the accident scene, showing the location of all evidence essential to understanding the accident situation. Distances involved should be measured and recorded on the sketch. Also, record any DOT station numbers if the diagram is of an accident site on a road that is under construction.

H. *Examine the evidence.* Anything that will provide information about what happened, how it happened, and why it happened must be identified and examined.

I. *Interview and obtain statements from all witnesses.* All persons who may be able to contribute information about the accident should be interviewed as soon as possible after the accident.

J. *Prepare an accident report.* A written report must be prepared for all serious accidents and a copy mailed to the risk management department as well as a copy to your insurance carrier and broker.

1. Securing the accident site

   To prevent any alteration of the scene prior to completion of the investigation, the site must be secured immediately after the accident. The site should be maintained as it was at the time of the accident so that it can be examined and preserved on film. Securing the site should also be considered if aspects of the accident could develop an adverse reaction from the public or other employees.

   The method used to secure the site will depend on the conditions and circumstances involved. Several methods are:

   Roping off the area using barricade, tape, string, or rope.

   Closing a walkway or stairway leading to the area.

   Using company employees to prohibit access to the area.

   Hiring security guards to patrol the area and prohibit access by unauthorized persons.

   Whatever methods are used, the following procedures should be followed:

   a. Nothing should be removed from the scene without the approval of the person in charge. Preserve all physical evidence.

b. All guards (employees) should be instructed not to touch, move, or mark anything at the accident site unless authorized.
c. An entry point should be provided to control the number of people going in and out.
d. Security guards (employees) should be provided with a list of personnel authorized to enter the area.
e. Security guards used should be closely supervised.
2. Diagramming accident sites
   Diagrams should be prepared if they contribute to the accident report. They should be cross-referenced in the report unless they are self-explanatory. Only information essential to understanding the accident situation should be included on accident diagrams. They should be easy to understand, accurate, and well-drawn. Viewers should be able to easily orient themselves and place critical items in the drawing.

   All relevant distances, measurements, angles, DOT station numbers, etc., should be included in the diagrams of the accident sites. Many times these distances, measurements, etc., may not seem important but can have a very significant bearing on the outcome of the payment of a claim. The plan view (looking down on a scene) is common. A side view should be prepared if it will assist in clarifying the situation. Persons trained in preparing diagrams, such as surveyors, draftspersons, and engineers, can be of great assistance in preparing diagrams.
3. Interview
   a. *Interviews with witnesses.* To determine all facts and circumstances surrounding an accident, witnesses must be interviewed and the information gained from them documented. The information obtained will be used as a basis for preventing future accidents and will also be beneficial to your company if a lawsuit follows.

   All persons who can contribute information about the accident should be interviewed. It is important to question witnesses as soon as possible, while the event is fresh in their minds. The longer the timespan between the accident and the interview, the more details are forgotten. Witnesses may see or hear things that revise their thinking; they may unconsciously adjust their observations to fit what they hear from others. The longer the timespan, the more opportunity people have to establish alibis or stories. Promptness in interviewing is the best solution.

b. *Interview location.*   Interviews should be held either at the scene of the accident or in a private location. An interview at the scene of the accident is advantageous if the extent and nature of damages, noise level, and degree of privacy will not interfere with the interview. At the site, interviews usually result in better recall of details as the witness has physical reminders to stimulate his or her memory. There is also more accurate positioning of relevant items and people involved than there is by reference to a diagram.

c. *The interviewer.*   Before beginning to interview witnesses, it is advisable for the interviewer to survey the accident scene and environment to get the big picture of the accident. A brief visual orientation of the situation helps put the investigator on the same plane as the witness. Awareness of what the area looked like and what machines or materials were involved makes it easier for the investigator to absorb and rank information, to see meaningful data from witnesses and to sift out information that is irrelevant or in conflict with the physical facts of the accident.

Several qualifications possessed by a good interviewer are:

- Good listener—all useful information will come from the witness. Let the story be told in full without interruption before asking specific questions.

- Open mind—even though the interviewer may have some preconceived ideas about the event, he or she must attempt to erase these notions and not jump to conclusions during the interview.

- Self-control—any loss of temper during the interview will cause an immediate breakdown of the interview and destroy its effectiveness.

- Courteous—if the interviewer becomes antagonistic, the effectiveness of the interview will, in most cases, end at that point.

- Thorough—at the completion of the witness's statement, the interviewer must ask questions to clear up any gray areas and obtain a more complete story.

An interviewer must assume the level of the person being interviewed. When the interviewer talks down to the witness, the effectiveness of the interview may be lost.

d. *The interview.*   The success or failure of an interview is often determined at the opening point. The witness should be put at ease as much as possible by explaining that the object of the interview is to obtain all information available so that problems can be identified and future similar accidents avoided. The witness should be told that the interview is for *fact-finding*, not *fault-finding*.

e. *Witness's version.* The actual interview should be started by asking witnesses to relate, *in their own words*, what they know about the accident, what they saw, what they heard, what they felt, and what they did after the accident. Each witness should be allowed to relate events and observations of the story chronologically or in any other manner. The interviewer must be a good listener during this time in order to structure the evidence, as well as determine the areas which need in-depth expansion.

The witness should be allowed periods of silence to collect his or her thoughts and explore memories. Neutral questions such as, "Then what did you do or see?" can help keep the witness talking. Witnesses should be allowed to talk but not ramble. If necessary, the interviewer should interrupt and turn the witness back to the subject and keep the pace moving along.

4. Written statements

Written statements should be obtained from all key witnesses (see Fig. 15.1). These can help to provide a clear, detailed description of an accident and can be used in legal hearings. If more than one page is used, the witness should initial each page below the bottom line to prevent allegations that comments were added without the witness's knowledge. If the witness refuses to sign the statement, the interviewer should write a comment to that effect and sign the statement alone.

The interviewer should construct the statement in the general language of the witness. The statement should be organized with facts and points in logical order. After the statement is written, it should be read aloud to the witness, point by point. The witness should comment on each point and be given the chance to correct any words and writing above them. The witness should sign at the end of the statement and initial beside each change made.

5. Accident report

A written report should be completed for all serious accidents. The report should be completed by the person(s) who investigate the accident and should be completed as soon as possible after the investigation is completed. The report should contain the following:

WRITTEN STATEMENT FORMAT

NAME _____   EMPLOYER _____

ADDRESS _____   POSITION _____

_____   THIS STATEMENT IS IN REFERENCE TO

_____   _____

TEL. NO. _____   _____

_____

DATE _____

DESCRIBE THE EVENTS THAT LED TO THE ACCIDENT, HOW THE ACCIDENT
HAPPENED AND ANY OTHER INFORMATION WHICH MAY BE HELPFUL.

_____

_____

_____

_____

_____

<u>NAME OF OTHERS WITH KNOWLEDGE OF THE ACCIDENT</u>

1. _____   The foregoing statement is true to the best
                                        of my knowledge and memory.
2. _____

3. _____   _____

4. _____   Signature of Witness

                                        _____

                                        Signature of Interviewer

Figure 15.1   Written statement format.

a. Detailed description of the accident including answers to
   the following:
   - What happened?
   - Who (individuals and companies) was involved?
   - When did the accident occur?
   - Why did it happen?
   - What injuries/property damage resulted?

b. List of who was notified—owner, insurance company,
   OSHA, MSHA, etc., and when they were notified

c. List of who investigated the accident—owner, insurance company, OSHA, etc.
d. Photographs taken
e. Diagrams made
f. Witnesses' statements
g. Contract documents involved—rental agreement, hold-harmless clause, etc.

A copy of the completed report should be sent to the risk management department and your insurance carrier and broker.

Managing Safety for Profit

Steven D. Davis, CPCU, ARM
Willis Corroon Construction

16.1 Program Administration

16.1.1 Corporate safety and health policy/mission statement

It is very important that a construction company convey to its employees and to the industry its philosophy. This can be accomplished through either a mission statement or a safety policy statement.

Often the success of the company's efforts depends upon a thorough understanding and acceptance of safety and health principles. Primarily, this philosophy should encompass the belief that all personal injuries can be prevented and health risks contained. A commitment to safety is a commitment to doing things right *the first time*. If a company truly believes this, what you gain, ultimately, is the elimination of *all* accidents. Accidents do not have to happen.

16.1.2 Yearly goals and objectives

A company must establish criteria to promote improvement. This can best be achieved through the establishment of quantifiable yearly goals and objectives. The objectives should be analyzed for measurability and examined for attainability, and a determination should be made as to whether or not they are rewarding and challenging.

16.1.3 Commitment by top management

Within an industry where competition is the byword, safety management has become an extremely important component. The safest construction companies are those which recognize the benefits of safety and provide a culture in which the employees see safety as a given, companies in which safety is an acknowledged priority. These companies create a culture that values safety.

Planning, organizing, and managing the safety effort in a manner consistent with overall management of quality, production, schedule, and cost is key to the success of the construction company. It is not enough for management to just approve of a safety program; management must ensure the safety process is working. Top management must become involved on a day-to-day basis. Employees, both superintendents and craftsworkers, have to see evidence of management desires and commitment concerning safety.

Although top managers are not present daily on projects to supervise construction work, if managers talk to superintendents about production, projects generally finish ahead of schedule. If cost is emphasized, the unit rates tend to improve. Where improvement of safety is the topic of conversation, safer patterns of behavior will result. Where safety is not specifically included as a focus of management, superintendents and workers alike pick up on the subtle difference in approach and respond accordingly.

16.1.4 Supervisory responsibility and accountability

In too many companies and especially in construction, using safety equipment and following safety procedures has been more or less the option of the employee. When a supervisor is queried about a safety problem, the frequent response is "I can't get the employees to use the equipment or to follow the procedure." In all other phases of operation, employees follow procedures. Why is safety different? Why should a supervisor or superintendent gamble with the profits or the future competitive position of a company? Tell employees what you want and then hold them accountable.

Where employees are to be held accountable, there must be performance standards. Companies need a safety and health procedures manual which sets forth, in substantial detail, the method and manner by which the company intends for its supervisors and craftsworkers to comply with each of the hundreds of industry-accepted safety practices and regulatory requirements. The safety manual should provide directions to workers for avoiding injuries and illnesses as well as damage to equipment and property. It is the blueprint for safety success.

Being held accountable monetarily can modify behavior to yield significant results. Examples of how this may be accomplished include: charging accident costs to the project by charging claim costs or by including accident cost in the profit and loss statement, prorating insurance premiums, putting safety into the supervisor's appraisal, and establishing safety rewards and penalties that affect the supervisor's income.

Setting goals and plans such as the establishment of a safety action plan cooperatively with the supervisor reinforces top management's commitment to safety.

16.1.5 The safety professional

Those companies that are large enough and choose to employ a safety professional need to be cognizant of the role this person should play in the safety process. The fact that safety is part of his or her title allows for misunderstanding. It must be understood that the safety professional has no responsibility for the safety record or the results. The safety professional's many activities include: consulting with management on the technical and organizational aspects of safety, training managers so that they may fulfill their safety responsibilities, introducing measurement systems to management by which supervisors can be held accountable for safety performance, assisting in monitoring safety performance of company projects, developing safety training and orientation materials, introducing safety considerations into planning at all levels, and keeping organizations up-to-date on safety changes. One of the most important aspects in utilizing the expertise of a safety professional effectively is to place that person in a position where he or she reports to someone who can affect change.

16.2 Program Content

16.2.1 Developing a safety action plan

Just as each construction company develops a game plan for each job and addresses issues such as construction techniques, optimum locations for material storage, amount and type of equipment, staffing levels, detailed schedules, items to be subcontracted, etc., to ensure success, planning for safety and health concerns is also essential. Planning reduces the potential for injuries and unwanted losses through identification of potential hazards and problem operations. Once these hazards are identified, they are documented. Managers and superintendents can then devise strategies for eliminating or controlling the hazards which are also documented. A plan that is unique to each job can be given to superintendents and workers. It clarifies for each person what is expected to be done. Keep in mind that a safety action plan is dynamic, ever changing to accommodate the needs of the job as the work progresses.

It is our belief that progressive contractors will benefit from the development and implementation of new and nontraditional approaches to workplace safety. The core of this effort is increased worker involvement in the daily work planning and increased worker commitment to error-free performance.

We suggest that substantial improvements in overall job performance can be achieved by conducting daily planning meetings between superintendents and supervisors and between supervisors and crew. Specific items should be on the daily agenda. The superintendent/supervisors meetings should cover: (1) work to be performed next shift; (2) methods, techniques, and equipment to be used and appropriate safety reminders; (3) any actual or potential production or safety problems; and (4) an assessment of work in progress as well as future work with immediate feedback if any potential interference with safety or production is perceived. The supervisor/crew meeting should cover: (1) work to be performed next shift; (2) methods, techniques, and equipment to be used and appropriate safety reminders to be given with each new work assignment; (3) any actual or potential production or safety problems; (4) an assessment of work in progress as well as future work with immediate feedback if any potential interference with safety or production is perceived; (5) any substandard practice or substandard condition witnessed that was not immediately corrected; and (6) any near-miss experiences or injuries incurred. If a substandard practice or condition was witnessed, supervisor should take notes and handle or pass up the line; if any near-miss experience or injury was incurred, supervisor should take notes, pass up, and participate in review and investigation.

It is our further belief that companies can dispense with the weekly toolbox safety meeting and move to a proven communication technique, the daily safety contact, wherein each supervisor has a one-minute safety meeting with each individual in the crew every day. The purpose of this contact is to convey a safety message and compliment positive behavior, although unsafe behavior must be dealt with promptly and effectively also. Brief notes should be kept by each supervisor as to employee name, topic of discussion, date, and time of day.

We believe that by increasing employee participation in the work assignment process and by discussing safety as a function of production management, you will obtain increased employee commitment to your goals in the areas of lowest possible cost, maintaining and improving schedule, and highest quality, error-free performance.

16.2.2   Accident investigation policy

An accident investigation policy should be written to establish what should be done following an injury, reinforcing the purpose and stressing promptness. A serious accident should be investigated immediately after its occurrence. The less time intervening between an accident and investigation, the more accurate the information that can be obtained.

Facts are more accurate because people have not had time to be biased by the opinions of others, memories are clearer, and more details are remembered.

Conditions at an accident scene are the only things that change faster than opinions when there is a delay. The contact phase of an accident is brief and initiates a wide spectrum of activity. People responding to an accident generally react rather than respond and, unless well trained, their reactions are not always rational. Injured people are moved about or removed for treatment. Equipment and items are moved about to treat the injured, provide passage, or restore operational work. Many times, things are stolen as the opportunity presents itself. Prompt arrival at the scene allows the investigator to observe evidence before it has been removed or altered.

16.2.3 Employee selection and orientation

Every company has its own way of determining how to hire employees. The company that feels it has to rush its applicants into employment usually does so because of one or more of the following: (1) the company does not have personnel to check the applicant's background and work history, (2) the company cannot afford the expense of preemployment physicals, (3) the worker is needed right away, (4) the applicant won't wait, or (5) a combination of the above.

The urgency to meet these operational needs can undermine your loss prevention efforts. Consider the benefits of a more thorough preemployment qualification process in assisting a company to match applicants to jobs they can safely perform. Verification of an applicant's good driving record before placing the applicant in a delivery position can be done through background checking. A preemployment physical may protect the safety of an applicant with a back problem.

Prior to employment the applicant should read and sign all of the company's rules of conduct, safety regulations, and procedures concerning drug testing. Each company should conduct formal and informal orientation programs for a new hire. The orientation should reinforce the safety program, train through coaching, and encourage involvement or worker participation.

Each employee should have the benefit of a brief but straightforward explanation of precisely what top management will do to achieve worker safety and to protect equipment and property. Employees should be told that they are never to undertake any task which they suspect could result in an injury to themselves or to fellow employees. They should also be instructed to immediately report to their supervisor regarding any substandard work practices or any tools or equipment which are not operating in accordance with manufacturer's specifications and OSHA and EPA requirements.

Depending upon the scope of the operation, a 10-minute video tape featuring the CEO giving the general introduction and welcoming the new employee to the project team, emphasizing that the employee's safety is a primary concern and that there is an expectation that he or she will conform with all safety requirements, provides an excellent vehicle of communication. This underscores management involvement, telling employees exactly what is wanted and what is required. The video may also be used to discuss key safety and health issues and advise how to report a safety or health concern.

An employee safety handbook should be prepared. It should identify key safety and health issues and a code of safe practices. Each supervisor should review the book with each new employee and both should sign a tear-out sheet documenting that the orientation was completed. That sheet must be placed in the employee's personnel file.

16.2.4   Ongoing safety training program

Companies should provide ongoing safety training for managers, superintendents and supervisors, employees, and safety staff. Safety-related work procedures, practices, and rules help prevent injuries; therefore, training should be an important part of the safety program. Accident prevention training is vital to an effective safety program. Experts estimate that up to 90 percent of all accidents result from human error, whether it be from the worker, the supervisor, the superintendent, or from management. To help prevent accidents, *doing it safely* must become synonymous with *doing it well* and *doing it right*.

How to train can be categorized in two ways: training for specific tasks and general training in accident avoidance and prevention. Equally important as knowing how to train is knowing when to train. Training should occur when a person is new to the company, new to a particular jobsite, new to a particular task or process, or new to a particular crew. Persons who supervise hazardous operations, e.g., trenching and excavation, and persons who are using or exposed to hazardous materials require training. Some off-the-job training may be necessary which, when completed, results in a certificate of competency.

Safety should be introduced as an integrated part of the project. Each time a supervisor provides instructions on how to perform a task, related safety aspects should be included. This process takes only minutes and helps employees to think about safety while performing the task.

16.2.5   Modified return-to-work program

Jobsite accidents have a costly impact on the construction industry. Work-related injuries and illnesses, including fatalities, in construction occur at a higher rate than the rate for all other industries, making it one of the most hazardous occupations. Contractors are finding that it is cheaper to pay employees their full pay for restricted work (modified duty) temporarily while they heal than it is to have lost time accident benefits paid and suffer the adverse effect these have on a company's insurance premium and also on employee morale. The lack of a comprehensive return-to-work (modified duty) program to assist the employee and employer in adjusting to the many variables in the workers' compensation process has resulted in an alarmingly small percentage of injured skilled employees returning to work at full duty after serious injury.

The primary objective of a modified duty program is to return employees to their general craft area in an efficient and positive manner within the restrictions prescribed by the treating physician. It should be noted that the program should not be a creation of makework or light duty positions, but must be viewed by all employees as productive.

Improved construction safety and the resulting cost benefits requires more awareness and understanding by company management. Construction management should recognize that the principles of management control commonly applied to cost, schedule, quality, and productivity are equally applicable to safety, and that if used, will improve the bottom line.

Construction management should attempt wherever possible to facilitate the return of employees suffering occupational injuries to their crafts. Each injured employee should be assisted on an individual basis by all of the following: qualified rehabilitation consultant, physician, insurance adjuster, and employer. The modified duty program is an individualized approach, tailoring a program to meet the individual employee's needs. The program should be based on a team approach which will provide a far more comprehensive service and a higher percentage of success. Primary methods used in tailoring the return-to-work programs are counseling (both personal and vocational), job analysis, job restructuring, job modification, and extensive follow-up. The combination of these areas forms a sound base for the return of employees to their general craft areas.

16.3 Program Assessment

16.3.1 Detailed record keeping and performance measurement

Research shows that when project injury and illness data are recorded and reported throughout the company, safety performance improves. Modern management gurus put it in more simple terms: what gets measured gets done. Injury costs now affect the bottom line much more directly than before. Therefore, it is important to focus on losses and the causal factors which contributed to the loss in order to learn how to prevent reoccurrences. This requires good investigations and the gathering of pertinent facts. Management has a responsibility to see that the record keeping system includes maintenance of incident statistics, enforcement procedures, and performance reviews.

Various measurement tools may be employed. Several of those tools are described accompanied by diagrammatic examples. A company may measure loss costs on a cost-per-hour-worked basis. The costs paid by insurance carriers, including reserves and the expenses paid out by the project, should be included. (See Fig. 16.1.)

OSHA lost workday injury/illness and recordable injury/illness cases should also be reported to allow internal and external comparisons. (See Fig. 16.2a and b.)

Workers' compensation and general liability losses as a percentage of covered payroll provide another measuring tool. (See Fig. 16.3.)

The report should be structured to develop the data for each project, each operating entity, and the company as a whole. Data should be displayed for the past month, the year-to-date, and project inception to date.

Performance goals which represent meaningful improvement over past performance should be set for each entity. Superintendents, project managers, and division or entity managers should be held accountable for achieving those goals.

Liability and auto losses should be reported and recorded on a job-by-job basis.

Use of this detailed reporting system enables management to identify trends and to initiate timely actions. It also creates peer pressure among the superintendents. Management may learn that some of the superintendents may not be as profitable as had been assumed. The bottom line is that detailed record keeping and performance measurement, when used effectively by management, forces safety excellence.

16.3.2 Subcontractor management

There is a clear need to manage those subcontractor operations which impact jobsite safety and health in a more structured and effective manner. A contractor cannot accept subcontractor practices and conditions which either pose a potential hazard to the contractor or set a poor example for employees. To do so creates the potential for litigation. The courts, with increasing frequency, are holding general contractors responsible for their subcontractors' safety failures. OSHA frequently has difficulty in identifying the responsible employer and tends to cite the general contractor.

Direct Costs

Policy Year	A Direct Subject Losses	B Frictional Costs	A+B=C Direct Claims Costs	D Hours Worked	C./D=E Direct Claim Cost/Hour Worked	Historical Payroll
Totals & Averages						

+ Indirect Costs

Policy Year	A Direct Subject Losses	F Indirect Loss Cost Factor	AxF=G Indirect Loss Cost	D Hours Worked	G./F=H Indirect Loss Cost/Hour Worked
Totals & Averages					

= Total Workers' Compensation Claims Cost

Policy Year	C Direct Claims Cost	G Indirect Loss Cost	C+G=I Total Claims Costs	D Hours Worked	I/D=J Indirect Claim Cost/Hour Worked
Totals & Averages					

Figure 16.1   Claims cost recap.

The subcontractor management process must begin at the prebid stage when the contractors give prospective subcontractors a document which details the minimum safety and health practices which will be acceptable on the jobsite. At that time, the contractor should require from each prospective subcontractor the following information from the previous three years: OSHA injury/illness rates, OSHA citations, and a recap of all workers' compensation and general liability claims in excess of $50,000.00.

Direct Costs

	A	B	A+B=C	D	C./D=E	
Policy Year	Direct Subject Losses	Frictional Costs	Direct Claims Costs	Hours Worked	Direct Claim Cost/Hour Worked	Historical Payroll
6/88-5/89	1,320,043	318,538	1,638,581	1,044,298	$1.57	11,136,074
6/89-5/90	575,282	268,036	843,318	1,223,531	$0.69	11,423,393
6/90-5/91	330,044	180,775	510,819	1,025,802	$0.59	9,658,831
Totals & Averages	$2,225,370	$767,349	$2,992,718	$3,293,631	$0.91	$32,218,298

+ Indirect Costs

	A	F	AxF=G	D	G./F=H
Policy Year	Direct Subject Losses	Indirect Loss Cost Factor	Indirect Loss Cost	Hours Worked	Indirect Loss Cost/Hour Worked
6/88-5/89	1,320,043	2.0	2,640,086	1,044,298	$2.53
6/89-5/90	575,282	2.0	1,150,564	1,223,531	$0.94
6/90-5/91	330,044	2.0	669,088	1,025,802	$0.64
Totals & Averages	$2,225,370		$4,450,739	$3,293,631	$1.35

= Total Workers' Compensation Claims Cost

	C	G	C+G=I	D	I/D=J
Policy Year	Direct Claims Cost	Indirect Loss Cost	Total Claims Costs	Hours Worked	Indirect Claim Cost/Hour Worked
6/88-5/89	1,638,581	2,640,086	4,278,667	1,044,298	$4.10
6/89-5/90	843,318	1,150,564	1,993,882	1,223,531	$1.63
6/90-5/91	510,810	668,088	1,170,907	1,025,802	$1.14
Totals & Averages	$2,992,718	$4,450,739	$7,443,458	$3,293,631	$2.26

Figure 16.1 *(Continued)*

The contractor should inform the subcontractor in writing that the contractor will hold the subcontractor responsible for attaining and maintaining the contractual level of safe work practices and working conditions. Further, the subcontractor's management must be informed that if the subcontractor supervisor cannot or will not maintain the desired level of performance from each one of its employees, then the contractor will direct the removal of that supervisor. If safety problems continue with subsequent supervisors, the contractor may elect to terminate the subcontract.

National Average
Construction 6.8

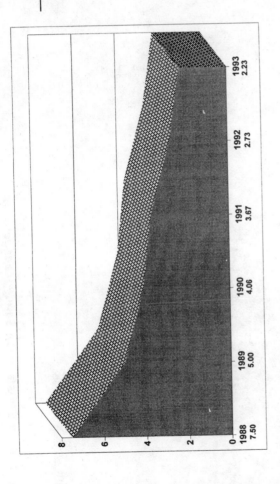

Injuries / Illnesses x 200,000
Hours Worked   = Rate

Pursuing Excellence in Construction Safety Management

Figure 16.2a Lost workday injury/illness rates (per 200,000 hours worked).

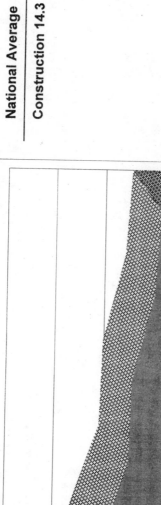

National Average

Construction 14.3

	1988	1989	1990	1991	1992	1993
	18.30	13.36	10.34	10.08	7.92	7.32

$$\underline{\frac{\text{Injuries / Illnesses x 200,000}}{\text{Hours Worked}}} = \text{Rate}$$

Pursuing Excellence in Construction Safety Management

Figure 16.2b   OSHA recordable case rate (per 200,000 hours worked).

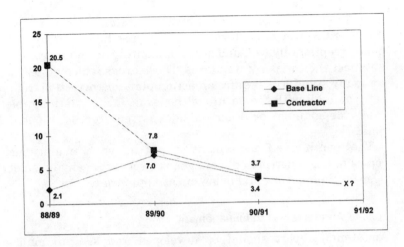

Pursuing Excellence in Construction Safety Management

Figure 16.3   Workers' compensation and general liability losses as a percentage of covered payroll.

It is important to note that, except in an imminent danger situation, all communication will be directed to the subcontractor supervisor. Otherwise, there is a risk of being identified as a coemployer.

16.3.3 Work area safety evaluation

Historically, in most companies the safety department inspects the jobsites to identify deficiencies and then prepares a report to management with copies to the job superintendent, construction manager, project manager, et al. However, the identification of safety and health concerns is generally dependent upon the skills of the safety staff, the relationship which they may or may not have with the superintendent or project manager, and the availability of the right people to provide input to the safety specialist. That traditional approach to problem identification and problem solving is one part of the equation which has kept the construction industry at the bottom of the rankings, i.e., with the highest injury and illness rate of any of the industrial classifications.

Several of the more successful construction companies have changed their approach to jobsite safety surveys and are using a self-assessment approach by the jobsite staff in lieu of the routine home office safety inspection.

The assessment is performed by two or more supervisory personnel and covers all facets of the company's safety program OSHA compliance. In each category a brief narrative statement is made reflecting the overall level of performance with specific comments on things done exceptionally well and also commenting on those areas which do not meet the company standards. Those areas which do not meet the company standards require an action plan for correction, commensurate in detail with the nature of the issue, including timetable. If the safety department or other outsider is required, it is noted in the plan.

The monthly self assessment is sent to senior management with copies to the construction manager, project manager, and safety manager for their review and follow-up as appropriate.

16.3.4   Annual report to management

An annual report should be developed and sent to management reviewing the last year's accomplishments, determining obligations for the upcoming years and providing an outline of what is needed from management.

16.4   Program Support

16.4.1   MVR screening

If a company has drivers and/or operators of equipment, it is the company's responsibility to check the applicant's record. The company should conduct a review of equipment operated previously and develop procedures for rewards and penalties for the individual once he or she becomes an employee.

16.4.2   Substance abuse program

Not only should the use of illegal drugs and the problems they have caused in the workplace be addressed, attention must be given to alcohol and the problems associated with the work to be performed. A substance abuse program should include preemployment screening, periodic testing, reasonable testing, and post-accident testing.

16.4.3   Preemployment workers' compensation screening

Today more than ever fraud has become a major area of cost to the construction industry. Only if we take the time to check can we control this problem. A company must develop a program procedure to verify applicant information, match applicant to job requirements, and provide guidance to the physician performing a physical examination.

16.5   Employee Commitment or Buy-in

Each of the safety program elements which have been reviewed is an essential part of structuring the company safety effort so that no one in the company can ignore or bypass the fundamental safety concerns. If a company operates with safety as a primary concern, frequently expressing a desire to avoid employee injury and the need to eliminate losses, the employees will respond. They need to be reminded frequently, and management needs to remind itself frequently, that safety—the control of potential losses—is important to a company's success.

Some construction companies use safety recognition awards for employees as a means of maintaining visibility and recognizing accomplishment. These range from baseball caps, T-shirts, sweatshirts, and jackets with company logos, to stainless steel thermoses and igloo lunch pails with appropriate labels, to cast brass belt buckles. Some companies use monetary rewards, sharing a portion of the savings derived from injury rates and/or accident loss costs which bettered the company goals for the project.

Regardless of the means chosen to communicate safety commitment, the important issue is that management create an environment where the status of safety and health are continuously monitored by every employee, craftsworker, supervisor, and superintendent alike. Substandard work practices must be red flagged and corrected before they result in a loss. When you achieve employee commitment, your safety effort will move to a higher level of intensity. Safety is directed from the top and driven from the bottom. Because safety is expected, it gets done. It is a win-win situation.

Life Insurance and Succession Planning for Contractors

Douglas R. McPherson
McPherson Enterprises Limited

According to several knowledgeable sources, only 35 percent of successful construction firms continue to exist following the founder's death. Less than 20 percent survive into the third generation. Obviously, there are a number of factors that contribute to this failure, such as competition, family conflict, regulatory issues, or technological change. However, a common cause for failure is a lack of funds to adequately accomplish the succession plan. If properly arranged, life insurance can be a superb vehicle for assuring that funds will be available to accomplish the contractor's goals, purchase the corporate equity, provide for family income, indemnify the surety, create a friendly family bank for future generations, and pay estate taxes, all at a reasonable cost. Life insurance proceeds can significantly contribute to the continuation of a construction company by providing the funding mechanism for a well-designed succession plan.

17.1 The Two Succession Plan Components

There are two major components of any succession plan: the *management component,* or who will run the company, and the *equity component,* or who will own the stock. The management component deals with people issues and the equity component refers to those who will benefit economically from the stock ownership. The shareholders who own the voting stock elect the directors of the company. The directors elect the officers. The officers, who are the senior management, appoint the lower levels of management. (See Fig. 17.1.) In many cases, the contractor owns all of the stock, both voting and nonvoting, if any; sits on the board of directors, totally comprising family members; and is the

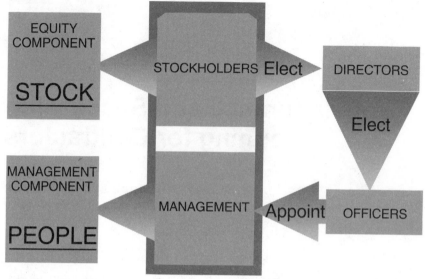

Figure 17.1   Equity/management equation.

president. The contractor essentially runs the whole show! It is difficult for him or her to see that the two components can be separated and
transferred at different times and in different proportions. Successor
management can be in place without any transfer of equity. In many
situations some transfer of stock, frequently nonvoting, occurs with
management succession.

Family succession plans.   When beginning to plan for management succession the contractor must first identify suitable candidates to carry
on the business. This process can be very difficult. Are family members
in succession positions just because they are related? If they were not
related, would they even be company employees? Frequently, one or
two nonfamily employees are extremely qualified to run the business.
A candid appraisal of the players and their capabilities is needed.
When the company is to be preserved for the family, it is imperative to
bring the spouse and other major interests into the planning process.
The contractor often leaves the spouse out of the planning loop. Many
times when a plan is completed and the documents are all ready for
signatures, the spouse is presented with an involved arrangement with
no prior discussion of the mechanics of the plan, which could even
involve the future pledging of marital assets to the surety. This lack of
involvement frequently creates tension between the parties. Leaving
family members out of this planning process can easily derail the best
of plans. The people expected to carry out the contractor's future plans

should be part of the planning process at every stage that affects them so that the ultimate result is acceptable to everyone.

Succession plans with partners.   In the situation where there are two or more shareholders active in the business, the management component does not appear to be such a pressing issue; however, the equity component can become much more complicated. The contractor could end up in business with a deceased partner's spouse or family. There is no more critical issue in the future than dealing with the surviving spouse. Typically, that spouse does not understand the contractor's problems. After all, the deceased partner owned half the business, was responsible for most of its success, and is not here any longer. The surviving spouse suddenly is face-to-face with the surviving partner who explains that the business just can't afford to continue to pay at the former salary level. Substantial accounts may have been lost, the bank has called the note, little of the money that is due is coming in right now, and business is tough. What does the contractor say to the surviving spouse who says, "No problem. I'll be in Monday to help take over."?

The family construction business with children.   The contractor is married, has three children and owns the entire company, and wants the oldest child to have the business. The surviving spouse will be adequately provided for, but the contractor does not know how to get the business to the oldest child and still provide for the other children. If the contractor leaves the business to the spouse and expects the oldest child to run it, there will most likely be significant problems. The spouse will not understand the situation that the child will face in attempting to grow the business. In addition, any contracting company where the principal has just died and the child has never run a business before and is undercapitalized is not considered to be a great investment for a surviving spouse, who will want income from that business and will become more conservative as the child becomes more aggressive. On the other hand, the contractor does not want to give the business outright to the child. What this contractor probably needs is a buy-sell agreement that is funded with life insurance so that the spouse will get the money and the child will have the business free and clear, and the other children will receive their shares at the surviving spouse's subsequent death.

The unfunded agreement, or why you may wish you were not the survivor. If the buyout formula provides that funds will be paid out over a period of years, the contractor may not know that those payments will not be deductible to the corporation. In making those payments to the estate or to the surviving spouse, the contractor discovers that a bigger income in a higher tax bracket must be earned; the contractor will get only what's left. In addition to making the interest payments on the funds, the costs are like a mortgage and may very well double the actual purchase price. The cost of life insurance would have been considerably less than the cost of borrowing money from the bank. Even though the interest cost is deductible, every cent of the principal has to be repaid in addition to the interest payments. The annual premium for a life insurance policy for a 40-year-old is about 2 percent of the face amount and guarantees the future use of the money. If the worst happens, the plan is self-completing and the contractor will have the cash to give to the surviving spouse. Then the spouse will have money and the contractor will own the business. The cost of doing it this way is *very* much less than the cost of doing it any other way!

17.2   Integrating Life Insurance into the Succession Plan

In order to integrate the purchase of life insurance into the succession plan, it must first be determined where dollars will be needed in the succession plan to fulfill the plan objectives. Depending on the elements of the plan, cash may be needed when a single shareholder dies, when a multiple shareholder dies, or when a surviving spouse dies. The desired timing will indicate whether an individual life, survivor first-to-die, or survivor second-to-die product structure will be needed.

The second step is to determine the amount of dollars that will be needed to fulfill the succession plan objectives. Sophisticated planning techniques exist to prevent the life insurance proceeds from being taxed in the contractor's estate, usually at rates of 55 to 60 percent. There should be a very compelling reason if the life insurance is not to be removed from the estate! When determining the cash needs of the plan, all potential needs at death should be identified to avoid falling short when the need arises. Funding for stock purchases, keyman indemnification, creditor satisfaction, estate tax, surety requirements, family income, estate conservation, and charitable goals may all contribute to the successful succession plan.

The third step is to determine an acceptable level of cost to fulfill the succession plan objectives. Insurance acquisition will be more costly at older ages and when the insured is in less-than-perfect health. However, the cost can be mitigated through the use of a split-dollar plan or through survivor life insurance, as long as a spouse is alive at the point of acquisition. In addition, even where higher rates are caused by poor health, the cost is significantly lower than borrowing when the need arises, because the principal insurance amount is not a debt to be repaid.

The fourth step is to determine the acceptable level of risk in the selection of insurance carrier and product type. At the younger ages (under age 55) and where survivor second-to-die insurance is to be acquired, it becomes extremely important to evaluate the carrier stability and potential insurance product volatility. This will provide a greater assurance that the policy will be in force far in the future when the need arises.

17.3 Business Insurance Planning Techniques

A number of business insurance techniques exist that are frequently implemented by construction company owners. Some of these techniques are listed along with a short summary of how each could be used in continuation and succession structures.

17.3.1 Keyman insurance

An important element in succession planning is to provide funds to replace the value of a key employee of the corporation. The purpose is to provide funds to find, hire, train, and pay a replacement employee, enhance the corporation's balance sheet, replace lost profits and/or comfort the surety. This person may not be a shareholder, but the loss of this employee's services would cause a significant risk to the corporate succession plan. Because the insurance premiums are paid with corporate after-tax dollars, the cash that the corporation receives at the key employee's death is generally income tax free. If the construction company is a C corporation, there may be an alternative minimum tax due on a portion of the death benefits when received.

There are several methods of determining the value of an individual to the business. The amounts determined by the following formulas will only satisfy the financial indemnification to the company in the event of death and will not address any other needs, such as family income, corporate or estate liquidity, or disability. However, the process of defining a specific individual's value to the corporation can provide indicators that would help define value in the event of disability or premature death.

The three formulas are the *human life value formula,* the *contribution to earnings formula,* and the *weighted keyman formula.*

The *human life value formula:* Computes the key individual's value by estimating the amount that annual corporate earnings would be reduced as a result of this employee's loss and multiplying that figure by the number of years until the income can be replaced. This amount may be discounted by applying an interest factor based on the number of years to replacement.

The *contribution to earnings formula:* Calculated by dividing corporate pretax income by book value to produce an average return. The average return minus fair return if invested elsewhere equals the percent of pretax income that is attributable to management. This amount is then allocated between the selected key individuals on an arbitrary basis as needed, and then multiplied by the number of years needed to prepare the replacement.

The *weighted keyman formula:* Takes into consideration the replacement's effectiveness, the time needed for the replacement to become fully effective, and the cost of the replacement. It requires the evaluation of the time required and the percentage of effectiveness of the key executive's replacement. This method usually produces the highest keyman value.

Although the values produced by these formulas are approximate, it is better to evaluate the potential loss due to death or disability of a construction company owner or key employee than to ignore it. Once an evaluation has been made and the areas of potential loss identified, steps can be taken to provide the means to mitigate those losses.

17.3.2 Salary continuation plan

The existence of well-documented corporate plans for salary and corporate continuation in the event of the corporate owner's disability form a solid base for a well-rounded financial plan. In order for a salary continuation plan to be tax deductible, it must be a plan for employees. However, it may limit benefits to selected groups, such as officers, supervisors, etc. In addition, the plan should be in writing and be adopted by means of a corporate resolution. It should be obvious that few companies can afford to continue long-term salary payments to disabled or nonproductive employees, whether or not they are owners. By utilizing a type of insurance designed to replace income in the event of disability, an indefinite future liability may be converted into a deductible business expense that has a fixed cost. The ownership of the disability income policy can be arranged in several ways:

1. The owner/employee may pay the premium and own the policy. The premiums are not tax-deductible, but any benefits are received income tax free to the employee.

2. The corporation may pay the premium and the owner/employee may own the policy. The premiums would be tax-deductible to the corporation, but not taxable income to the employee. However, any disability benefits received by the employee would be taxable as income.

3. The corporation may pay the premium and own the policy. The premiums would not be tax-deductible to the corporation, but the proceeds would be received income tax free by the corporation.

This information is only applicable where a regular C corporation exists, not to an S corporation, because of the differences in tax structure.

17.3.3 Section 162 bonus plan

This technique provides a means for the corporation to pay for personal life insurance coverage for key employees and/or owners. However, the full amount of the bonus is taxable income to the employee or contractor and the corporation usually has no ownership rights in the policy. Because this insurance is usually arranged to be included in the insured's estate, it is not always the most efficient use of premium dollars. This planning technique is frequently used to supplement the price arranged for the purchase of corporate stock at the contractor's death.

17.3.4 Deferred compensation plan

In the succession context, when a contractor is disabled, dies or retires, it becomes very difficult for the construction company to pay a substantial salary on a tax-deductible basis. We have developed a structure to provide the greatest possible reasonable compensation to an employee shareholder at disability or retirement and those payments can be continued to the spouse following the contractor's death. The plan provides these payments in part as compensation for being underpaid in the past. It may be executed at any time during the contractor's lifetime, as desired. The plan comprises three separate elements: a *postretirement consulting contract*, a *deferred compensation agreement*, and a *deferred bonus earn-out*. Although the total compensation provided by all three elements would be subject to the reasonable compensation test by the IRS, each element would have its own justification. Using corporate minutes for documentation and consistency in adhering to the terms of the agreements in both profitable and nonprofitable years has helped many taxpayers when the IRS has challenged their compensation as unreasonable. The ability to withdraw money from the construction company on a favorable *tax-deductible* basis encourages the implementation of the contractor's succession plan while he or she can enjoy the fruits of the work.

17.3.5 Stock redemption

When a succession plan results in a corporate obligation to redeem a shareholder's stock at his death, it makes good sense to provide an assurance that the funds will, in fact, be available. Even though there may appear to be sufficient corporate liquidity to accomplish a buyout, future continuation of liquidity is not guaranteed. If the death of a shareholder should coincide with a business downturn cycle, a problem job, a need for corporate borrowing, or any number of adverse business conditions, liquidity levels could be reduced. Funding this type of plan with life insurance is cost-effective and provides a timely delivery of cash where it is most needed. Because the corporation owns the life insurance, pays the premiums, and ultimately receives the death benefit, the policy values are exposed to corporate creditors. In addition, in the case of "C" corporations (not Subchapter S corporations), a portion of the build-up and/or the death benefit proceeds may be exposed to corporate alternative minimum taxes. These are areas that should be quantified before finalizing the succession plan.

17.3.6　Cross-purchase

When a succession plan results in a agreement by the shareholders to individually purchase stock at the death of one of the shareholders, personal liquidity becomes the issue. This type of plan is often recommended so that the surviving shareholders will experience an increase in the cost basis of the stock that they acquire. This issue is not usually as important in Subchapter S corporation situations. A significant factor in selecting this type of plan is the number of shareholders involved. If a life insurance policy is owned by each shareholder on every other shareholder, a large number of policies may result. When the first shareholder dies, a significant problem may arise with the policies owned by that deceased shareholder on the surviving shareholders. The purchase or transfer of those policies directly to those other shareholders may be considered a "transfer for value" which would cause the subsequent death benefits to become taxable income to the beneficiaries. The arrangement of the cross purchase policies should be reviewed to avoid possible exposure to potential income taxation.

Although the policies funding a cross-purchase agreement are personally owned by the respective shareholders, it is usually possible to use some form of split-dollar arrangement to fund the premiums on the policies. The majority shareholder rules discussed later should be reviewed for their present or future application to split-dollar funding of policies intended to fund a cross-purchase agreement.

17.3.7　Partial redemption–section 303

This technique permits the partial redemption of a deceased shareholder's interest without adverse income taxation consequences and is frequently used to permit corporate dollars to fund the payment of estate expenses and taxes for a major shareholder. The comments relative to a complete redemption apply here.

17.3.8 Split-dollar plan

This technique is a funding method for life insurance where the corporation provides premium payments for policies owned by a shareholder or a third party. Third-party ownership is advantageous in that it removes the value of the life insurance death benefit from the insured's estate. Children of the insured, a properly structured irrevocable trust, or a properly structured partnership are all appropriate third-party owners. The construction company pays the premiums on the policy but the contractor only reports the amount of the economic benefit for income tax purposes. Depending on the age of the insured and the structure of the policy, this imputed income, which is called *PS-58 cost,* can be significantly less than the actual premium paid. The actual amount is derived from a government table, but if the term rates of the issuing company are lower, they may be used. This PS-58 cost is currently taxable as income, unless it is separately contributed by the insured. A disadvantage of this type of plan is that the premiums advanced by the construction company must be repaid at the insured's death. Another disadvantage arises when the insured is a majority shareholder of the construction company; the insured's majority ownership could cause the insurance death benefit to be included in his or her estate. In order to avoid this result, the construction company may not have any rights in the policy during the insured's lifetime. (See Fig. 17.2.)

There are two methods of arranging split-dollar plans depending on where the control of the policy is desired. They are called the *endorsement* method and the *collateral assignment* method. With the endorsement method, ownership and control of the policy is retained by the corporation. With the collateral assignment method, ownership and control of the policy is retained by the third-party owner, such as a trust, a child, or a partnership. Significant creditor exposure of the policy's living and death values can be avoided through properly structured ownership arrangements.

17.3.9 Reverse split-dollar

This technique reverses the traditional split-dollar policy owner and beneficiary; the insured shareholder/employee owns the policy instead, and the corporation receives the death benefit. The PS-58 cost is paid by the corporation. The higher government table amount is usually used to maximize the premium without causing income to the insured.

EMPLOYER—TRUST/SHARED PREMIUM

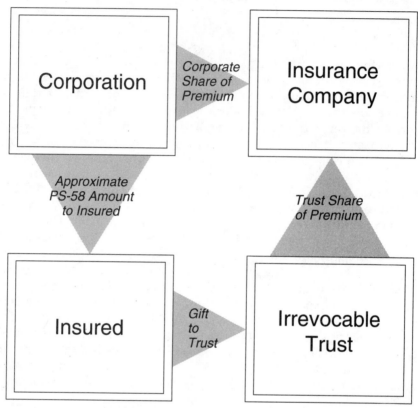

Figure 17.2 Majority shareholder split-dollar.

Any premium deposited in excess of the PS-58 cost is charged as income to the owner/insured. The reverse split-dollar technique can provide the company with a significant death benefit in the event of the premature death of the insured shareholder/employee. It can also help alleviate any Section 531 accumulation of earnings problems a corporation may have. At the same time, a substantial personal benefit is created for the insured—the policy is personally-owned with the potential for significant cash-value accumulation. This cash value is not a corporate asset and is not subject to corporate creditors. If death occurred before the reverse split-dollar arrangement was terminated, the amount paid to the corporation would be included in the insured/owner's estate, so care must be used in the estate planning process to accommodate that possibility. Use of various ownership structures may be a part of this planning.

This can be a very effective method for transferring corporate dollars to the shareholder on a tax-favored basis. Reasonable parameters of design should be observed so that the IRS will not contend that no split-dollar arrangement exists, which would cause the premiums to be entirely taxable to the shareholder. Since relative tax rates between corporate and personal income fluctuate, consideration should be given to a flexible premium product. This technique is generally not suitable for use with S corporation shareholders.

17.4 The Life Insurance Selection Process

The following are items that the buyer of life insurance should consider when acquiring life insurance vehicles. Products can be selected to accommodate price parameters and risk tolerance.

Carrier financial strength. One of the major considerations when purchasing a life insurance policy is the financial stability of the insurance carrier issuing the policy. Since policy death benefits will generally not be paid until the contractor's life expectancy or beyond, it is important to choose a carrier with excellent financial ratings. There are four major rating companies providing financial strength ratings for life insurance companies: A.M. Best, Standard & Poor's, Moody's, and Duff & Phelps. In addition, financial information on any insurer is public information and may be obtained from any state insurance department where an insurer is licensed to do business. The rating categories are all designated using letters, but are not consistent as to quality designation. For example, an A+ rating is A.M. Best's second highest rating, but it is the fifth highest rating for Standard & Poor's. Therefore, it is important to know where a specific rating from a specific company falls in relation to that company's rating spectrum. In addition to the letter ratings, extensive details are available from the rating companies in their full reports.

A.M. Best rates most of the life insurance companies. There are 15 levels of ratings, from A++ to F. Less than 10 percent of the companies reviewed received A.M. Best's highest rating of A++.

Duff & Phelps provides claims-paying ability ratings upon an insurer's request only. There are 19 levels of letter ratings including + and – modifiers, ranging from A+ to F.

Moody's will assign an insurer financial strength rating upon request only. Nineteen levels of letter ratings with number modifiers are used.

Standard & Poor's will assign a claims-paying ability rating to an insurer upon request. There are 17 levels of rating ranging from AAA to CCC, including + and – modifiers.

All companies are not rated by every rating company. Most financial advisers suggest that selection of life insurance carriers be limited to companies with top ratings from at least two of the four rating services and that the carrier have no rating below investment grade.

Types of life insurance.   There are only two basic types of life insurance: *term insurance,* in which the entire face amount is at risk, and *permanent insurance,* which accumulates a cash value. Term insurance is appropriate for situations of short-term need or where a large amount of insurance is needed at a low cost. Rates generally increase on a scheduled basis, yearly, every 5 years, or every 10 years, etc. Often term policies are offered with an option to requalify for initial rates after a specified period of years. Many times the requalification requirements are much more stringent than new underwriting requirements and the anticipated lower insurance rates are not in fact obtained.

Whole life and universal life are variations of permanent insurance, and these variations represent a spectrum of costs, guarantees, and risks. Whole life offers the fewest risks in that the mortality costs are guaranteed, the face amount is guaranteed, and the cash value is guaranteed. Policy charges are guaranteed, but are not disclosed. Universal life offers the fewest guarantees in that the mortality costs may fluctuate between minimum and maximum levels, the face amount is guaranteed only as long as there is sufficient cash value to pay the mortality costs, and the cash value has a minimum crediting rate of interest and receives an excess interest crediting. Policy charges are not guaranteed, and are disclosed.

There are almost an infinite number of policy variations and combinations of risk elements present in the marketplace today, such as modified life, adjustable life, flexible life, etc. Another variation in permanent insurance is in the investment element of the policy. Whole life policies are invested in an insurance company's bond and mortgage general account. Universal life policies are invested in an insurance company's short-term guaranteed interest general account. Variable life, which may have either whole life or universal life structure, is invested in separate accounts comprising selected stock, bond or money market mutual funds, or guaranteed interest accounts. This permits the policyholder to participate in appreciation in the cash value account instead of, or in addition to, interest accumulation.

Another aspect of the life insurance product portfolio is the availability of a single policy on two or more insured individuals. A policy which pays a death benefit at the second death of two insureds is called *second-to-die* or *survivor* life insurance. It is frequently used in the estate planning context where a significant tax burden exists at the second death. A policy that pays a death benefit at the first death of two or more insureds is called a *first-to-die* policy. This type of policy is frequently used in the business continuation and succession situation. Both types of policies are less expensive to purchase than single life policies on each individual. Not all life insurance companies issue these products, nor are they available in every format, such as term, whole life, universal life, or variable life.

Comparison of life insurance illustrations. When reviewing life insurance projections, it is important to be aware that certain columns represent guarantees and other columns represent projections based on assumptions that may or may not be achievable. These areas are not always discernible at a glance, but the illustrations should be closely reviewed so as to be clear about what is guaranteed and what is not. Another area to be aware of is the use of marketing terms, such as vanishing premium, to describe the payment of premiums from excess cash value or dividend accumulation. In order for any policy to remain in force, the mortality charges must be paid. Universal life and some whole life policies deduct these charges from the cash value on a monthly basis. Because cash value interest accumulates income tax free, the mortality charges in a policy may be paid with before-tax dollars, thereby lowering the ultimate policy cost. Because of this benefit, there are limitations on the amounts of cash that can be deposited into a life insurance policy and penalties for exceeding those amounts.

When comparing whole life and universal life illustrations, you should be aware of the risk involved in the lower projected premium for the universal life policy. Life insurance should not be purchased entirely on the basis of the lowest premium or highest cash value illustrated. There are several acceptable methods for comparing different premium patterns, cash value accumulations, and death benefit projections, such as interest adjusted payment, interest adjusted surrender cost, internal rate of return at death, internal rate of return on surrender, and net clear death benefit. All of these methods quantify premiums, cash values, and death benefits with some form of cost-to-benefit yield calculation at a given point of time in the future.

17.4.1 Considerations in the selection of insurance companies and products

Most contractors do not have all of their cash in one bank or all of their investment dollars in a single product. When substantial amounts of life insurance are to be acquired, it makes sense to diversify that purchase in several ways.

- *By carrier.* There are stock companies and mutual companies, companies domiciled in the U.S. and companies domiciled in Canada, big companies and small companies, and old companies and new companies. You should review illustrations from top-rated companies in several selected categories.

- *By product type.* Because of the different risk levels in the different types of policies discussed previously, a life insurance portfolio should reflect low risk, moderate risk and high risk product designs in a ratio that is comfortable for the contractor.

- *By investment mix.* Purchasing universal life or variable life policies from two or more different carriers will not necessarily diversify the insurance portfolio investment mix. The underlying investments should be reviewed and evaluated for this purpose.

Important considerations. It may be important to the contractor to be able to vary the timing and amount of premium deposits to the policy. This is accomplished through loans in the traditional whole life format in which interest is accrued until the loan is paid back. The universal life format easily accommodates flexible premium payments except at very low levels of cash value.

Another area of flexibility that may be desired is in the death benefit structure. Fixed-level death benefits and death benefit increases through dividends are available in the whole life structure, and reductions in face amount are accomplished through loans that accrue interest. The universal life format provides fixed-level death benefits which include cash value accumulation, or death benefits which are in addition to the cash value. Additions may be added to the original face amount, usually requiring underwriting. The face amount may be lowered through loans, withdrawals, or reductions. (See Fig. 17.3.)

Ownership/beneficiary structure. While the ownership and beneficiary designation of an insurance policy would appear to be a simple matter, important tax and other consequences may be involved. Although the death benefit of a life insurance policy is income tax free, the death benefit could be subject to estate tax in the estate of the owner.

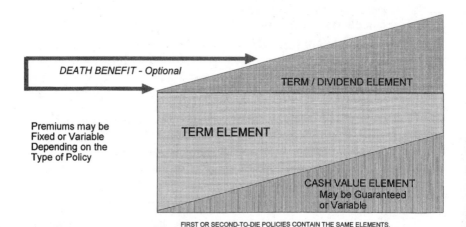

Figure 17.3 Life insurance policy structure.

It is important to have knowledgeable advisers evaluate the ultimate effects of insurance ownership arrangements.

Other owner/beneficiary issues.   Control of policy values during life and at death is another aspect of the ownership decision-making process. Some contractors want the insurance to be reflected as a corporate asset and others do not. A policy owned by a construction company is subject to the creditors of that company. State laws vary as to the creditor protection afforded to death benefits. Policy death proceeds may be used to satisfy agreements, to satisfy creditors, or for surety needs. (See Fig. 17.4.) If ownership is properly structured, the life insurance death benefit may be immune from creditors.

When policy ownership is changed, the policy proceeds are subjected to the *Transfer For Value* rule. This provides that the entire death benefit will be taxable as income to the beneficiary if the policy is transferred "for a valuable consideration." Valuable consideration encompasses some transactions that do not appear to involve a sale, with three exceptions to this rule, which are briefly: (1) a transfer to the insured, (2) a transfer to a partner or partnership of the insured, or to a corporation in which the insured is a officer or shareholder, and (3) where the income tax basis in the policy is carried over from the transferor to the transferee. This potential problem of "transfer for value" is often neglected, but it can seriously upset the best of plans and should be reviewed whenever an ownership change is contemplated.

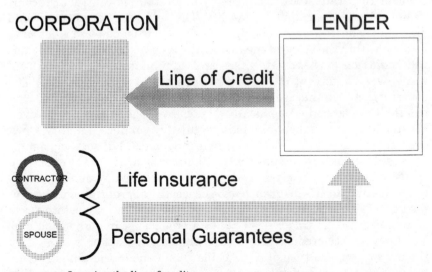

Figure 17.4   Insuring the line of credit.

Whenever split-dollar agreements are under consideration, to prevent controlling shareholder issues, it is important to evaluate whether the insured is a majority shareholder, which causes the possibility of the death benefit being included in the insured's estate.

17.5   The Acquisition Process

No premium quote reviewed up to this point is final until you make a formal application to an insurance company and they have made an offer. Before the offer is made, you will have taken a routine physical exam. Depending on the amount of insurance applied for, there will be a personal inspection and some financial information will most likely have to be disclosed. The company then reviews this information and makes the offer. If the contractor's health is not excellent, he or she may be offered what is called a *rated* policy. This means additional mortality costs are needed to fund the policy. It is in this area that broker expertise is very important, since insurers vary greatly in the amount and type of information they require and the types of health risks they are willing to assume.

17.6   Solving the Contractor's Estate Tax Problem with Survivor Life Insurance

Transferring the contracting business from the contractor to his or her children has become increasingly difficult. The continued passage of antifamily business legislation by Congress makes it very expensive to transfer corporate stock to children while retaining control or income rights in the corporation during life. The rules are very complicated and a fairly narrow path is being defined as to what can and what can't be done within the scope of the law. At the contractor's death, if no marital deduction exists, estate taxes and administration costs can consume over 55 percent of the contractor's estate. (See Fig. 17.5.) The construction firm that is frequently the major estate asset may be considerably weakened economically. This percentage of loss can be even greater if a forced sale of assets is required to produce liquidity. However, as long as estate assets qualify for the marital deduction, significant estate taxes are not payable until after the death of the surviving spouse. Uncertainty relative to which spouse will die last causes second-to-die or survivor life insurance to be especially appropriate for the production of liquidity at the same time as the need to pay estate taxes occurs. Proceeds of this type of policy are normally immune from personal and surety credit relationships and can be further insulated by the use of a special type of irrevocable trust. Due to the unique structure of this type of policy, the cost is usually much less than single life policies. In addition, either the husband or wife may be uninsurable as long as the other is reasonably healthy.

Structuring the ownership and premium payment arrangements for this type of policy requires considerable attention to detail. Because there are no benefits payable at the first death, the policy provides no income to the surviving spouse. If the proceeds of a $1,000,000 policy were included in the second estate at an estate tax rate of 55 percent, the taxes would consume $550,000. Since this is clearly unacceptable, someone other than the contractors and their spouses should own the policy. The children are a logical first choice. However, several factors reduce the effectiveness of this choice: coordination requirements relative to assuring the annual exclusion for gifts of the premiums; the loss of control over the cash values of the policy; and the possibility that the death proceeds will not be used as intended by one or more of the children to the detriment of the other(s).

Possibly a better choice is an irrevocable trust specifically structured to own a survivor life policy. The contractor's children and/or grandchildren can be beneficiaries of the trust and the contractor can make gifts to the trust which will qualify for the annual gift tax exclusion. In order to assure that the proceeds will remain out of both insured spouses' estates for tax purposes, neither the contractor nor the spouse can serve as trustee of this trust. In addition, the premium payments

HOW DOES IT WORK?

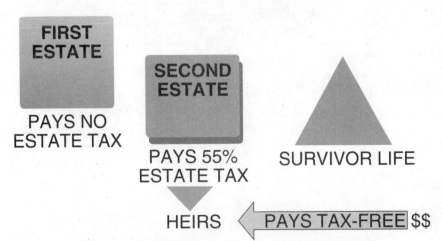

Ownership and Premium arrangements may affect this result.
No Family Income Benefits are derived from this arrangement.

Figure 17.5 Using survivor life to pay estate taxes.

can be structured on a split-dollar basis, where the corporation pays the premium and the contractor pays the cost of the joint economic benefit which is quite small in the case of survivor life. This joint economic benefit, also known as PS-38 cost, is considered to be the amount of the gift to the trust; this considerably reduces the exposure of those gifts to gift tax requirements. After the death of the first insured, the joint economic benefit used to calculate the income to the contractor and the gift to the trust is changed to the individual split-dollar economic benefit, or PS-58 cost. Planning is required to accommodate this future adjustment.

Survivor life insurance, in combination with a properly drafted irrevocable trust, can significantly enhance the contractor's ability to pass the business and other assets to his or her heirs at the lowest possible cost.

17.7   Summary

The acquisition of life insurance in the business succession planning process can be extremely complex. It is important that the individual selected by the contractor to implement this process understands the alternatives and trade-offs involved in the selection of an appropriate product to satisfy the contractor's stated goals.

A

Insurance and Bonding Terminology

actual cash value   Cost of replacing damaged or destroyed property with comparable new property, minus depreciation and obsolescence.

additional insured   Another person, firm, or other entity enjoying the same protection as the named insured.

agent   Individual who sells and services insurance policies and represents the insured.

alien insurer   Insurance company formed according to the legal requirements of a foreign country.

all-risk   Insurance that covers each and every loss except for those specifically excluded. If the insurance company does not specifically exclude a particular loss, it is automatically covered. This is the broadest type of property policy that can be purchased.

anti-rebate law   Statute that makes it illegal in most states for an agent to rebate (return) any portion of his or her commission as an inducement for an applicant to purchase insurance from the agent.

assigned risk   Statutory requirement, whereby high-risk contractors are guaranteed the availability to purchase insurance. This applies to insurance that is required by statute, such as workers' compensation and automobile liability.

assigned risk plan   Coverage in which individuals who cannot obtain conventional automobile liability insurance, usually because of adverse or low experience, are placed in a residual insurance market. Insurance companies are assigned to write insurance at higher rates, in proportion to the premiums written in a particular state. These plans protect motorists who suffer injury or property damage through the negligence of two drivers who otherwise would not have insurance.

bid bond   Bond required of a contractor submitting the lowest bid on a project. If the contractor then refuses to undertake the project, the bid bond assures that the developer will be paid the difference between the lowest and next lowest bid. The bid bond encourages contractors to make realistic bids and live up to their obligations.

binder   Temporary insurance contract providing coverage until a permanent policy is issued. In property and casualty insurance, some agents have authority to bind the insurance company to cover until a policy can be issued. For example, the purchaser of an automobile can call the agent, who can then bind the insurance company to temporary coverage.

bond   Form of suretyship.

builder's risks forms   Types of contracts that insure building contractors for damage to property under construction.

business crime insurance   Protection for the assets of a business (including merchandise for sale, real property, money, and securities) in the event of robbery, burglary, larceny, forgery, and embezzlement.

business interruption   A break in commercial activities due to the occurrence of a peril. Coverage against business interruption by various named perils can be obtained through insurance.

business interruption insurance   Indemnification for the loss of profits and the continuing fixed expenses due to a business interruption.

business liability insurance   Coverage for liability exposure resulting from the activities of a business.

businessowners' policy (BOP)   Policy combining property, liability, contractors, and business interruption coverages. It is usually written to small and medium-size businesses to cover expenses resulting from damage or destruction of business's property, or when actions or nonactions of the business's representatives result in bodily injury or property damage to one or more individuals.

buy-sell agreement   Approach used for sole proprietorships, partnerships, and close corporations in which the business interests of a deceased or disabled proprietor, partner, or shareholder are sold according to a predetermined formula to the remaining member(s) of the business. For example, a partnership has three principals. Upon the death of one, the two survivors have agreed to purchase, and the deceased partner's estate has agreed to sell, the interest of that partner according to a predetermined formula for valuing the partnership to the survivors. Funds for buying out the deceased partner's interest are usually provided by life insurance policies, with each partner purchasing a policy on the other partners. Each is the owner and beneficiary of the policies purchased on the other partners.

When a sole proprietor dies, usually a key employee is the buyer and successor. Under the entity plan, the sole proprietorship, partnership, and close corporation can buy and own life insurance policies on the proprietor, partner, or shareholder and achieve the same result as when an individual buys and owns the policies.

captive insurance company   Insurance company formed to insure the risks of its parent or reinsure losses fronted by a licensed insurer.

care, custody, and control   Phrase in most liability insurance policies which eliminates from coverage damage or destruction to property under the care, custody, and control of an insured. Such coverage is excluded from liability policies because the insured either has some ownership interest in such property (better covered through property, not liability, insurance) or is a bailor of the property (and can better cover this bailment exposure through an appropriate bailee policy).

cash flow plans   Method of payment of an insurance premium which allows an insured to regulate the amount and frequency of the premium payments in accordance with cash flow over a stipulated period of time. This enables the insured to maintain control over the funds for a longer period of time and thus reap benefits from their earnings.

Central Index Bureau (CIB)   Insurance bureau, now called the Index System, that keeps data on employee's previous workers' compensation claims and claimants' general liability or tort claims.

civil damages   Sums payable to the winning plaintiff by the losing defendant in a court of law; can take any or all of these forms: *general, punitive,* and *special.*

civil liability   Negligent acts and/or omissions, other than breach of contract, normally independent of moral obligations for which a remedy can be provided in a court of law.

coinsurance   Defines the amount of each loss that the insurance company pays according to the following relationship:

$$\frac{\text{Amount of insurance carried}}{\text{Amount of insurance required}} \times \text{amount of loss} = \text{insurance company payment}$$

where Amount of insurance required = value of property insured × coinsurance clause percentage

collision   Physical contact of an automobile with another inanimate object resulting in damage to the insured car. Insurance coverage is available to provide protection against this occurrence.

collision damage waiver   Special property damage coverage purchased by an individual renting an automobile where the rental company waives any right to recover property damage to the automobile from that individual, regardless of who is at fault. A significant fee is paid by the individual to the rental company for this waiver for coverage that may already be provided by a personal automobile policy (PAP). Some states require personal automobile insurance policies to cover the majority of damage to a rented vehicle.

combined single limit   Bodily injury liability and property damage liability expressed as a single sum of coverage. A $500,000 limit of liability for bodily injury and property damage is an example of a CSL.

commercial blanket bond   Coverage of the employer for all employees on a *blanket* basis, where the maximum limit of coverage is not applied to any one loss regardless of the number of employees involved. Both commercial and position blanket bonds work the same way if only one employee causes the loss, or if the guilty employee(s) cannot be identified.

commercial general liability insurance (CGL)   Coverage against all liability exposures of a contractor unless specifically excluded. Coverage includes products, completed operations, premises and operations, elevators, and independent contractors.

comparative negligence   Principle of tort law in some states providing that in the event of an accident, each party's negligence is based on that party's contribution to the accident. For example, if in an auto accident both parties fail to obey the yield sign, their negligence would be equal and neither would collect legal damages from the other.

completed operations insurance   Coverage for a contractor's liability for injuries or property damage suffered by third parties as the result of the contractor completing an operation. The contractor must take reasonable care in rendering a project safe and free from all reasonable hazards.

completion bond   Protection for a mortgagee guaranteeing that the mortgagor will complete construction. The mortgagee lends money to the mortgagor in order to pay the contractor who is physically building the project. Upon completion, the project then serves to secure the loan. Should the project not be completed, the mortgage is protected through the completion bond.

constructive total loss   Partial loss of such significance that the cost of restoring damaged property would exceed its value after restoration. For example, an automobile is so badly damaged by fire that fixing it would cost more than the restored vehicle would be worth.

contract bond   A guarantee of the performance of a contractor. In general, contract bonds are used to guarantee that the contractor will perform according to the specifications of the construction contract. If the contractor fails to perform according to contract, the insurance company is responsible to the insured for payment, up to the limit of the bond, which is usually an amount equal to the cost of the construction project. The insurance company then has recourse against the contractor for reimbursement.

contractor's equipment floater   Form of marine insurance that covers mobile equipment of a contractor, including road building machinery, steam shovels, hoists, and derricks used on the job by builders of structures, roads, bridges, dams, tunnels, and mines. Coverage is provided on a specified peril of an all-risk basis, subject to exclusion of wear and tear, work, and nuclear disaster.

contributory negligence   Principle of law recognizing that injured persons may have contributed to their own injuries. For example, by not observing the *don't walk* sign at a crosswalk, pedestrians may cause accidents in which they are injured.

date of loss (D/O/L)   The date the injury occurred, not the date the claim is reported to the insurance company.

demolition insurance Coverage that will indemnify the insured up to the limits of the policy for the expenses incurred if a building is damaged by peril such as fire, and if zoning and/or building codes require that the building be demolished.

depreciation Actual or accounting recognition of the decrease in the value of a hard asset (property) over a period of time, according to a predetermined schedule such as *straight line* depreciation.

directors' and officers' liability insurance Coverage when a director or officer of a company commits a negligent act, omission, misstatement, or misleading statement, and a successful libel suit is brought against the company as a result. Usually a large deductible is required. The policy provides coverage for directors' and officers' liability exposure if they are sued as individuals. Coverage is also provided for the costs of defense such as legal fees and other court costs.

drive-other-car insurance (DOC) Endorsement to an automobile insurance policy which protects an insured in either or both of two circumstances when driving a nonowned car:

1. *Business endorsements.* If the insured's negligent acts or omissions result in bodily injury or property damage to a third party while the insured is driving a nonowned car for business activities

2. *Personal endorsements.* If the insured's negligent acts or omissions result in bodily injury or property damage to a third party while the insured is driving a nonowned car for nonbusiness activities

employer's liability Claims that involve employment but cannot be settled under the scheduled benefits provisions of workers' compensation laws. These are settled in tort although administered under the workers' compensation policy.

experience modification Adjustment of premiums resulting from the use of experience ratings. Experience rating plans take the form of *retrospective* or *prospective* plans. Under retrospective plans, premiums are modified after the fact. That is, once the policy period ends, premiums are adjusted to reflect actual loss experience of an insured. In contrast, under prospective plans, an insured's past experience (usually for the three years immediately preceding) is used to determine the premium for the current year of coverage.

exposure The payrolls, number of employees, number of plants or deports, number of trucks, sales, and other proxies for measuring exposure to legal liabilities and fortuitous losses. In workers' compensation, the exposure unit is every $100 of payroll.

extra expense insurance Form that covers exposures associated with efforts to operate a business that is damaged by a peril such as fire. For example, a special electrical generator may have to be purchased in the event of a long-range loss of electricity if the business is to continue to operate.

extraterritoriality Provision in workers' compensation insurance under which an employee who incurs an injury in another state, and elects to come under the law of the home state, will retain coverage under the workers' compensation policy.

finite insurance The transfer of a fixed, agreed amount of liabilities to a specialty insurance company for a premium. For such a transaction to be attractive, the discounted value of the loss reserves transferred should be less than the premium. Ideally, the agreed amount of liability—or the limits of liability in the insurance policy—should exceed the loss reserves transferred or "sold." Provided there is a cash payment of the premium and the limits exceed the losses, the premiums paid should be tax-deductible. Premiums earned by the finite insurer that are not needed to pay claims earn interest in an experience account which is governed by a trust agreement with a financial institution. At any time the insured may *commute* or take back the agreed amount of liabilities and get back whatever funds remain in the experience account.

fire legal liability insurance Coverage for property loss liability as the result of negligent acts and/or omission of the insured which allows a spreading fire to damage others' property. Negligent acts and omissions can result in fire legal liability. For example, an insured, through negligence, allows a fire to spread to a neighbor's property. The neighbor then brings suit against the insured for negligence. In another example, a tenant occupying another party's property causes serious fire damage to the property through negligence.

frequency The number of claims, usually expressed as a rate per unit of exposure. Workers' compensation frequency rates are expressed as the number of losses per x amount of dollars in payroll.

hold-harmless agreement Assumption of liability through contractual agreement by one party, thereby eliminating liability on the part of another party. An example is a railroad sidetrack agreement with a contracting company under which the contractor is held harmless for damage to railroad equipment and tracks.

incidence rate The number of fortuitous events per unit of exposure. An incident does not always result in lost time or a financial loss, e.g., a near miss.

incurred losses Paid losses plus reserves, including paid allocated loss adjustment expenses.

indemnity bond Coverage for loss of an obligee in the event that the principal fails to perform according to standards agreed upon between the obligee and the principal.

independent agency system Means of selling and servicing property and casualty insurance through agents who represent different insurance companies.

lex loci The jurisdiction in which a claim or suit can be brought against the contractor.

liability insurance Coverage for all sums which the insured becomes legally obligated to pay because of bodily injury or property damage, and sometimes other wrongs to which an insurance policy applies.

loss prevention   The practice of avoiding or reducing overall claims costs by containing the frequency of claims, incorporating engineering and behavior modification techniques. Loss prevention is the cornerstone to controlling workers' compensation costs.

loss reduction   Claims management practice of reducing losses after they occur. Investigations and effective medical case management are loss reduction techniques.

lost time claim   A workers' compensation injury claim which, in addition to medical expenses, involves a period of time where the employee does not report to work. Under state laws there are waiting periods until the worker can collect indemnity benefits for lost time.

maximum medical improvement date   Time at which injured workers will not heal any further and thus be eligible for permanent awards.

monopolistic state fund   Quasi-government state-operated insurance funds used in workers' compensation insurance in some states where the risks are so great that the commercial insurance companies cannot operate at affordable rates.

Monte Carlo simulation   A computer-based actuarial technique used to measure or quantify uncertainty. After projecting ultimate loss costs and calculating loss severities, a consulting actuary sorts the loss severities into size intervals. The distribution is then fit to the most appropriate statistical distribution. The Monte Carlo simulation model then is used to make a number of trials based on the expected number of claims. For example, if after 100 simulations, 75 percent of the losses are less than $10 million, then one can be 75 percent confident that losses will be equal to or less than $10 million.

named insured   Person, business, or organization specified as the insured(s) in a property or liability insurance policy. In some instances, a policy provides broader coverage to persons other than those named in the policy if they have the insured's permission to use the property that is insured.

National Council on Compensation Insurance (NCCI)   Membership organization of insurance companies that write workers' compensation insurance. The organization collects statistics on the frequency and severity of job-related injuries to establish a rate structure for member companies, files rate plans with insurance commissioners' offices for member companies, and generates forms and policies for member companies.

negligence   In tort law, the failure to provide the duty of care which a reasonable, prudent organization or individual would provide.

no-fault insurance   Coverage afforded without regard to fault or negligence. Workers' compensation insurance is a type of no-fault insurance.

nonscheduled loss   Losses which involve the body as a whole rather than specific members. The basic theory in awarding compensation is that the injury has resulted in a diminution of earning capacity. It is not necessary for an employee to demonstrate a specific loss of earnings or earning capacity in order to justify an award for partial permanent disability in many states. The mere fact that the worker can establish functional disability as a result of the injury will justify an award of compensation.

obligor   Individual or other entity who has promised to perform a certain act. For example, an insurance company promises to pay a death benefit if a life insurance policy is in force at the time of the death of an insured.

Occupational Safety and Health Act (OSHA)   Federal legislation regulating safety standards for employers and imposing fines for violation of safety regulations.

occurrence   An accident which results in a financial loss. There can be more than one claim, such as would occur if more than one employee was injured in the same accident.

paid-loss retrospectively rated program   A version of retrospectively rated insurance in which the premium payments are tied to the loss payment schedule of the insured and not reserves established by the insurance company.

permanency benefits   Indemnity payments paid to the permanently disabled employee.

permanent partial disability (PPD)   An award for a permanent disability that is partial in nature. For example, some states use AMA guidelines to calculate the percentage loss of use of a limb as a result of an accident. Many states have scheduled awards for PPD.

permanent total disability (PTD)   A permanent disability which entitles the employee to the maximum awards.

personal injury   Wrongful conduct causing false arrest, invasion of privacy, libel, slander, defamation of character, and bodily injury. The injury is against the person, in contrast to property damage or destruction.

pollution exclusion   Liability insurance exception for pollution coverage that is not both sudden and accidental from the contractor's perspective. As a result of the damage suits from such incidents as the chemical pollution at Love Canal, insurance companies began to modify pollution coverage in their liability policies in the 1970s. First, companies changed coverage to apply only if pollution was sudden and accidental rather than gradual.

preexisting condition   A medical condition that the employee had prior to the injury being claimed. In some states reinjury costs get transferred to a second injury fund established to encourage employers to hire workers who were injured in previous accidents.

renewal certificate   Form showing notification that an insurance policy has been renewed with the same provisions, clauses, and benefits of the previous policy.

residual market loads (RMLs)   Deficits arising out of a state's assigned-risk workers' compensation acts, in all states except Alaska, Florida, Kentucky, Maine, Minnesota, Montana, and Nevada. Benefits are paid pursuant to a statutory list or schedule for specific injuries. In general, scheduled losses are limited to specific members of the body, such as the arm, leg, hand, etc. In New

York, backs are scheduled, while in other states they are rated as nonscheduled injuries.

retrospective rating Method of establishing rates in which the current year's premium is calculated to reflect the actual current year's loss experience. An initial premium is charged and then adjusted at the end of the policy year to reflect the actual loss experience of the business.

risk management The five-step process of (1) identifying fortuitous risks; (2) measuring the exposure to loss; (3) selecting risk treatment techniques; (4) selecting risk control techniques; and (5) allocating unreimbursed losses, premiums, and other *costs of risk* (COR).

Risk Management Information System (RMIS) Computer information system designed to track losses and payment history.

second injury fund A specific fund arranged by most states and funded by insurers and self-insurers whereby the employer can recover a workers' compensation loss if the employee is reinjured. The injury sustained must be the same as the original injury and the employer must have prior knowledge of the original injury at the time employment was offered.

self-insurance Protecting against loss by setting aside one's own money. This can be done on a mathematical basis by establishing a separate fund into which funds are deposited on a periodic basis. Through self-insurance it is possible to protect against high-frequency, low-severity losses.

severity The average dollar size of the loss.

special investigation units (SIUs) The insurance industry's dedicated investigative units that investigate employees suspected of currently engaging in workers' compensation fraud.

strict liability In tort law, liability imposed without regard to degree of negligence. Strict liability applies to inherent dangerous activities such as blasting or scaffolding.

structured settlement Periodic payments to an injured person or survivor for a determined number of years or for life typically in settlement of a claim under a liability policy. Terms may include immediate reimbursement for medical and legal expenses and rehabilitation, and long-term payments for loss of income or as compensation for other injuries.

subrogation The policy-given right of an insurer to take legal action against a third party responsible for a loss, subsequent to the insurer's payment of such loss to its insured.

subrogation clause Section of property insurance and liability insurance policies giving an insurer the right to take legal action against a third party responsible for a loss to an insured for which a claim has been paid.

temporary partial disability (TPD) A temporary disability which prevents the employee from resuming the full level of his or her duties, for a limited period of time.

temporary total disability (TTD) A temporary disability which prevents the employee from working at all, for a limited period of time.

third party Individual other than the insured or insurer who has incurred a loss or is entitled to receive a benefit payment as the result of the acts or omissions of the insured.

tort law Legislation governing wrongful acts, other than breaches of contract by one person against another or the property of another, for which civil action can be brought. Tort law and contract law define civil liability exposures. The four areas of torts are negligence, intentional interference, absolute liability, and strict liability. For example, the owner of a decrepit boat dock that collapses while people are standing on it might be liable under negligence. The owner of a poisonous snake that bites someone could be liable for injury under strict liability, even if the owner did not intend to harm anyone. The maker of a defective product that harms the buyer might be held liable under strict liability.

umbrella liability insurance Excess liability coverage above the limits of a basic business liability insurance policy. For example, if a basic policy has a limit of $1,000,000, and it is exhausted by claims, the umbrella will pay the excess above $1,000,000 up to the limit of the umbrella policy, which may be as high as $10 to $25 million or more. The umbrella policy also fills gaps in coverage under basic liability policies.

Unfair Claims Practice Acts The National Association of Insurance Commissioners (NAIC) has developed model legislation requiring that claims be handled fairly and that there be free communication between insured and insurance company. Many states have adopted these unfair claims practice laws.

valuation (1) Method of determining the worth of property to be insured, or of property that has been lost or damaged, or (2) method of setting insurance company reserves to pay future claims.

workers' compensation benefits Income, medical, rehabilitation, death, and survivor payments to workers injured on the job. State workers' compensation laws, which date from early in the twentieth century, provide that employers take responsibility for on-the-job injuries. Each state defines the benefit level for employers in that state. Although these benefits were designed to be the final obligation of employers to their employees, there has been some erosion of this concept since the early 1970s; workers have been allowed by the courts to sue employers for various on-the-job injuries in addition to receiving workers' compensation benefits. Because workers' compensation benefits are a routine and fairly predictable risk, many employers use self-insurance. Some states mandate that employers buy workers' compensation insurance from a state fund, but some offer a choice of a state fund, self-insurance, or commercial insurance.

workers' compensation insurance A no-fault insurance scheme imposed upon employers by statute in mandated limits, providing coverage for employees injured on the job in exchange for the employee relinquishing the right to sue the employer in tort.

wrap-up insurance A jumbo insurance program insuring various entities at a construction site, usually including workers' compensation, employer's liability, and general liability exposures. Insured entities usually include the project owners, construction manager, general contractor, and subcontractors.

B

Examples of Insurance Policies

- Commercial General Liability Policy
- Workers Compensation and Employers Liability Insurance Policy
- Commercial and Business Auto Policy

COMMON POLICY DECLARATIONS

POLICY NO.

RENEWAL OF:
1. NATIONAL UNION FIRE INSURANCE COMPANY OF PITTSBURGH,PA.
2. AMERICAN HOME ASSURANCE COMPANY
3. THE INSURANCE COMPANY OF THE STATE OF PENNSYLVANIA

MEMBERS OF THE
AMERICAN INTERNATIONAL GROUP, INC
EXECUTIVE OFFICES
70 PINE STREET
NEW YORK, N.Y.

COVERAGE IS PROVIDED IN THE
COMPANY DESIGNATED BY NUMBER
A STOCK INSURANCE COMPANY
(HEREIN CALLED THE COMPANY)

NAMED INSURED

MAILING ADDRESS

POLICY PERIOD: From   **To**   **At**
12:01 A.M. Standard Time at your mailing address shown above

BUSINESS DESCRIPTION

IN RETURN FOR THE PAYMENT OF THE PREMIUM, AND SUBJECT TO ALL TERMS OF THIS POLICY, WE AGREE WITH YOU TO PROVIDE THE INSURANCE AS STATED IN THIS POLICY.

THIS POLICY CONSISTS OF THE FOLLOWING COVERAGE PARTS FOR WHICH A PREMIUM IS INDICATED. THIS PREMIUM MAY BE SUBJECT TO ADJUSTMENT

	PREMIUM
Boiler and Machinery Coverage Part	$ NOT COVERED
Commercial Auto Coverage Part	$ NOT COVERED
Commercial Crime Coverage Part	$ NOT COVERED
Commercial General Liability Coverage Part	$
Commercial Inland Marine Coverage Part	$ NOT COVERED
Commercial Property Coverage Part	$ NOT COVERED
Farm Coverage Part	$ NOT COVERED

TOTAL $

Premium shown is payable: $ at inception.

Forms applicable to all Coverage Parts:

COUNTERSIGNED _____ BY _____
                    (Date)   (Authorized Signature)

In Witness Whereof, we have caused this policy to be executed and attested, and, if required by state law, this policy shall not be valid unless countersigned by our authorized representative.

Elizabeth M. Tuck
Secretary
National Union Fire Insurance Company of Pittsburgh PA.
American Home Assurance Company
The Insurance Company Of The
State Of Pennsylvania

William D Smith
President
National Union Fire Insurance
Company of Pittsburgh, PA.

1. NATIONAL UNION FIRE INSURANCE
   COMPANY OF PITTSBURGH
   A STOCK COMPANY

2. AMERICAN HOME ASSURANCE COMPANY
   A STOCK COMPANY

   AIG Member Companies of
   American International Group

   EXECUTIVE OFFICES

   70 PINE STREET, NEW YORK, N.Y. 10270

3. THE INSURANCE COMPANY OF THE
   STATE OF PENNSYLVANIA
   A STOCK COMPANY

   COVERAGE IS PROVIDED IN THE
   COMPANY DESIGNATED BY NUMBER
   (HEREIN CALLED THE COMPANY).

COMMERCIAL GENERAL LIABILITY DECLARATIONS

POLICY NO.

NAMED INSURED

MAILING ADDRESS

POLICY PERIOD: From   to   at

   12:01 A.M. Standard Time at your mailing address shown above

IN RETURN FOR THE PAYMENT OF THE PREMIUM, AND SUBJECT TO ALL TERMS OF THIS POLICY,
WE AGREE WITH YOU TO PROVIDE THE INSURANCE AS STATED IN THIS POLICY.

LIMITS OF INSURANCE

GENERAL AGGREGATE LIMIT (Other Than Prod–Comp Operations)	$
PRODUCTS–COMPLETED OPERATIONS AGGREGATE LIMIT	$
PERSONAL & ADVERTISING INJURY LIMIT	$
EACH OCCURRENCE LIMIT	$
FIRE DAMAGE LIMIT	$ Any One Fire
MEDICAL EXPENSE LIMIT	$ Any One Person

Forms Of Business: ☐ Individual   ☐ Partnership   ☐ Joint Venture   ☐ Organization
(Other than Partnership
or Joint Venture)

Business Description:

Location Of All Premises You Own, Rent or Occupy:

				ADVANCE PREMIUM	
CLASSIFICATION	CODE NO.	PREMIUM BASIS	RATE.	PR/CO	ALL OTHER

Premium shown is payable: $ at inception. | TOTAL: $ |

ENDORSEMENTS ATTACHED TO THIS POLICY:_____

COUNTERSIGNED _____ BY _____
                        (Date)   (Authorized Representative)

COMMON POLICY CONDITIONS

All Coverage Parts included in this policy are subject to the following conditions.

A. CANCELLATION

1. The first Named Insured shown in the Declarations may cancel this policy by mailing or delivering to us advance written notice of cancellation.

2. We may cancel this policy by mailing or delivering to the first Named Insured written notice of cancellation at least:

   a. 10 days before the effective date of cancellation if we cancel for nonpayment of premium; or

   b. 30 days before the effective date of cancellation if we cancel for any other reason.-

3. We will mail or deliver our notice to the first Named Insured's last mailing address known to us.

4. Notice of cancellation will state the effective date of cancellation. The policy period will end on that date.

5. If this policy is cancelled, we will send the first Named Insured any premium refund due. If we cancel, the refund will be pro rata. If the first Named Insured cancels, the refund may be less than pro rata. The cancellation will be effective even if we have not made or offered a refund.

6. If notice is mailed, proof of mailing will be sufficient proof of notice.

B. CHANGES

This policy contains all the agreements between you and us concerning the insurance afforded. The first Named Insured shown in the Declarations is authorized to make changes in the terms of this policy with our consent. This policy's terms can be amended or waived only by endorsement issued by us and made a part of this policy.

C. EXAMINATION OF YOUR BOOKS AND RECORDS

We may examine and audit your books and records as they relate to this policy at any time during the policy period and up to three years afterward.

D. INSPECTIONS AND SURVEYS

We have the right but are not obligated to:

1. Make inspections and surveys at any time;

2. Give you reports on the conditions we find; and

3. Recommend changes.

Any inspections, surveys, reports or recommendations relate only to insurability and the premiums to be charged. We do not make safety inspections. We do not undertake to perform the duty of any person or organization to provide for the health or safety of workers or the public. And we do not warrant that conditions:

1. Are safe or healthful; or

2. Comply with laws, regulations, codes or standards.

This condition applies not only to us, but also to any rating, advisory, rate service or similar organization which makes insurance inspections, surveys, reports or recommendations.

E. PREMIUMS

The first Named Insured shown in the Declarations:

1. Is responsible for the payment of all premiums; and

2. Will be the payee for any return premiums we pay.

F. TRANSFER OF YOUR RIGHTS AND DUTIES UNDER THIS POLICY

Your rights and duties under this policy may not be transferred without our written consent except in the case of death of an individual named insured.

If you die, your rights and duties will be transferred to your legal representative but only while acting within the scope of duties as your legal representative. Until your legal representative is appointed, anyone having proper temporary custody of your property will have your rights and duties but only with respect to that property.

COMMERCIAL GENERAL LIABILITY COVERAGE FORM

Various provisions in this policy restrict coverage. Read the entire policy carefully to determine rights, duties and what is and is not covered.

Throughout this policy the words "you" and "your" refer to the Named Insured shown in the Declarations, and any other person or organization qualifying as a Named Insured under this policy. The words "we", "us" and "our" refer to the company providing this insurance.

The word "insured" means any person or organization qualifying as such under WHO IS AN INSURED (SECTION II).

Other words and phrases that appear in quotation marks have special meaning. Refer to DEFINITIONS (SECTION V).

SECTION I - COVERAGES

COVERAGE A. BODILY INJURY AND PROPERTY DAMAGE LIABILITY

1. **Insuring Agreement.**
   a. We will pay those sums that the insured becomes legally obligated to pay as damages because of "bodily injury" or "property damage" to which this insurance applies. We will have the right and duty to defend any "suit" seeking those damages. We may at our discretion investigate any "occurrence" and settle any claim or "suit" that may result. But
      (1) The amount we will pay for damages is limited as described in LIMITS OF INSURANCE (SECTION III); and
      (2) Our right and duty to defend end when we have used up the applicable limit of insurance in the payment of judgments or settlements under Coverages A or B or medical expenses under Coverage C.

      No other obligation or liability to pay sums or perform acts or services is covered unless explicitly provided for under SUPPLEMENTARY PAYMENTS – COVERAGES A AND B.

   b. This insurance applies to "bodily injury" and "property damage" only if:
      (1) The "bodily injury" or "property damage" is caused by an "occurrence" that takes place in the "coverage territory"; and

      (2) The "bodily injury" or "property damage" occurs during the policy period.
   c. Damages because of "bodily injury" include damages claimed by any person or organization for care, loss of services or death resulting at any time from the "bodily injury".

2. **Exclusions.**

   This insurance does not apply to:

   a. **Expected or Intended Injury**

      "Bodily injury" or "property damage" expected or intended from the standpoint of the insured. This exclusion does not apply to "bodily injury" resulting from the use of reasonable force to protect persons or property.

   b. **Contractual Liability**

      "Bodily injury" or "property damage" for which the insured is obligated to pay damages by reason of the assumption of liability in a contract or agreement. This exclusion does not apply to liability for damages:
      (1) Assumed in a contract or agreement that is an "insured contract", provided the "bodily injury" or "property damage" occurs subsequent to the execution of the contract or agreement; or
      (2) That the insured would have in the absence of the contract or agreement.

   c. **Liquor Liability**

      "Bodily injury" or "property damage" for which any insured may be held liable by reason of:
      (1) Causing or contributing to the intoxication of any person;
      (2) The furnishing of alcoholic beverages to a person under the legal drinking age or under the influence of alcohol; or
      (3) Any statute, ordinance or regulation relating to the sale, gift, distribution or use of alcoholic beverages.

      This exclusion applies only if you are in the business of manufacturing, distributing, selling, serving or furnishing alcoholic beverages.

d. Workers Compensation and Similar Laws

Any obligation of the insured under a workers compensation, disability benefits or unemployment compensation law or any similar law.

e. Employer's Liability

"Bodily injury" to:

(1) An "employee" of the insured arising out of and in the course of:

  (a) Employment by the insured; or

  (b) Performing duties related to the conduct of the insured's business; or

(2) The spouse, child, parent, brother or sister of that "employee" as a consequence of paragraph (1) above.

This exclusion applies:

(1) Whether the insured may be liable as an employer or in any other capacity; and

(2) To any obligation to share damages with or repay someone else who must pay damages because of the injury.

This exclusion does not apply to liability assumed by the insured under an "insured contract".

f. Pollution

(1) "Bodily injury" or "property damage" arising out of the actual, alleged or threatened discharge, dispersal, seepage, migration, release or escape of pollutants:

  (a) At or from any premises, site or location which is or was at any time owned or occupied by, or rented or loaned to, any insured;

  (b) At or from any premises, site or location which is or was at any time used by or for any insured or others for the handling, storage, disposal, processing or treatment of waste;

  (c) Which are or were at any time transported, handled, stored, treated, disposed of, or processed as waste by or for any insured or any person or organization for whom you may be legally responsible; or

(d) At or from any premises, site or location on which any insured or any contractors or subcontractors working directly or indirectly on any insured's behalf are performing operations:

  (i) If the pollutants are brought on or to the premises, site or location in connection with such operations by such insured, contractor or subcontractor; or

  (ii) If the operations are to test for, monitor, clean up, remove, contain, treat, detoxify or neutralize, or in any way respond to, or assess the effects of pollutants.

Subparagraphs (a) and (d)(i) do not apply to "bodily injury" or "property damage" arising out of heat, smoke or fumes from a hostile fire.

As used in this exclusion, a hostile fire means one which becomes uncontrollable or breaks out from where it was intended to be.

(2) Any loss, cost or expense arising out of any:

  (a) Request, demand or order that any insured or others test for, monitor, clean up, remove, contain, treat, detoxify or neutralize, or in any way respond to, or assess the effects of pollutants; or

  (b) Claim or suit by or on behalf of a governmental authority for damages because of testing for, monitoring, cleaning up, removing, containing, treating, detoxifying or neutralizing, or in any way responding to, or assessing the effects of pollutants.

Pollutants means any solid, liquid, gaseous or thermal irritant or contaminant, including smoke, vapor, soot, fumes, acids, alkalis, chemicals and waste. Waste includes materials to be recycled, reconditioned or reclaimed.

g. Aircraft, Auto or Watercraft

"Bodily injury" or "property damage" arising out of the ownership, maintenance, use or entrustment to others of any aircraft, "auto" or watercraft owned or operated by or rented or loaned to any insured. Use includes operation and "loading or unloading".

This exclusion does not apply to:

(1) A watercraft while ashore on premises you own or rent;

(2) A watercraft you do not own that is:

    (a) Less than 26 feet long; and

    (b) Not being used to carry persons or property for a charge;

(3) Parking an "auto" on, or on the ways next to, premises you own or rent, provided the "auto" is not owned by or rented or loaned to you or the insured;

(4) Liability assumed under any "insured contract" for the ownership, maintenance or use of aircraft or watercraft; or

(5) "Bodily injury" or "property damage" arising out of the operation of any of the equipment listed in paragraph f.(2) or f.(3) of the definition of "mobile equipment".

h. **Mobile Equipment**

"Bodily injury" or "property damage" arising out of:

(1) The transportation of "mobile equipment" by an "auto" owned or operated by or rented or loaned to any insured; or

(2) The use of "mobile equipment" in, or while in practice for, or while being prepared for, any prearranged racing, speed, demolition, or stunting activity.

i. **War**

"Bodily injury" or "property damage" due to war, whether or not declared, or any act or condition incident to war. War includes civil war, insurrection, rebellion or revolution. This exclusion applies only to liability assumed under a contract or agreement.

j. **Damage to Property**

"Property damage" to:

(1) Property you own, rent, or occupy;

(2) Premises you sell, give away or abandon, if the "property damage" arises out of any part of those premises;

(3) Property loaned to you;

(4) Personal property in the care, custody or control of the insured;

(5) That particular part of real property on which you or any contractors or subcontractors working directly or indirectly on your behalf are performing operations, if the "property damage" arises out of those operations; or

(6) That particular part of any property that must be restored, repaired or replaced because "your work" was incorrectly performed on it.

Paragraph (2) of this exclusion does not apply if the premises are "your work" and were never occupied, rented or held for rental by you.

Paragraphs (3), (4), (5) and (6) of this exclusion do not apply to liability assumed under a sidetrack agreement.

Paragraph (6) of this exclusion does not apply to "property damage" included in the "products-completed operations hazard".

k. **Damage to Your Product**

"Property damage" to "your product" arising out of it or any part of it.

l. **Damage to Your Work**

"Property damage" to "your work" arising out of it or any part of it and included in the "products-completed operations hazard".

This exclusion does not apply if the damaged work or the work out of which the damage arises was performed on your behalf by a subcontractor.

m. **Damage to Impaired Property or Property Not Physically Injured**

"Property damage" to "impaired property" or property that has not been physically injured, arising out of:

(1) A defect, deficiency, inadequacy or dangerous condition in "your product" or "your work"; or

(2) A delay or failure by you or anyone acting on your behalf to perform a contract or agreement in accordance with its terms.

This exclusion does not apply to the loss of use of other property arising out of sudden and accidental physical injury to "your product" or "your work" after it has been put to its intended use.

n. Recall of Products, Work or Impaired Property

Damages claimed for any loss, cost or expense incurred by you or others for the loss of use, withdrawal, recall, inspection, repair, replacement, adjustment, removal or disposal of:

(1) "Your product";

(2) "Your work"; or

(3) "Impaired property";

if such product, work, or property is withdrawn or recalled from the market or from use by any person or organization because of a known or suspected defect, deficiency, inadequacy or dangerous condition in it.

Exclusions c. through n. do not apply to damage by fire to premises while rented to you or temporarily occupied by you with permission of the owner. A separate limit of insurance applies to this coverage as described in LIMITS OF INSURANCE (Section III).

COVERAGE B. PERSONAL AND ADVERTISING INJURY LIABILITY

1. Insuring Agreement.

a. We will pay those sums that the insured becomes legally obligated to pay as damages because of "personal injury" or "advertising injury" to which this insurance applies. We will have the right and duty to defend any "suit" seeking those damages. We may at our discretion investigate any "occurrence" or offense and settle any claim or "suit" that may result. But:

(1) The amount we will pay for damages is limited as described in LIMITS OF INSURANCE (SECTION III); and

(2) Our right and duty to defend end when we have used up the applicable limit of insurance in the payment of judgments or settlements under Coverage A or B or medical expenses under Coverage C.

No other obligation or liability to pay sums or perform acts or services is covered unless explicitly provided for under SUPPLEMENTARY PAYMENTS COVERAGES A AND B.

b. This insurance applies to:

(1) "Personal injury" caused by an offense arising out of your business, excluding advertising, publishing, broadcasting or telecasting done by or for you;

(2) "Advertising injury" caused by an offense committed in the course of advertising your goods, products or services;

but only if the offense was committed in the "coverage territory" during the policy period.

2. Exclusions.

This insurance does not apply to:

a. "Personal injury" or "advertising injury":

(1) Arising out of oral or written publication of material, if done by or at the direction of the insured with knowledge of its falsity;

(2) Arising out of oral or written publication of material whose first publication took place before the beginning of the policy period;

(3) Arising out of the willful violation of a penal statute or ordinance committed by or with the consent of the insured; or

(4) For which the insured has assumed liability in a contract or agreement. This exclusion does not apply to liability for damages that the insured would have in the absence of the contract or agreement.

b. "Advertising injury" arising out of:

(1) Breach of contract, other than misappropriation of advertising ideas under an implied contract;

(2) The failure of goods, products or services to conform with advertised quality or performance;

(3) The wrong description of the price of goods, products or services; or

(4) An offense committed by an insured whose business is advertising, broadcasting, publishing or telecasting.

COVERAGE C. MEDICAL PAYMENTS

1. Insuring Agreement.

a. We will pay medical expenses as described below for "bodily injury" caused by an accident:

   (1) On premises you own or rent;

   (2) On ways next to premises you own or rent; or

   (3) Because of your operations;

   provided that:

   (1) The accident takes place in the "coverage territory" and during the policy period;

   (2) The expenses are incurred and reported to us within one year of the date of the accident; and

   (3) The injured person submits to examination, at our expense, by physicians of our choice as often as we reasonably require.

b. We will make these payments regardless of fault. These payments will not exceed the applicable limit of insurance. We will pay reasonable expenses for:

   (1) First aid administered at the time of an accident;

   (2) Necessary medical, surgical, x-ray and dental services, including prosthetic devices; and

   (3) Necessary ambulance, hospital, professional nursing and funeral services.

2. Exclusions.

We will not pay expenses for "bodily injury":

a. To any insured.

b. To a person hired to do work for or on behalf of any insured or a tenant of any insured.

c. To a person injured on that part of premises you own or rent that the person normally occupies.

d. To a person, whether or not an "employee" of any insured, if benefits for the "bodily injury" are payable or must be provided under a workers compensation or disability benefits law or a similar law.

e. To a person injured while taking part in athletics.

f. Included within the "products-completed operations hazard".

g. Excluded under Coverage A.

h. Due to war, whether or not declared, or any act or condition incident to war. War includes civil war, insurrection, rebellion or revolution.

SUPPLEMENTARY PAYMENTS - COVERAGES A AND B

We will pay, with respect to any claim or "suit" we defend:

1. All expenses we incur.

2. Up to $250 for cost of bail bonds required because of accidents or traffic law violations arising out of the use of any vehicle to which the Bodily Injury Liability Coverage applies. We do not have to furnish these bonds.

3. The cost of bonds to release attachments, but only for bond amounts within the applicable limit of insurance. We do not have to furnish these bonds.

4. All reasonable expenses incurred by the insured at our request to assist us in the investigation or defense of the claim or "suit", including actual loss of earnings up to $100 a day because of time off from work.

5. All costs taxed against the insured in the "suit".

6. Prejudgment interest awarded against the insured on that part of the judgment we pay. If we make an offer to pay the applicable limit of insurance, we will not pay any prejudgment interest based on that period of time after the offer.

7. All interest on the full amount of any judgment that accrues after entry of the judgment and before we have paid, offered to
pay, or deposited in court the part of the
judgment that is within the applicable limit
of insurance.

These payments will not reduce the limits of
insurance.

SECTION II - WHO IS AN INSURED

1. If you are designated in the Declarations as:

   a. An individual, you and your spouse are
   insureds, but only with respect to the
   conduct of a business of which you are
   the sole owner.

   b. A partnership or joint venture, you are
   an insured. Your partners, your partners,
   and their spouses are also insureds, but
   only with respect to the conduct of your
   business.

   c. An organization other than a partnership
   or joint venture, you are an insured. Your
   "executive officers" and directors are insureds, but only with respect to their
   duties as your officers or directors. Your
   stockholders are also insureds, but only
   with respect to their liability as stockholders.

2. Each of the following is also an insured:

   a. Your "employees", other than your
   "executive officers", but only for acts
   within the scope of their employment by
   you or while performing duties related
   to the conduct of your business. However, no "employee" is an insured for:

      (1) "Bodily injury" or "personal injury":

         (a) To you, to your partners or mem-
         bers (if you are a partnership or
         joint venture), or to a co-
         "employee" while in the course of
         his or her employment or while
         performing duties related to the
         conduct of your business;

         (b) To the spouse, child, parent,
         brother or sister of that co-
         "employee" as a consequence of
         paragraph (1)(a) above;

         (c) For which there is any obligation
         to share damages with or repay
         someone else who must pay dam-
         ages because of the injury de-
         scribed in paragraphs (1)(a) or (b)
         above; or

         (d) Arising out of his or her providing
         or failing to provide professional
         health care services.

      (2) "Property damage" to property:

         (a) Owned, occupied or used by,

         (b) Rented to, in the care, custody or
         control of, or over which physical
         control is being exercised for any
         purpose by

         you, any of your "employees" or, if
         you are a partnership or joint venture,
         by any partner or member.

   b. Any person (other than your "employee"),
   or any organization while acting as your
   real estate manager.

   c. Any person or organization having proper
   temporary custody of your property if
   you die, but only:

      (1) With respect to liability arising out of
      the maintenance or use of that prop-
      erty; and

      (2) Until your legal representative has been
      appointed.

   d. Your legal representative if you die, but
   only with respect to duties as such. That
   representative will have all your rights and
   duties under this Coverage Part.

3. With respect to "mobile equipment" registered in your name under any motor vehicle
registration law, any person is an insured
while driving such equipment along a public
highway with your permission. Any other
person or organization responsible for the
conduct of such person is also an insured,
but only with respect to liability arising out
of the operation of the equipment, and only
if no other insurance of any kind is available
to that person or organization for this liability. However, no person or organization
is an insured with respect to:

   a. "Bodily injury" to a co-"employee" of the
   person driving the equipment; or

   b. "Property damage" to property owned
   by, rented to, in the charge of or occupied by you or the employer of any
   person who is an insured under this
   provision.

4. Any organization you newly acquire or form,
other than a partnership or joint venture, and
over which you maintain ownership or majority interest, will qualify as a Named Insured if there is no other similar insurance
available to that organization. However:

   a. Coverage under this provision is afforded
   only until the 90th day after you acquire
   or form the organization or the end of
   the policy period, whichever is earlier;

b. Coverage A does not apply to "bodily injury" or "property damage" that occurred before you acquired or formed the organization; and

c. Coverage B does not apply to "personal injury" or "advertising injury" arising out of an offense committed before you acquired or formed the organization.

No person or organization is an insured with respect to the conduct of any current or past partnership or joint venture that is not shown as a Named Insured in the Declarations.

SECTION III - LIMITS OF INSURANCE

1. The Limits of Insurance shown in the Declarations and the rules below fix the most we will pay regardless of the number of:

   a. Insureds;

   b. Claims made or "suits" brought; or

   c. Persons or organizations making claims or bringing "suits".

2. The General Aggregate Limit is the most we will pay for the sum of:

   a. Medical expenses under Coverage C;

   b. Damages under Coverage A, except damages because of "bodily injury" or "property damage" included in the "products-completed operations hazard"; and

   c. Damages under Coverage B.

3. The Products-Completed Operations Aggregate Limit is the most we will pay under Coverage A for damages because of "bodily injury" and "property damage" included in the "products-completed operations hazard".

4. Subject to 2. above, the Personal and Advertising Injury Limit is the most we will pay under Coverage B for the sum of all damages because of all "personal injury" and all "advertising injury" sustained by any one person or organization.

5. Subject to 2. or 3. above, whichever applies, the Each Occurrence Limit is the most we will pay for the sum of:

   a. Damages under Coverage A; and

   b. Medical expenses under Coverage C

   because of all "bodily injury" and "property damage" arising out of any one "occurrence".

6. Subject to 5. above, the Fire Damage Limit is the most we will pay under Coverage A for damages because of "property damage" to premises, while rented to you or temporarily occupied by you with permission of the owner, arising out of any one fire.

7. Subject to 5. above, the Medical Expense Limit is the most we will pay under Coverage C for all medical expenses because of "bodily injury" sustained by any one person.

The Limits of Insurance of this Coverage Part apply separately to each consecutive annual period and to any remaining period of less than 12 months, starting with the beginning of the policy period shown in the Declarations, unless the policy period is extended after issuance for an additional period of less than 12 months. In that case, the additional period will be deemed part of the last preceding period for purposes of determining the Limits of Insurance.

SECTION IV - COMMERCIAL GENERAL LIABILITY CONDITIONS

1. Bankruptcy.

   Bankruptcy or insolvency of the insured or of the insured's estate will not relieve us of our obligations under this Coverage Part.

2. Duties In The Event Of Occurrence, Offense, Claim Or Suit.

   a. You must see to it that we are notified as soon as practicable of an "occurrence" or an offense which may result in a claim. To the extent possible, notice should include:

      (1) How, when and where the "occurrence" or offense took place;

      (2) The names and addresses of any injured persons and witnesses; and

      (3) The nature and location of any injury or damage arising out of the "occurrence" or offense.

   b. If a claim is made or "suit" is brought against any insured, you must:

      (1) Immediately record the specifics of the claim or "suit" and the date received; and

      (2) Notify us as soon as practicable.

      You must see to it that we receive written notice of the claim or "suit" as soon as practicable.

c. You and any other involved insured must

(1) Immediately send us copies of any demands, notices, summonses or legal papers received in connection with the claim or "suit";

(2) Authorize us to obtain records and other information;

(3) Cooperate with us in the investigation, settlement or defense of the claim or "suit"; and

(4) Assist us, upon our request, in the enforcement of any right against any person or organization which may be liable to the insured because of injury or damage to which this insurance may also apply.

d. No insureds will, except at their own cost, voluntarily make a payment, assume any obligation, or incur any expense, other than for first aid, without our consent.

3. Legal Action Against Us.

No person or organization has a right under this Coverage Part:

a. To join us as a party or otherwise bring us into a "suit" asking for damages from an insured; or

b. To sue us on this Coverage Part unless all of its terms have been fully complied with.

A person or organization may sue us to recover on an agreed settlement or on a final judgment against an insured obtained after an actual trial; but we will not be liable for damages that are not payable under the terms of this Coverage Part or that are in excess of the applicable limit of insurance. An agreed settlement means a settlement and release of liability signed by us, the insured and the claimant or the claimant's legal representative.

4. Other Insurance.

If other valid and collectible insurance is available to the insured for a loss we cover under Coverages A or B of this Coverage Part, our obligations are limited as follows:

a. Primary Insurance

This insurance is primary except when b. below applies. If this insurance is primary, our obligations are not affected unless any of the other insurance is also primary. Then, we will share with all that other insurance by the method described in c. below.

b. Excess Insurance

This insurance is excess over any of the other insurance, whether primary, excess, contingent or on any other basis:

(1) That is Fire, Extended Coverage, Builder's Risk, Installation Risk or similar coverage for "your work";

(2) That is Fire insurance for premises rented to you; or

(3) If the loss arises out of the maintenance or use of aircraft, "autos" or watercraft to the extent not subject to Exclusion g. of Coverage A (Section I).

When this insurance is excess, we will have no duty under Coverage A or B to defend any claim or "suit" that any other insurer has a duty to defend. If no other insurer defends, we will undertake to do so, but we will be entitled to the insured's rights against all those other insurers.

When this insurance is excess over other insurance, we will pay only our share of the amount of the loss, if any, that exceeds the sum of:

(1) The total amount that all such other insurance would pay for the loss in the absence of this insurance; and

(2) The total of all deductible and self-insured amounts under all that other insurance.

We will share the remaining loss, if any, with any other insurance that is not described in this Excess Insurance provision and was not bought specifically to apply in excess of the Limits of Insurance shown in the Declarations of this Coverage Part.

c. Method of Sharing

If all of the other insurance permits contribution by equal shares, we will follow this method also. Under this approach each insurer contributes equal amounts until it has paid its applicable limit of insurance or none of the loss remains, whichever comes first.

If any of the other insurance does not permit contribution by equal shares, we will contribute by limits. Under this method, each insurer's share is based on the ratio of its applicable limit of insurance to the total applicable limits of insurance of all insurers.

5. **Premium Audit.**
   a. We will compute all premiums for this Coverage Part in accordance with our rules and rates.
   b. Premium shown in this Coverage Part as advance premium is a deposit premium only. At the close of each audit period we will compute the earned premium for that period. Audit premiums are due and payable on notice to the first Named Insured. If the sum of the advance and audit premiums paid for the policy period is greater than the earned premium, we will return the excess to the first Named Insured.
   c. The first Named Insured must keep records of the information we need for premium computation, and send us copies at such times as we may request.

6. **Representations.**
   By accepting this policy, you agree:
   a. The statements in the Declarations are accurate and complete;
   b. Those statements are based upon representations you made to us; and
   c. We have issued this policy in reliance upon your representations.

7. **Separation Of Insureds.**
   Except with respect to the Limits of Insurance, and any rights or duties specifically assigned in this Coverage Part to the first Named Insured, this insurance applies:
   a. As if each Named Insured were the only Named Insured; and
   b. Separately to each insured against whom claim is made or "suit" is brought.

8. **Transfer Of Rights Of Recovery Against Others To Us.**
   If the insured has rights to recover all or part of any payment we have made under this Coverage Part, those rights are transferred to us. The insured must do nothing after loss to impair them. At our request, the insured will bring "suit" or transfer those rights to us and help us enforce them.

9. **When We Do Not Renew.**
   If we decide not to renew this Coverage Part, we will mail or deliver to the first Named Insured shown in the Declarations written notice of the nonrenewal not less than 30 days before the expiration date.
   If notice is mailed, proof of mailing will be sufficient proof of notice.

SECTION V - DEFINITIONS

1. "Advertising injury" means injury arising out of one or more of the following offenses:
   a. Oral or written publication of material that slanders or libels a person or organization or disparages a person's or organization's goods, products or services;
   b. Oral or written publication of material that violates a person's right of privacy;
   c. Misappropriation of advertising ideas or style of doing business; or
   d. Infringement of copyright, title or slogan.

2. "Auto" means a land motor vehicle, trailer or semitrailer designed for travel on public roads, including any attached machinery or equipment. But "auto" does not include "mobile equipment".

3. "Bodily injury" means bodily injury, sickness or disease sustained by a person, including death resulting from any of these at any time.

4. "Coverage territory" means:
   a. The United States of America (including its territories and possessions), Puerto Rico and Canada;
   b. International waters or airspace, provided the injury or damage does not occur in the course of travel or transportation to or from any place not included in **a.** above; or
   c. All parts of the world if:
      (1) The injury or damage arises out of:
          (a) Goods or products made or sold by you in the territory described in **a.** above; or

(b) The activities of a person whose home is in the territory described in a. above, but is away for a short time on your business; and

(2) The insured's responsibility to pay damages is determined in a "suit" on the merits, in the territory described in a. above or in a settlement we agree to.

5. "Employee" includes a "leased worker". "Employee" does not include a "temporary worker".

6. "Executive officer" means a person holding any of the officer positions created by your charter, constitution, by-laws or any other similar governing document.

7. "Impaired property" means tangible property, other than "your product" or "your work", that cannot be used or is less useful because:

   a. It incorporates "your product" or "your work" that is known or thought to be defective, deficient, inadequate or dangerous; or

   b. You have failed to fulfill the terms of a contract or agreement;

   if such property can be restored to use by:

   a. The repair, replacement, adjustment or removal of "your product" or "your work"; or

   b. Your fulfilling the terms of the contract or agreement.

8. "Insured contract" means:

   a. A contract for a lease of premises. However, that portion of the contract for a lease of premises that indemnifies any person or organization for damage by fire to premises while rented to you or temporarily occupied by you with permission of the owner is not an "insured contract";

   b. A sidetrack agreement;

   c. Any easement or license agreement, except in connection with construction or demolition operations on or within 50 feet of a railroad;

   d. An obligation, as required by ordinance, to indemnify a municipality, except in connection with work for a municipality;

   e. An elevator maintenance agreement;

f. That part of any other contract or agreement pertaining to your business (including an indemnification of a municipality in connection with work performed for a municipality) under which you assume the tort liability of another party to pay for "bodily injury" or "property damage" to a third person or organization. Tort liability means a liability that would be imposed by law in the absence of any contract or agreement.

Paragraph f. does not include that part of any contract or agreement:

(1) That indemnifies a railroad for "bodily injury" or "property damage" arising out of construction or demolition operations, within 50 feet of any railroad property and affecting any railroad bridge or trestle, tracks, road-beds, tunnel, underpass or crossing;

(2) That indemnifies an architect, engineer or surveyor for injury or damage arising out of:

   (a) Preparing, approving or failing to prepare or approve maps, drawings, opinions, reports, surveys, change orders, designs or specifications; or

   (b) Giving directions or instructions, or failing to give them, if that is the primary cause of the injury or damage; or

(3) Under which the insured, if an architect, engineer or surveyor, assumes liability for an injury or damage arising out of the insured's rendering or failure to render professional services, including those listed in (2) above and supervisory, inspection or engineering services.

9. "Leased worker" means a person leased to you by a labor leasing firm under an agreement between you and the labor leasing firm, to perform duties related to the conduct of your business. "Leased worker" does not include a "temporary worker".

10. "Loading or unloading" means the handling of property:

    a. After it is moved from the place where it is accepted for movement into or onto an aircraft, watercraft or "auto";

b. While it is in or on an aircraft, watercraft or "auto"; or

c. While it is being moved from an aircraft, watercraft or "auto" to the place where it is finally delivered;

but "loading or unloading" does not include the movement of property by means of a mechanical device, other than a hand truck, that is not attached to the aircraft, watercraft or "auto".

11. "Mobile equipment" means any of the following types of land vehicles, including any attached machinery or equipment:

a. Bulldozers, farm machinery, forklifts and other vehicles designed for use principally off public roads;

b. Vehicles maintained for use solely on or next to premises you own or rent;

c. Vehicles that travel on crawler treads;

d. Vehicles, whether self-propelled or not, maintained primarily to provide mobility to permanently mounted:

(1) Power cranes, shovels, loaders, diggers or drills; or

(2) Road construction or resurfacing equipment such as graders, scrapers or rollers;

e. Vehicles not described in a., b., c. or d. above that are not self-propelled and are maintained primarily to provide mobility to permanently attached equipment of the following types:

(1) Air compressors, pumps and generators, including spraying, welding, building cleaning, geophysical exploration, lighting and well servicing equipment; or

(2) Cherry pickers and similar devices used to raise or lower workers;

f. Vehicles not described in a., b., c. or d. above maintained primarily for purposes other than the transportation of persons or cargo.

However, self-propelled vehicles with the following types of permanently attached equipment are not "mobile equipment" but will be considered "autos":

(1) Equipment designed primarily for:

(a) Snow removal;

(b) Road maintenance, but not construction or resurfacing; or

(c) Street cleaning;

(2) Cherry pickers and similar devices mounted on automobile or truck chassis and used to raise or lower workers;

(3) Air compressors, pumps and generators, including spraying, welding, building cleaning, geophysical exploration, lighting and well servicing equipment.

12. "Occurrence" means an accident, including continuous or repeated exposure to substantially the same general harmful conditions.

13. "Personal injury" means injury, other than "bodily injury", arising out of one or more of the following offenses:

a. False arrest, detention or imprisonment;

b. Malicious prosecution;

c. The wrongful eviction from, wrongful entry into, or invasion of the right of private occupancy of a room, dwelling or premises that a person occupies by or on behalf of its owner, landlord or lessor;

d. Oral or written publication of material that slanders or libels a person or organization or disparages a person's or organization's goods, products or services; or

e. Oral or written publication of material that violates a person's right of privacy.

14.a. "Products-completed operations hazard" includes all "bodily injury" and "property damage" occurring away from premises you own or rent and arising out of "your product" or "your work" except:

(1) Products that are still in your physical possession; or

(2) Work that has not yet been completed or abandoned.

b. "Your work" will be deemed completed at the earliest of the following times:

(1) When all of the work called for in your contract has been completed.

(2) When all of the work to be done at the site has been completed if your contract calls for work at more than one site.

(3) When that part of the work done at a job site has been put to its intended use by any person or organization other than another contractor or subcontractor working on the same project.

Work that may need service, maintenance, correction, repair or replacement, but which is otherwise complete, will be treated as completed.

c. This hazard does not include "bodily injury" or "property damage" arising out of:

    (1) The transportation of property, unless the injury or damage arises out of a condition in or on a vehicle created by the "loading or unloading" of it;

    (2) The existence of tools, uninstalled equipment or abandoned or unused materials; or

    (3) Products or operations for which the classification in this Coverage Part or in our manual of rules includes products or completed operations.

15. "Property damage" means:

a. Physical injury to tangible property, including all resulting loss of use of that property. All such loss of use shall be deemed to occur at the time of the physical injury that caused it; or

b. Loss of use of tangible property that is not physically injured. All such loss of use shall be deemed to occur at the time of the "occurrence" that caused it.

16. "Suit" means a civil proceeding in which damages because of "bodily injury", "property damage", "personal injury" or "advertising injury" to which this insurance applies are alleged. "Suit" includes:

a. An arbitration proceeding in which such damages are claimed and to which you must submit or do submit with our consent; or

b. Any other alternative dispute resolution proceeding in which such damages are claimed and to which you submit with our consent.

17. "Your product" means:

a. Any goods or products, other than real property, manufactured, sold, handled, distributed or disposed of by:

    (1) You;

    (2) Others trading under your name; or

    (3) A person or organization whose business or assets you have acquired; and

b. Containers (other than vehicles), materials, parts or equipment furnished in connection with such goods or products.

"Your product" includes:

a. Warranties or representations made at any time with respect to the fitness, quality, durability, performance or use of "your product"; and

b. The providing of or failure to provide warnings or instructions.

"Your product" does not include vending machines or other property rented to or located for the use of others but not sold.

18. "Temporary worker" means a person who is furnished to you to substitute for a permanent "employee" on leave or to meet seasonal or short-term workload conditions.

19. "Your work" means:

a. Work or operations performed by you or on your behalf; and

b. Materials, parts or equipment furnished in connection with such work or operations.

"Your work" includes:

a. Warranties or representations made at any time with respect to the fitness, quality, durability, performance or use of "your work"; and

b. The providing of or failure to provide warnings or instructions.

THIS ENDORSEMENT CHANGES THE POLICY. PLEASE READ IT CAREFULLY.

AMENDMENT–AGGREGATE LIMITS OF INSURANCE PER PROJECT

This endorsement modifies insurance provided under the following:

COMMERCIAL GENERAL LIABILITY COVERAGE PART.

The General Aggregate Limit under LIMITS OF INSURANCE (SECTION III) applies separately to each of your projects away from premises owned by or rented to you.

COMMERCIAL GENERAL LIABILITY

THIS ENDORSEMENT CHANGES THE POLICY. PLEASE READ IT CAREFULLY.

AMENDMENT-AGGREGATE LIMITS OF INSURANCE PER LOCATION

This endorsement modifies insurance provided under the following:

COMMERCIAL GENERAL LIABILITY COVERAGE PART.

The General Aggregate Limit under LIMITS OF INSURANCE (Section III) applies separately to each of your "locations" owned by or rented to you.

"Location" means premises involving the same or connecting lots, or premises whose connection is interrupted only by a street, roadway, waterway or right-of-way of a railroad.

POLICY NUMBER: COMMERCIAL GENERAL LIABILITY
                                                       CG 20 10 10 93

THIS ENDORSEMENT CHANGES THE POLICY. PLEASE READ IT CAREFULLY.

ADDITIONAL INSURED - OWNERS, LESSEES OR CONTRACTORS (FORM B)

This endorsement modifies insurance provided under the following:

COMMERCIAL GENERAL LIABILITY COVERAGE PART

                                SCHEDULE
Name of Person or Organization:

(If no entry appears above, information required to complete this endorsement will be shown in the Declarations as applicable to this endorsement.)

WHO IS AN INSURED (Section II) is amended to include as an insured the person or organization shown in the Schedule, but only with respect to liability arising out of your ongoing operations performed for that insured.

POLICY NUMBER: COMMERCIAL GENERAL LIABILITY

THIS ENDORSEMENT CHANGES THE POLICY. PLEASE READ IT CAREFULLY.

AMENDMENT OF CONTRACTUAL LIABILITY EXCLUSION FOR PERSONAL INJURY

This endorsement modifies insurance provided under the following:

COMMERCIAL GENERAL LIABILITY COVERAGE PART

SCHEDULE

Designated contract or agreement:

(If no entry appears above, information required to complete this endorsement will be shown in the Declarations applicable to this endorsement.)

With respect to the contract or agreement designated in the Schedule above, paragraph a.(4) of Exclusion 2. of Coverage B (Section I) is replaced by the following:

a. (4) For which the insured has assumed liability in a contract or agreement. This exclusion does not apply to:

 (a) liability for damages that the insured would have in the absence of the contract or agreement; or

 (b) liability for "personal injury," arising out of the offenses of false arrest, detention or imprisonment, undertaken in that part of the contract or agreement pertaining to your business shown in the schedule in which you assume the tort liability of another. The contract or agreement must be made prior to the offense. Tort liability means a liability that would be imposed by law in the absence of any contract or agreement.

ENDORSEMENT MS #

This endorsement, effective   **12:01 A.M.**   forms a part of

policy No.:   issued to

By:

O.C.I.P. AND C.C.I.P. - WRAP UP EXCLUSION

This endorsement modifies insurance provided under the following:
COMMERCIAL GENERAL LIABILITY COVERAGE PART

It is agreed that this policy does not apply to any work performed by or on behalf of you under any Owner Controlled Insurance Program (O.C.I.P.) or Contractor Controlled Insurance Program (C.C.I.P.), otherwise referred to as Wrap Up Program, that you enter into except as respects excess coverage for the Products - completed operations hazard for "your work."

All other terms, conditions, and exclusions of this policy remain unchanged.

_____
Authorized Representative

ENDORSEMENT MS #

This endorsement, effective   **12:01 A.M.**   forms a part of

policy No.   issued to

By:

CANCELLATION BY US

This endorsement modifies insurance provided under the following:
COMMERCIAL GENERAL LIABILITY COVERAGE PART

SCHEDULE

Number Of Days   60

Paragraph 2 of CANCELLATION (Common Policy Conditions) is replaced by the following:

2.   We may cancel or non-renew this Coverage Part by mailing or delivering to the first Named Insured written notice of cancellation or non-renewal at least:

   a.   10 days before the effective date of cancellation if we cancel for non-payment of premium; or

   b.   The number of days shown in the Schedule before the effective date of cancellation or non-renewal if we cancel or non-renew for any other reason.

All other terms, conditions, and exclusions of this policy remain unchanged.

_____
Authorized Representative

ENDORSEMENT MS #

This endorsement, effective   **12:01 A.M.**   forms a part of

policy No.   issued to

By:

THIS ENDORSEMENT CHANGES THE POLICY. PLEASE READ IT CAREFULLY
JOINT VENTURE EXCLUSION ENDORSEMENT

THIS ENDORSEMENT MODIFIES INSURANCE PROVIDED UNDER THE FOLLOWING:
COMMERCIAL GENERAL LIABILITY COVERAGE PART

It is agreed that all of your Joint Ventures that you enter into are excluded except, for damages arising out of the "Products/Completed Operations Hazards." The insurance for such damages shall be excess over any other insurance whether primary, excess, contingent or on any other basis.

Coverage as provided by this endorsement, will be afforded you in accordance with your interest in the Joint Venture as stipulated by the Joint Venture Agreement. In the absence of a formal joint venture agreement, no coverage will apply.

This exclusion does not apply to those Joint Ventures specifically listed on this policy's named insured endorsement.

All Other Terms, Conditions, and Exclusions Of This Policy Remain Unchanged.

Authorized Representative

ENDORSEMENT MS #

This endorsement, effective   **12:01 A.M.**   forms a part of

policy No.   issued to

By:

THIS ENDORSEMENT CHANGES THE POLICY. PLEASE READ IT CAREFULLY.
AMENDMENT OF CONDITION: NOTICE TO THE COMPANY

This endorsement modifies insurance provided under the following:
COMMERCIAL GENERAL LIABILITY COVERAGE FORM
Section IV - Commercial General Liability Conditions, paragraph 2 - Duties
In The Event Of Occurrence, Offense, Claim Or Suit

It is agreed that when the Insured reports the occurrence of any accident to the insurance company that provides its Workers' Compensation coverage and such loss later develops into a claim involving liability that is covered by the policy to which this endorsement is attached, the failure by the Insured to report such an accident to us at the time of the occurrence shall not be deemed to be in violation of the General Condition entitled Duties In The Event Of Occurrence, Offense, Claim, or Suit. However, the Insured shall give notification to this Company as soon as the Insured is made aware of the fact that the particular loss is a liability case rather than a compensation case.

All other terms, conditions, and exclusions of this policy remain unchanged.

_____
Authorized Representative

WORKERS COMPENSATION AND EMPLOYERS LIABILITY

INSURANCE POLICY

National Union Fire Insurance
Company of Pittsburgh, Pa.

American Home Assurance Company

The Insurance Company of
The State of Pennsylvania

Birmingham Fire Insurance Company
of Pennsylvania

Commerce and Industry
Insurance Company

Member Companies of
American International Group, Inc.
EXECUTIVE OFFICES
70 PINE STREET
NEW YORK, N.Y. 10270

Coverage is provided by the Company designated on the Information Page
A Stock Insurance Company

**WORKERS COMPENSATION AND EMPLOYERS LIABILITY INSURANCE POLICY
QUICK REFERENCE**

QUICK REFERENCE – CONTINUED

IMPORTANT: This Quick Reference is **not** part of the Workers Compensation and Employers Liability Policy and does **not** provide coverage. Refer to the Workers Compensation and Employers Liability Policy itself for actual contractual provisions.

PLEASE READ THE WORKERS COMPENSATION AND EMPLOYERS LIABILITY POLICY CAREFULLY

ATTACH FORM AND ENDORSEMENTS (IF ANY) HERE

WORKERS COMPENSATION AND EMPLOYERS LIABILITY INSURANCE POLICY

In return for the payment of the premium and subject to all terms of this policy, we agree with you as follows.

GENERAL SECTION

A. The Policy

This policy includes at its effective date the Information Page and all endorsements and schedules listed there. It is a contract of insurance between you (the employer named in Item 1 of the Information Page) and us (the insurer named on the Information Page). The only agreements relating to this insurance are stated in this policy. The terms of this policy may not be changed or waived except by endorsement issued by us to be part of this policy.

B. Who Is Insured

You are insured if you are an employer named in Item 1 of the Information Page. If that employer is a partnership, and if you are one of its partners, you are insured, but only in your capacity as an employer of the partnership's employees.

C. Workers Compensation Law

Workers Compensation Law means the workers or workmen's compensation law and occupational disease law of each state or territory named in Item 3.A. of the Information Page. It includes any amendments to that law which are in effect during the policy period. It does not include any federal workers or workmen's compensation law, any federal occupational disease law or the provisions of any law that provide nonoccupational disability benefits.

D. State

State means any state of the United States of America, and the District of Columbia.

E. Locations

This policy covers all of your workplaces listed in Items 1 or 4 of the Information Page; and it covers all other workplaces in Item 3.A states unless you have other insurance or are self-insured for such workplaces.

PART ONE · WORKERS COMPENSATION INSURANCE

A. How This Insurance Applies

This workers compensation insurance applies to bodily injury by accident or bodily injury by disease. Bodily injury includes resulting death.

1. Bodily injury by accident must occur during the policy period.

2. Bodily injury by disease must be caused or aggravated by the conditions of your employment. The employee's last day of last exposure to the conditions causing or aggravating such bodily injury by disease must occur during the policy period.

B. We Will Pay

We will pay promptly when due the benefits required of you by the workers compensation law.

C. We Will Defend

We have the right and duty to defend at our expense any claim, proceeding or suit against you for benefits payable by this insurance. We have the right to investigate and settle these claims, proceedings or suits.

We have no duty to defend a claim, proceeding or suit that is not covered by this insurance.

D. We Will Also Pay

We will also pay these costs, in addition to other amounts payable under this insurance, as part of any claim, proceeding or suit we defend: .

1. reasonable expenses incurred at our request, but not loss of earnings;

2. premiums for bonds to release attachments and for appeal bonds in bond amounts up to the amount payable under this insurance;

3. litigation costs taxed against you;

4. interest on a judgment as required by law until we offer the amount due under this insurance; and

5. expenses we incur.

E. Other Insurance

We will not pay more than our share of benefits and costs covered by this insurance and other insurance or self-insurance. Subject to any limits of liability that may apply, all shares will be equal until the loss is paid. If any insurance or self-insurance is exhausted, the shares of all remaining insurance will be equal until the loss is paid.

F. Payments You Must Make

You are responsible for any payments in excess of the benefits regularly provided by the workers compensation law including those required because:

1. of your serious and willful misconduct;

2. you knowingly employ an employee in violation of law;

3. you fail to comply with a health or safety law or regulation; or

4. you discharge, coerce or otherwise discriminate against any employee in violation of the workers compensation law.

If we make any payments in excess of the benefits regularly provided by the workers compensation law on your behalf, you will reimburse us promptly.

G. Recovery From Others

We have your rights, and the rights of persons entitled to the benefits of this insurance, to recover our payments from anyone liable for the injury. You will do everything necessary to protect those rights for us and to help us enforce them.

H. Statutory Provisions

These statements apply where they are required by law.

1. As between an injured worker and us, we have notice of the injury when you have notice.

2. Your default or the bankruptcy or insolvency of you or your estate will not relieve us of our duties under this insurance after an injury occurs.

3. We are directly and primarily liable to any person entitled to the benefits payable by this insurance. Those persons may enforce our duties; so may an agency authorized by law. Enforcement may be against us or against you and us.

4. Jurisdiction over you is jurisdiction over us for purposes of the workers compensation law. We are bound by decisions against you under that law, subject to the provisions of this policy that are not in conflict with that law.

5. This insurance conforms to the parts of the workers compensation law that apply to:

    a. benefits payable by this insurance or;

    b. special taxes, payments into security or other special funds, and assessments payable by us under that law.

6. Terms of this insurance that conflict with the workers compensation law are changed by this statement to conform to that law.

Nothing in these paragraphs relieves you of your duties under this policy.

PART TWO - EMPLOYERS LIABILITY INSURANCE

A. How This Insurance Applies

This employers liability insurance applies to bodily injury by accident or bodily injury by disease. Bodily injury includes resulting death.

1. The bodily injury must arise out of and in the course of the injured employee's employment by you.

2. The employment must be necessary or incidental to your work in a state or territory listed in Item 3.A. of the Information Page.

3. Bodily injury by accident must occur during the policy period.

4. Bodily injury by disease must be caused or aggravated by the conditions of your employment. The employee's last day of last exposure to the conditions causing or aggravating such bodily injury by disease must occur during the policy period.

5. If you are sued, the original suit and any related legal actions for damages for bodily injury

by accident or by disease must be brought in the United States of America, its territories or possessions, or Canada.

B. **We Will Pay**

We will pay all sums you legally must pay as damages because of bodily injury to your employees, provided the bodily injury is covered by this Employers Liability Insurance.

The damages we will pay, where recovery is permitted by law, include damages:

1. for which you are liable to a third party by reason of a claim or suit against you by that third party to recover the damages claimed against such third party as a result of injury to your employee;

2. for care and loss of services; and

3. for consequential bodily injury to a spouse, child, parent, brother or sister of the injured employee;

provided that these damages are the direct consequence of bodily injury that arises out of and in the course of the injured employee's employment by you; and

4. because of bodily injury to your employee that arises out of and in the course of employment, claimed against you in a capacity other than as employer.

C. **Exclusions**

This insurance does not cover:

1. liability assumed under a contract. This exclusion does not apply to a warranty that your work will be done in a workmanlike manner;

2. punitive or exemplary damages because of bodily injury to an employee employed in violation of law;

3. bodily injury to an employee while employed in violation of law with your actual knowledge or the actual knowledge of any of your executive officers;

4. any obligation imposed by a workers compensation, occupational disease, unemployment compensation, or disability benefits law, or any similar law;

5. bodily injury intentionally caused or aggravated by you;

6. bodily injury occurring outside the United States of America, its territories or possessions, and Canada. This exclusion does not apply to bodily injury to a citizen or resident of the United States of America or Canada who is temporarily outside these countries;

7. damages arising out of coercion, criticism, demotion, evaluation, reassignment, discipline, defamation, harassment, humiliation, discrimination against or termination of any employee, or any personnel practices, policies, acts or omissions.

8. bodily injury to any person in work subject to the Longshore and Harbor Workers' Compensation Act (33 USC Sections 901-950), the Nonappropriated Fund Instrumentalities Act (5 USC Sections 8171-8173), the Outer Continental Shelf Lands Act (43 USC Sections 1331-1356), the Defense Base Act (42 USC Sections 1651-1654), the Federal Coal Mine Health and Safety Act of 1969 (30 USC Sections 901-942), any other federal workers or workmen's compensation law or other federal occupational disease law, or any amendments to these laws.

9. bodily injury to any person in work subject to the Federal Employers' Liability Act (45 USC Sections 51-60), any other federal laws obligating an employer to pay damages to an employee due to bodily injury arising out of or in the course of employment, or any amendments to those laws.

10. bodily injury to a master or member of the crew of any vessel.

11. fines or penalties imposed for violation of federal or state law.

12. damages payable under the Migrant and Seasonal Agricultural Worker Protection Act (29 USC Sections 1801-1872) and under any other federal law awarding damages for violation of those laws or regulations issued thereunder, and any amendments to those laws.

D. **We Will Defend**

We have the right and duty to defend, at our expense, any claim, proceeding or suit against you for damages payable by this insurance. We have the right to investigate and settle these claims, proceedings and suits.

We have no duty to defend a claim. proceeding or suit that is not covered by this insurance. We have no duty to defend or continue defending after we have paid our applicable limit of liability under this insurance.

E. **We Will Also Pay**

We will also pay these costs. in addition to other amounts payable under this insurance, as part of any claim proceeding, or suit we defend;

1. reasonable expenses incurred at our request; but not loss of earnings;

2. premiums for bonds to release attachments and for appeal bonds in bond amounts up to the limit of our liability under this insurance;

3. litigation costs taxed against you;

4. interest on a judgment as required by law until we offer the amount due under this insurance: and

5. expenses we incur.

F. **Other Insurance**

We will not pay more than our share of damages and costs covered by this insurance and other in- surance or self-insurance. Subject to any limits of liability that apply, all shares will be equal until the loss is paid. If any insurance or self-insurance is exhausted, the shares of all remaining insurance and self-insurance will be equal until the loss is paid.

G. **Limits of Liability**

Our liability to pay for damages is limited. Our limits of liability are shown in Item 3.B. of the Information Page. They apply as explained below.

1. Bodily Injury by Accident. The limit shown for "bodily injury by accident-each accident" is the most we will pay for all damages covered by this insurance because of bodily injury to one or more employees in any one accident.

A disease is not bodily injury by accident un- less it results directly from bodily injury by ac- cident.

2. Bodily Injury by Disease. The limit shown for "bodily injury by disease-policy limit" is the most we will pay for all damages covered by this insurance and arising out of bodily injury by disease, regardless of the number of em- ployees who sustain bodily injury by disease. The limit shown for "bodily injury by disease- each employee" is the most we will pay for all damages because of bodily injury by disease to any one employee.

Bodily injury by disease does not include dis- ease that results directly from a bodily injury by accident.

3. We will not pay any claims for damages after we have paid the applicable limit of our liability under this insurance.

H. **Recovery From Others**

We have your rights to recover our payment from anyone liable for an injury covered by this insur- ance. You will do everything necessary to protect those rights for us and to help us enforce them.

I. **Actions Against Us**

There will be no right of action against us under this insurance unless:

1. You have complied with all the terms of this policy; and

2. The amount you owe has been determined with our consent or by actual trial and final judgment.

This insurance does not give anyone the right to add us as a defendant in an action against you to determine your liability. The bankruptcy or insolvency of you or your estate will not relieve us of our obligations under this Part.

PART THREE - OTHER STATES INSURANCE

A. **How This Insurance Applies**

1. This other states insurance applies only if one or more states are shown in Item 3.C. of the Information Page.

2. If you begin work in any one of those states after the effective date of this policy and are not insured or are not self-insured for such work, all provisions of the policy will apply as

though that state were listed in Item 3.A. of the Information Page.

3. We will reimburse you for the benefits required by the workers compensation law of that state if we are not permitted to pay the benefits di- rectly to persons entitled to them.

4. If you have work on the effective date of this policy in any state not listed in Item 3.A. of the

Information Page. coverage will not be afforded for that state unless we are notified within thirty days.

B. Notice

Tell us at once if you begin work in any state listed in Item 3.C. of the Information Page.

PART FOUR - YOUR DUTIES IF INJURY OCCURS

Tell us at once if injury occurs that may be covered by this policy. Your other duties are listed here.

1. Provide for immediate medical and other services required by the workers compensation law.

2. Give us or our agent the names and addresses of the injured persons and of witnesses. and other information we may need.

3. Promptly give us all notices. demands and legal papers related to the injury. claim. proceeding or suit.

4. Cooperate with us and assist us. as we may request. in the investigation, settlement or defense of any claim. proceeding or suit.

5. Do nothing after an injury occurs that would interfere with our right to recover from others.

6. Do not voluntarily make payments. assume obligations or incur expenses, except at your own cost.

PART FIVE - PREMIUM

A. Our Manuals

All premium for this policy will be determined by our manuals of rules. rates. rating plans and classifications. We may change our manuals and apply the changes to this policy if authorized by law or a governmental agency regulating this insurance.

B. Classifications

Item 4 of the Information Page shows the rate and premium basis for certain business or work classifications. These classifications were assigned based on an estimate of the exposures you would have during the policy period. If your actual exposures are not properly described by those classifications. we will assign proper classifications. rates and premium basis by endorsement to this policy.

C. Remuneration

Premium for each work classification is determined by multiplying a rate times a premium basis. Remuneration is the most common premium basis. This premium basis includes payroll and all other remuneration paid or payable during the policy period for the services of:

1. All your officers and employees engaged in work covered by this policy; and

2. All other persons engaged in work that could make us liable under Part One (Workers Compensation Insurance) of this policy. If you do not have payroll records for these persons, the contract price for their services and materials may be used as the premium basis. This paragraph 2 will not apply if you give us proof

that the employers of these persons lawfully secured their workers compensation obligations.

D. Premium Payments

You will pay all premium when due. You will pay the premium even if part or all of a workers compensation law is not valid.

E. Final Premium

The premium shown on the Information Page. schedules. and endorsements is an estimate. The final premium will be determined after this policy ends by using the actual. not the estimated. premium basis and the proper classifications and rates that lawfully apply to the business and work covered by this policy. If the final premium is more than the premium you paid to us. you must pay us the balance. If it is less. we will refund the balance to you. The final premium will not be less than the highest minimum premium for the classifications covered by this policy.

If this policy is canceled, final premium will be determined in the following way unless our manuals provide otherwise.

1. If we cancel, final premium will be calculated pro rata based on the time this policy was in force. Final premium will not be less than the pro rata share of the minimum premium.

2. If you cancel. final premium will be more than pro rata; it will be based on the time this policy was in force. and increased by our short rate

cancellation table and procedure. Final premium will not be less than the minimum premium.

F. Records

You will keep records of information needed to compute premium. You will provide us with copies of those records when we ask for them.

G. Audit

You will let us examine and audit all your records that relate to this policy. These records include ledgers, journals, registers, vouchers, contracts, tax reports, payroll and disbursement records, and programs for storing and retrieving data. We may conduct the audits during regular business hours during the policy period and within three years after the policy period ends. Information developed by audit will be used to determine final premium. Insurance rate service organizations have the same rights we have under this provision.

PART SIX - CONDITIONS

A. Inspection

We have the right, but are not obliged to inspect your workplaces at any time. Our inspections are not safety inspections. They relate only to the insurability of the workplaces and the premiums to be charged. We may give you reports on the conditions we find. We may also recommend changes. While they may help reduce losses, we do not undertake to perform the duty of any person to provide for the health or safety of your employees or the public. We do not warrant that your workplaces are safe or healthful or that they comply with laws, regulations, codes or standards. Insurance rate service organizations have the same rights we have under this provision.

B. Long Term Policy

If the policy period is longer than one year and sixteen days, all provisions of this policy will apply as though a new policy were issued on each annual anniversary that this policy is in force.

C. Transfer of Your Rights and Duties

Your rights or duties under this policy may not be transferred without our written consent.

If you die and we receive notice within thirty days after your death, we will cover your legal representative as insured.

D. Cancellation

1. You may cancel this policy. You must mail or deliver advance written notice to us stating when the cancellation is to take effect.

2. We may cancel this policy. We must mail or deliver to you not less than ten days advance written notice stating when the cancellation is to take effect. Mailing that notice to you at your mailing address shown in Item 1 of the Information Page will be sufficient to prove notice.

3. The policy period will end on the day and hour stated in the cancellation notice.

4. Any of these provisions that conflicts with a law that controls the cancellation of the insurance in this policy is changed by this statement to comply with that law.

E. Sole Representative

The insured first named in Item 1 of the Information Page will act on behalf of all insureds to change this policy, receive return premium, and give or receive notice of cancellation.

In Witness Whereof, the company has caused this policy to be executed and attested, but this policy shall not be valid unless countersigned by a duly authorized representative of the company.

President
The Insurance Company
of The State of Pennsylvania

President
Birmingham Fire
Insurance Company of
Pennsylvania

President
Commerce and Industry
Insurance Company

President
National Union Fire Insurance
Company of Pittsburgh, PA

President
American Home
Assurance Company

Secretary
National Union Fire Insurance Company of Pittsburgh, PA
American Home Assurance Company
The Insurance Company of The State of Pennsylvania
Birmingham Fire Insurance Company of Pennsylvania
Commerce and Industry Insurance Company

ISSUED BY THE STOCK INSURANCE COMPANY HEREIN CALLED THE COMPANY

AGENT NUMBER   POLICY NUMBER

INCORPORATED UNDER THE LAWS OF
ITEM 1   NAMED INSURED   MAILING ADDRESS   IDENTIFICATION NO

Member Companies of American International Group
EXECUTIVE OFFICES:
70 PINE STREET, NEW YORK, N.Y. 10270

I.D.# _____

PRODUCERS NAME & MAILING ADDRESS

WORKERS COMPENSATION AND EMPLOYERS LIABILITY POLICY INFORMATION PAGE

INSURED IS   PREVIOUS POLICY NUMBER

OTHER WORKPLACES NOT SHOWN ABOVE

ITEM 2 POLICY PERIOD 12:01 A.M. standard time at the insured's mailing address
FROM   TO

ITEM 3 A. Workers Compensation Insurance: Part One of the policy applies to the Workers Compensation Law of the states listed here:

B. Employers Liability Insurance: Part Two of the policy applies to the work in each state listed in item 3.A. The limits of our liability under Part Two are:
Bodily Injury by Accident $_____ each accident
Bodily Injury by Disease $_____ policy limit
Bodily Injury by Disease $_____ each employee

C. Other States Insurance: Part Three of the policy applies to the states, if any, listed here:

ITEM 4 The premium for this policy will be determined by our Manuals of Rules, Classifications, Rates and Rating Plans. All information required below is subject to verification and change by audit.

Classifications	Code Number	Estimated Total Remuneration ☒ Annual ☐ 3 Year	Rate Per $100 OF Remuneration	Estimated Premium ☒ Annual ☐ 3 Year

EXPENSE CONSTANT (EXCEPT WHERE APPLICABLE BY STATE)

MINIMUM PREMIUM $ TOTAL ESTIMATED PREMIUM

If indicated below, interim adjustments of premium shall be made:
☐ Semi-Annually   ☐ Quarterly   ☐ Monthly

ENDORSEMENTS (FORM NUMBER)   DEPOSIT PREMIUM $

Issue Date   Issuing Office   Authorized Representative

LONGSHORE AND HARBOR WORKERS' COMPENSATION ACT COVERAGE ENDORSEMENT

This endorsement changes the policy to which it is attached effective on the inception date of the policy unless a different date is indicated below.

(The following "attaching clause" need be completed only when this endorsement is issued subsequent to preparation of the policy).

This endorsement, effective 12:01 AM   / /   forms a part of Policy No.   WC   – –

Issued to

By

This endorsement applies only to work subject to the Longshore and Harbor Workers' Compensation Act in a state shown in the Schedule. The policy applies to that work as though that state were listed in Item 3.A. of the Information Page.

General Section C. **Workers' Compensation Law** is replaced by the following:

C.   Workers' Compensation Law

Workers' Compensation Law means the workers or workmen's compensation law and occupational disease law of each state or territory named in Item 3.A. of the Information Page and the Longshore and Harbor Workers' Compensation Act (33 USC Sections 901-950). It includes any amendments to those laws that are in effect during the policy period. It does not include any other federal workers or workmen's compensation law, other federal occupational disease law or the provisions of any law that provide nonoccupational disability benefits.

Part Two (Employers Liability Insurance), C. Exclusions., exclusion 8, does not apply to work subject to the Longshore and Harbor Workers' Compensation Act.

This endorsement does not apply to work subject to the Defense Base Act, the Outer Continental Shelf Lands Act, or the Nonappropriated Fund Instrumentalities Act.

Schedule

State	Longshore and Harbor Workers' Compensation Act Coverage Percentage

The rates for classifications with code numbers not followed by the letter "F" are rates for work not ordinarily subject to the Longshore and Harbor Workers' Compensation Act. If this policy covers work under such classifications, and if the work is subject to the Longshore and Harbor Workers' Compensation Act, those non-F classification rates will be increased by the Longshore and Harbor Workers' Compensation Act Coverage Percentage shown in the Schedule.

Countersigned by _____

Authorized Representative

MARITIME COVERAGE ENDORSEMENT

This endorsement changes the policy to which it is attached effective on the inception date of the policy unless a different date is indicated below.

(The following "attaching clause" need be completed only when this endorsement is issued subsequent to preparation of the policy).

This endorsement, effective 12:01 AM   / /   forms a part of Policy No.   WC   - -

Issued to

By

This endorsement changes how insurance provided by Part Two (Employers Liability Insurance) applies to bodily injury to a master or member of the crew of any vessel.

A. **How This Insurance Applies** is replaced by the following:

    A. **How This Insurance Applies**

        This insurance applies to bodily injury by accident or bodily injury by disease. Bodily injury includes resulting death.

        1. The bodily injury must arise out of and in the course of the injured employee's employment by you.

        2. The employment must be necessary or incidental to work described in Item 1 of the Schedule of the Maritime Coverage Endorsement.

        3. The bodily injury must occur in the territorial limits of, or in the operation of a vessel sailing directly between the ports of, the continental United States of America, Alaska, Hawaii or Canada.

        4. Bodily injury by accident must occur during the policy period.

        5. Bodily injury by disease must be caused or aggravated by the conditions of your employment. The employee's last day of last exposure to the conditions causing or aggravating such bodily injury by disease must occur during the policy period.

        6. If you are sued, the original suit and any related legal actions for damages for bodily injury by accident or by disease must be brought in the United States of America, its territories or possessions, or Canada.

C. **Exclusions** is changed by removing exclusion 10 and by adding exclusions 13 and 14.

    This insurance does not cover:

    13. bodily injury covered by a Protection and Indemnity Policy or similar policy issued to you or for your benefit. This exclusion applies even if the other policy does not apply because of another insurance clause, deductible or limitation of liability clause, or any similar clause.

    14. your duty to provide transportation, wages, maintenance and cure. This exclusion does not apply if a premium entry is shown in Item 2 of the Schedule.

D. **We Will Defend** is changed by adding the following statement:

    We will treat a suit or other action in rem against a vessel owned or chartered by you as a suit against you.

G. **Limits of Liability**

    Our liability to pay for damages is limited. Our limits of liability are shown in the Schedule. They apply as explained below.

1. Bodily Injury by Accident. The limit shown for "bodily injury by accident-each accident" is the most we will pay for all damages covered by this insurance because of bodily injury to one or more employees in any one accident.

   A disease is not bodily injury by accident unless it results directly from bodily injury by accident.

2. Bodily Injury by Disease. The limit shown for "bodily injury by disease-aggregate" is the most we will pay for all damages covered by this insurance because of bodily injury by disease to one or more employees. The limit applies separately to bodily injury by disease arising out of work in each state shown in Item 3.A. of the Information Page. Bodily Injury by disease will be deemed to occur in the state of the vessel's home port.

   Bodily injury by disease does not include disease that results directly from a bodily injury by accident.

3. We will not pay any claims for damages after we have paid the applicable limit of our liability under this insurance.

Schedule

1. Description of work: See schedule of operations

2. Transportation, Wages, Maintenance and Cure Premium $

3. Limits of Liability

   Bodily Injury by Accident $ each accident
   Bodily Injury by Disease $ aggregate

Countersigned by _____

                                        Authorized Representative

SOLE PROPRIETORS, PARTNERS, OFFICERS AND OTHERS COVERAGE ENDORSEMENT

This endorsement changes the policy to which it is attached effective on the inception date of the policy unless a different date is indicated below.

(The following "attaching clause" need be completed only when this endorsement is issued subsequent to preparation of the policy).

This endorsement, effective 12:01 AM   / /   forms a part of Policy No.   WC   – –

Issued to

By

An election was made by or on behalf of each person described in the Schedule to be subject to the workers compensation law of the state named in the Schedule. The premium basis for the policy includes the remuneration of such persons.

Schedule

Persons   **State**

Sole Proprietor:

Partners:

Officers:

Others:

Countersigned by _____

Authorized Representative

VOLUNTARY COMPENSATION AND EMPLOYERS LIABILITY COVERAGE ENDORSEMENT

This endorsement changes the policy to which it is attached effective on the inception date of the policy unless a different date is indicated below.

(The following "attaching clause" need be completed only when this endorsement is issued subsequent to preparation of the policy).

This endorsement, effective 12:01 AM   /   /   forms a part of Policy No.   WC   -   -

Issued to

By

This endorsement adds Voluntary Compensation Insurance to the policy.

A. How This Insurance Applies
This insurance applies to bodily injury by accident or bodily injury by disease. Bodily injury includes resulting death.

1. The bodily injury must be sustained by an employee included in the group of employees described in the Schedule.

2. The bodily injury must arise out of and in the course of employment necessary or incidental to work in a state listed in the Schedule.

3. The bodily injury must occur in the United States of America, its territories or possessions, or Canada, and may occur elsewhere if the employee is a United States or Canadian citizen temporarily away from those places.

4. Bodily injury by accident must occur during the policy period.

5. Bodily injury by disease must be caused or aggravated by the conditions of your employment. The employee's last day of last exposure to the conditions causing or aggravating such bodily injury by disease must occur during the policy period.

B. We Will Pay
We will pay an amount equal to the benefits that would be required of you if you and your employees described in the Schedule were subject to the workers compensation law shown in the Schedule. We will pay those amounts to the persons who would be entitled to them under the law.

C. Exclusions
This insurance does not cover:

1. any obligation imposed by a workers compensation or occupational disease law, or any similar law.

2. bodily injury intentionally caused or aggravated by you.

D. Before We Pay
Before we pay benefits to the persons entitled to them, they must:

1. Release you and us, in writing, of all responsibility for the injury or death.

2. Transfer to us their right to recover from others who may be responsible for the injury or death.

3. Cooperate with us and do everything necessary to enable us to enforce the right to recover from others.

If the persons entitled to the benefits of this insurance fail to do those things, our duty to pay ends at once. If they claim damages from you or from us for the injury or death, our duty to pay ends at once.

WAIVER OF OUR RIGHT TO RECOVER FROM OTHERS ENDORSEMENT

This endorsement changes the policy to which it is attached effective on inception date of the policy unless a different date is indicated below.

(The following "attaching clause" need be completed only when this endorsement is issued subsequent to preparation of the policy).

This endorsement, effective 12:01 AM   / /   forms a part of Policy No. WC – –

Issued to

By

Premium

We have the right to recover our payments from anyone liable for an injury covered by this policy. We will not enforce our right against the person or organization named in the Schedule. This agreement applies only to the extent that you perform work under a written contract that requires you to obtain this agreement from us.

This agreement shall not operate directly or indirectly to benefit any one not named in the Schedule.

Schedule

Countersigned by _____

Authorized Representative

COMMERCIAL AUTO COVERAGE PART
BUSINESS AUTO DECLARATIONS

Renewal of No.
Policy No.

COVERAGE IS PROVIDED IN THE
COMPANY DESIGNATED BY NUMBER
A STOCK INSURANCE COMPANY
(HEREIN CALLED THE COMPANY)

1. NATIONAL UNION FIRE INSURANCE
   COMPANY OF PITTSBURGH,PA.
2. AMERICAN HOME ASSURANCE COMPANY
3. THE INSURANCE COMPANY OF THE
   STATE OF PENNSYLVANIA
4. THE BIRMINGHAM FIRE INSURANCE
   COMPANY OF PENNSYLVANIA
5. COMMERCE AND INDUSTRY
   INSURANCE COMPANY

ITEM ONE NAMED INSURED & MAILING ADDRESS

PRODUCER'S NAME & MAILING ADDRESS

**MEMBERS OF THE
AMERICAN INTERNATIONAL GROUP,INC.**
EXECUTIVE OFFICES
70 PINE STREET
NEW YORK, N.Y. 10270

FORM OF BUSINESS: ☐CORPORATION ☐PARTNERSHIP☐INDIVIDUAL OR ☐OTHER
POLICY PERIOD: Policy covers FROM   at 12:01 A.M. Standard Time at your mailing address shown above.
IN RETURN FOR THE PAYMENT OF THE PREMIUM, AND SUBJECT TO ALL THE TERMS OF THIS POLICY, WE AGREE WITH YOU TO
PROVIDE THE INSURANCE AS STATED IN THIS POLICY.

ITEM TWO–SCHEDULE OF COVERAGES AND COVERED AUTOS
This policy provides only those coverages where a charge is shown in the premium column below. Each of these coverages will apply only to those "autos" shown as covered
"autos." "Autos" are shown as covered "autos" for a particular coverage by the entry of one or more of the symbols from the COVERED AUTO Section of the Business Auto
Coverage Form next to the name of the coverage.

COVERAGES	COVERED AUTOS (Entry of one or more of the symbols from the COVERED AUTOS Section of the Business Auto Coverage Form shows which autos are covered autos)	LIMIT THE MOST WE WILL PAY FOR ANY ONE ACCIDENT OR LOSS	PREMIUM
LIABILITY		$	$
PERSONAL INJURY PROTECTION (P.I.P.) (or equivalent No-fault cov)		SEPARATELY STATED IN EACH P.I.P ENDORSEMENT MINUS $ Deductible	$
ADDED P.I.P(or equivalent added No-fault cov)		SEPARATELY STATED IN EACH ADDED P.I.P ENDORSEMENT	$
PROPERTY PROTECTION INS. (P.P.I.) (Michigan only)		SEPARATELY STATED IN THE P.P.I. ENDORSEMENT MINUS $ Deductible FOR EACH ACCIDENT	$
AUTO MEDICAL PAYMENTS		$	$
UNINSURED MOTORISTS (UM)		$	$
UNDERINSURED MOTORISTS (when not included in UM Cov.)		$	$
PHYSICAL DAMAGE — COMPREHENSIVE COVERAGE		ACTUAL CASH VALUE $ Ded. FOR EACH COVERED AUTO, BUT NO DED. OR COST OF APPLIES TO LOSS CAUSED BY FIRE OR LIGHTNING.	$
PHYSICAL DAMAGE — SPECIFIED CAUSES OF LOSS COVERAGE		REPAIR, $25 Deductible FOR EACH COVERED AUTO FOR LOSS WHICHEVER CAUSED BY MISCHIEF OR VANDALISM	$
PHYSICAL DAMAGE — COLLISION COVERAGE		IS LESS MINUS $ Deductible FOR EACH COVERED AUTO	$
PHYSICAL DAMAGE — TOWING AND LABOR (Not Available in California)		$ for each disablement of a private passenger auto	$

FORMS AND ENDORSEMENTS APPLYING TO THIS COVERAGE PART AND MADE PART OF THIS POLICY AT TIME OF ISSUE :

PREMIUM FOR ENDORSEMENTS $
ESTIMATED TOTAL PREMIUM $

ITEM THREE–SCHEDULE OF COVERED AUTOS YOU OWN   See ITEM FOUR for hired or borrowed "autos".

Covered Auto No.	DESCRIPTION Year Model; Trade Name; Body Type Serial Number (S); Vehicle Identification Number (VIN)		PURCHASED			TERRITORY: Town & State Where the Covered AUTO will be principally garaged
		Original Cost New	Actual Cost &	NEW (N) USED (U)		
1						
2						
3						
4						

Covered Auto No.	CLASSIFICATION								Except for towing all physical damage loss is payable to you and the loss payee named below as interests may appear at the time of the loss
	Radius of Operation (In Miles)	Business use s=service r=retail c=comm'l	Size. GVW GCW or Vehicle Seating Capacity	Age Group	Primary Rating Factor Liab.	Primary Rating Factor Phy Damage	Secondary Rating Factor	Code	
1									
2									
3									
4									

Countersigned: By _____
                                        Authorized Representative

THESE DECLARATIONS AND THE COMMON POLICY DECLARATIONS, IF APPLICABLE, TOGETHER WITH THE COMMON POLICY CONDITIONS, COVERAGE
FORMS, AND FORMS AND ENDORSEMENTS IF ANY ISSUED TO FORM A PART THEREOF COMPLETE THE ABOVE NUMBERED POLICY

POLICY NUMBER:

CA 00 03 01 87   PART 2
BUSINESS AUTO DECLARATIONS (Continued)

ITEM THREE (Cont'd)

COVERAGES–PREMIUMS, LIMITS AND DEDUCTIBLES (Absence of a deductible or limit entry in any column below means that the limit or deductible entry in the corresponding ITEM TWO column applies instead)

Cov-ered Auto No.	Limit (in Thou-sands)	Premium	Limit* minus deductible shown below	Pre-mium	Limit*	Premium	Limit* minus deductible shown below	Pre-mium	Limit** (in Thou sands)	Pre-mium	Limit** minus deductible shown below	Pre-mium	Limit** Premium	Limit* minus deductible shown below	Premium	Limit** per dis-ablement	Premium
	LIABILITY		P.I.P.		ADDED P.I.P.	P.P.I. (Mich. only)			AUTO. MED. PAY.		COMPREHENSIVE		SPEC. CAUSES OF LOSS		COLLISION		TOWING & LABOR
1																	
2																	
3																	
4																	
Total Premium																	

Add'l Coverage(s)--Premium, Limit, Deductible: *Limit stated in each applicable P.I.P. or P.P.I. Endorsement. **Limit stated in ITEM TWO

ITEM FOUR

SCHEDULE OF HIRED OR BORROWED COVERED AUTO COVERAGE AND PREMIUM, LIABILITY COVERAGE–RATING BASIS, COST OF HIRE

STATE	ESITMATED COST OF HIRE FOR EACH STATE	RATE PER EACH $100 COST OF HIRE	FACTOR (IF LIAB. COV. IS PRIMARY)	PREMIUM
				$

Cost of hire means the total amount you incur for the hire of 'autos' you don't own (not including 'autos' you borrow or rent from your partners or employees or their family members). Cost of hire does not include charges for services performed by motor carriers of property or passengers.   **TOTAL PREMIUM** $

PHYSICAL DAMAGE COVERAGE

COVERAGES		LIMIT OF INSURANCE THE MOST WE WILL PAY. DEDUCTIBLE	RATE	PREMIUM
COMPREHENSIVE	ACTUAL CASH VALUE. COST OF REPAIRS OR	$ WHICHEVER IS LESS MINUS $ DEDUCTIBLE FOR EACH COVERED AUTO. BUT NO DEDUCTIBLE APPLIES TO LOSS CAUSED BY FIRE OR LIGHTNING		$
SPECIFIED CAUSES OF LOSS		$ WHICHEVER IS LESS MINUS $25 DEDUCTIBLE FOR EACH COVERED AUTO. FOR LOSS CAUSED BY MISCHIEF OR VANDALISM		$
COLLISION		$ WHICHEVER IS LESS MINUS $ DEDUCTIBLE FOR EACH COVERED AUTO		$

PHYSICAL DAMAGE COVERAGE for covered 'autos' you hire or borrow is excess unless indicated below by" ☐ ."   **TOTAL PREMIUM** $

☐   If this box is checked, PHYSICAL DAMAGE COVERAGE applies on a direct primary basis and for purposes of the condition entitled OTHER INSURANCE, any covered 'auto' you hire or borrow is deemed to be a covered 'auto' you own.

ITEM FIVE – SCHEDULE FOR NON-OWNERSHIP LIABILITY

NAMED INSURED'S BUSINESS	RATING BASIS	NUMBER	PREMIUM
Other than a Social Service Agency	Number of Employees		$
	Number of Partners		$
Social Service Agency	Number of Employees		$
	Number of Volunteers		$

ITEM SIX–SCHEDULE FOR GROSS RECEIPTS OR MILAGE BASIS–LIABILITY COVERAGE–PUBLIC AUTO OR LEASING RENTAL CONCERNS

Estimated Yearly	RATES		PREMIUMS	
☐ Gross Receipts ☐ Mileage	☐ Per $100 of Gross Receipts ☐ Per Mile			
	LIABILITY COVERAGE	AUTO MEDICAL PAYMENTS	LIABILITY COVERAGE	AUTO MEDICAL PAYMENTS
			$	$
			$	$
			$	$
			$	$

When used as a premium basis:

FOR PUBLIC AUTOS | TOTAL PREMIUMS $ | $ |
| MINIMUM PREMIUMS $ | $ |

Gross Receipts means the total amount to which you are entitled for transporting passengers, mail or merchandise during the policy period regardless of whether you or any other carrier originate the transportation. Gross Receipts does not include:
- A. Amounts you pay to railroads, steamship lines, airlines and other motor carrier operating under their own ICC or PUC permits.
- B. Advertising Revenue.
- C. Taxes which you collect as a separate item and remit directly to a governmental division.
- D. C.O.D. collections for cost of mail or merchandise including collection fees.

Mileage means the total live and dead mileage of all revenue producing units operated during the policy period.

FOR RENTAL OR LEASING CONCERNS

Gross Receipts means the total amount to which you are entitled for the leasing or rental of 'autos' during the policy period and includes taxes except those taxes which you collect as a separate item and remit directly to a governmental division.

Mileage means the total of all live and dead autos you leased or rented to others during the policy period.

In Witness Whereof, we have caused this policy to be executed and attested, and, if required by state law, this policy shall not be valid unless countersigned by our authorized representative.

President
National Union Fire Insurance
Company of Pittsburg, Pa.

President
American Home
Assurance Company

President
Commerce and Industry
Insurance Company

President
The Insurance Company
of The State of Pennsylvania
New Hampshire Insurance Group

President
The Birmingham Fire
Insurance Company of
Pennsylvania
AIU Insurance Company

Secretary
National Union Fire Insurance Company of Pittsburgh, Pa.
American Home Assurance Company
The Insurance Company of The State of Pennsylvania
The Birmingham Fire Insurance Company of Pennsylvania
Commerce and Industry Insurance Company
New Hampshire Insurance Co.
Granite State Insurance Co.
American Fidelity Company
Illinois National Insurance Co.

COMMON POLICY CONDITIONS

All Coverage Parts included in this policy are subject to the following conditions.

A. CANCELLATION

1. The first Named Insured shown in the Declarations may cancel this policy by mailing or delivering to us advance written notice of cancellation.

2. We may cancel this policy by mailing or delivering to the first Named Insured written notice of cancellation at least:

    a. 10 days before the effective date of cancellation if we cancel for nonpayment of premium; or

    b. 30 days before the effective date of cancellation if we cancel for any other reason.-

3. We will mail or deliver our notice to the first Named Insured's last mailing address known to us.

4. Notice of cancellation will state the effective date of cancellation. The policy period will end on that date.

5. If this policy is cancelled, we will send the first Named Insured any premium refund due. If we cancel, the refund will be pro rata. If the first Named Insured cancels, the refund may be less than pro rata. The cancellation will be effective even if we have not made or offered a refund.

6. If notice is mailed, proof of mailing will be sufficient proof of notice.

B. CHANGES

This policy contains all the agreements between you and us concerning the insurance afforded. The first Named Insured shown in the Declarations is authorized to make changes in the terms of this policy with our consent. This policy's terms can be amended or waived only by endorsement issued by us and made a part of this policy.

C. EXAMINATION OF YOUR BOOKS AND RECORDS

We may examine and audit your books and records as they relate to this policy at any time during the policy period and up to three years afterward.

D. INSPECTIONS AND SURVEYS

We have the right but are not obligated to:

1. Make inspections and surveys at any time:

2. Give you reports on the conditions we find; and

3. Recommend changes.

Any inspections, surveys, reports or recommendations relate only to insurability and the premiums to be charged. We do not make safety inspections. We do not undertake to perform the duty of any person or organization to provide for the health or safety of workers or the public. And we do not warrant that conditions:

1. Are safe or healthful; or

2. Comply with laws, regulations, codes or standards.

This condition applies not only to us, but also to any rating, advisory, rate service or similar organization which makes insurance inspections, surveys, reports or recommendations.

E. PREMIUMS

The first Named Insured shown in the Declarations:

1. Is responsible for the payment of all premiums; and

2. Will be the payee for any return premiums we pay.

F. TRANSFER OF YOUR RIGHTS AND DUTIES UNDER THIS POLICY

Your rights and duties under this policy may not be transferred without our written consent except in the case of death of an individual named insured.

If you die, your rights and duties will be transferred to your legal representative but only while acting within the scope of duties as your legal representative. Until your legal representative is appointed, anyone having proper temporary custody of your property will have your rights and duties but only with respect to that property.

COMMERCIAL AUTO
CA 00 01 12 93

BUSINESS AUTO COVERAGE FORM

Various provisions in this policy restrict coverage. Read the entire policy carefully to determine rights, duties and what is and is not covered.

Throughout this policy the words "you" and "your" refer to the Named Insured shown in the Declarations. The words "we", "us" and "our" refer to the Company providing this insurance.

Other words and phrases that appear in quotation marks have special meaning. Refer to SECTION V - DEFINITIONS.

SECTION I - COVERED AUTOS

ITEM TWO of the Declarations shows the "autos" that are covered "autos" for each of your coverages. The following numerical symbols describe the "autos" that may be covered "autos". The symbols entered next to a coverage on the Declarations designate the only "autos" that are covered "autos".

A. **DESCRIPTION OF COVERED AUTO DESIGNATION SYMBOLS**

SYMBOL	DESCRIPTION

1 = ANY "AUTO".

2 = OWNED "AUTOS" ONLY. Only those "autos" you own (and for Liability Coverage any "trailers" you don't own while attached to power units you own). This includes those "autos" you acquire ownership of after the policy begins.

3 = OWNED PRIVATE PASSENGER "AUTOS" ONLY. Only the private passenger "autos" you own. This includes those private passenger "autos" you acquire ownership of after the policy begins.

4 = OWNED "AUTOS" OTHER THAN PRIVATE PASSENGER "AUTOS" ONLY. Only those "autos" you own that are not of the private passenger type (and for Liability Coverage any "trailers" you don't own while attached to power units you own). This includes those "autos" not of the private passenger type you acquire ownership of after the policy begins.

5 = OWNED "AUTOS" SUBJECT TO NO-FAULT. Only those "autos" you own that are required to have No-Fault benefits in the state where they are licensed or principally garaged. This includes those "autos" you acquire ownership of after the policy begins provided they are required to have No-Fault benefits in the state where they are licensed or principally garaged.

6 = OWNED "AUTOS" SUBJECT TO A COMPULSORY UNINSURED MOTORISTS LAW. Only those "autos" you own that because of the law in the state where they are licensed or principally garaged are required to have and cannot reject Uninsured Motorists Coverage. This includes those "autos" you acquire ownership of after the policy begins provided they are subject to the same state uninsured motorists requirement.

7 = SPECIFICALLY DESCRIBED "AUTOS". Only those "autos" described in ITEM THREE of the Declarations for which a premium charge is shown (and for Liability Coverage any "trailers" you don't own while attached to any power unit described in ITEM THREE).

8 = HIRED "AUTOS" ONLY. Only those "autos" you lease, hire, rent or borrow. This does not include any "auto" you lease, hire, rent, or borrow from any of your employees or partners or members of their households.

9 = NONOWNED "AUTOS" ONLY. Only those "autos" you do not own, lease, hire, rent or borrow that are used in connection with your business. This includes "autos" owned by your employees or partners or members of their households but only while used in your business or your personal affairs.

B. **OWNED AUTOS YOU ACQUIRE AFTER THE POLICY BEGINS**

1. If symbols 1, 2, 3, 4, 5 or 6 are entered next to a coverage in ITEM TWO of the Declarations, then you have coverage for "autos" that you acquire of the type described for the remainder of the policy period.

2. But, if symbol **7** is entered next to a coverage in **ITEM TWO** of the Declarations, an "auto" you acquire will be a covered "auto" for that coverage only if:

a. We already cover all "autos" that you own for that coverage or it replaces an "auto" you previously owned that had that coverage; and

b. You tell us within 30 days after you acquire it that you want us to cover it for that coverage.

C. CERTAIN TRAILERS, MOBILE EQUIPMENT AND TEMPORARY SUBSTITUTE AUTOS

If Liability Coverage is provided by this Coverage Form, the following types of vehicles are also covered "autos" for Liability Coverage:

1. "Trailers" with a load capacity of 2,000 pounds or less designed primarily for travel on public roads.

2. "Mobile equipment" while being carried or towed by a covered "auto".

3. Any "auto" you do not own while used with the permission of its owner as a temporary substitute for a covered "auto" you own that is out of service because of its:

a. Breakdown;

b. Repair;

c. Servicing;

d. "Loss"; or

e. Destruction.

SECTION II - LIABILITY COVERAGE

A. COVERAGE

We will pay all sums an "insured" legally must pay as damages because of "bodily injury" or "property damage" to which this insurance applies, caused by an "accident" and resulting from the ownership, maintenance or use of a covered "auto".

We will also pay all sums an "insured" legally must pay as a "covered pollution cost or expense" to which this insurance applies, caused by an "accident" and resulting from the ownership, maintenance or use of covered "autos". However, we will only pay for the "covered pollution cost or expense" if there is either "bodily injury" or "property damage" to which this insurance applies that is caused by the same "accident".

We have the right and duty to defend any "insured" against a "suit" asking for such damages or a "covered pollution cost or expense". However, we have no duty to defend any "insured" against a "suit" seeking damages for "bodily injury" or "property damage" or a "covered pollution cost or expense" to which this insurance does not apply. We may investigate and settle any claim or "suit" as we consider appropriate. Our duty to defend or settle ends when the Liability Coverage Limit of Insurance has been exhausted by payment of judgments or settlements.

1. WHO IS AN INSURED

The following are "insureds":

a. You for any covered "auto".

b. Anyone else while using with your permission a covered "auto" you own, hire or borrow except:

(1) The owner or anyone else from whom you hire or borrow a covered "auto". This exception does not apply if the covered "auto" is a "trailer" connected to a covered "auto" you own.

(2) Your employee if the covered "auto" is owned by that employee or a member of his or her household.

(3) Someone using a covered "auto" while he or she is working in a business of selling, servicing, repairing, parking or storing "autos" unless that business is yours.

(4) Anyone other than your employees, partners, a lessee or borrower or any of their employees, while moving property to or from a covered "auto".

(5) A partner of yours for a covered "auto" owned by him or her or a member of his or her household.

c. Anyone liable for the conduct of an "insured" described above but only to the extent of that liability.

2. **COVERAGE EXTENSIONS**

a. Supplementary Payments. In addition to the Limit of Insurance, we will pay for the "insured":

(1) All expenses we incur.

(2) Up to $250 for cost of bail bonds (including bonds for related traffic law violations) required because of an "accident" we cover. We do not have to furnish these bonds.

(3) The cost of bonds to release attachments in any "suit" we defend, but only for bond amounts within our Limit of Insurance.

(4) All reasonable expenses incurred by the "insured" at our request, including actual loss of earning up to $100 a day because of time off from work.

(5) All costs taxed against the "insured" in any "suit" we defend.

(6) All interest on the full amount of any judgment that accrues after entry of the judgment in any "suit" we defend, but our duty to pay interest ends when we have paid, offered to pay or deposited in court the part of the judgment that is within our Limit of Insurance.

b. **Out-of-State Coverage Extensions.**

While a covered "auto" is away from the state where it is licensed we will:

(1) Increase the Limit of Insurance for Liability Coverage to meet the limits specified by a compulsory or financial responsibility law of the jurisdiction where the covered "auto" is being used. This extension does not apply to the limit or limits specified by any law governing motor carriers of passengers or property.

(2) Provide the minimum amounts and types of other coverages, such as no-fault, required of out-of-state vehicles by the jurisdiction where the covered "auto" is being used.

We will not pay anyone more than once for the same elements of loss because of these extensions.

B. **EXCLUSIONS**

This insurance does not apply to any of the following:

1. **EXPECTED OR INTENDED INJURY**

"Bodily injury" or "property damage" expected or intended from the standpoint of the "insured".

2. **CONTRACTUAL**

Liability assumed under any contract or agreement.

But this exclusion does not apply to liability for damages:

a. Assumed in a contract or agreement that is an "insured contract" provided the "bodily injury" or "property damage" occurs subsequent to the execution of the contract or agreement; or

b. That the "insured" would have in the absence of the contract or agreement.

3. **WORKERS' COMPENSATION**

Any obligation for which the "insured" or the "insured's" insurer may be held liable under any workers' compensation, disability benefits or unemployment compensation law or any similar law.

4. **EMPLOYEE INDEMNIFICATION AND EMPLOYER'S LIABILITY**

"Bodily injury" to:

a. An employee of the "insured" arising out of and in the course of employment by the "insured"; or

b. The spouse, child, parent, brother or sister of that employee as a consequence of paragraph a. above.

This exclusion applies:

(1) Whether the "insured" may be liable as an employer or in any other capacity; and

(2) To any obligation to share damages with or repay someone else who must pay damages because of the injury.

But this exclusion does not apply to "bodily injury" to domestic employees not entitled to workers' compensation benefits or to liability assumed by the "insured" under an "insured contract".

5. FELLOW EMPLOYEE

"Bodily injury" to any fellow employee of the "insured" arising out of and in the course of the fellow employee's employment.

6. CARE, CUSTODY OR CONTROL

"Property damage" to or "covered pollution cost or expense" involving property owned or transported by the "insured" or in the "insured's" care, custody or control. But this exclusion does not apply to liability assumed under a sidetrack agreement.

7. HANDLING OF PROPERTY

"Bodily injury" or "property damage" resulting from the handling of property:

a. Before it is moved from the place where it is accepted by the "insured" for movement into or onto the covered "auto"; or

b. After it is moved from the covered "auto" to the place where it is finally delivered by the "insured".

8. MOVEMENT OF PROPERTY BY MECHANICAL DEVICE

"Bodily injury" or "property damage" resulting from the movement of property by a mechanical device (other than a hand truck) unless the device is attached to the covered "auto".

9. OPERATIONS

"Bodily injury" or "property damage" arising out of the operation of any equipment listed in paragraphs 6.b. and 6.c. of the definition of "mobile equipment".

10. COMPLETED OPERATIONS

"Bodily injury" or "property damage" arising out of your work after that work has been completed or abandoned.

In this exclusion, your work means:

a. Work or operations performed by you or on your behalf; and

b. Materials, parts or equipment furnished in connection with such work or operations.

Your work includes warranties or representations made at any time with respect to the fitness, quality, durability or performance of any of the items included in paragraphs a. or b. above.

Your work will be deemed completed at the earliest of the following times:

(1) When all of the work called for in your contract has been completed.

(2) When all of the work to be done at the site has been completed if your contract calls for work at more than one site.

(3) When that part of the work done at a job site has been put to its intended use by any person or organization other than another contractor or subcontractor working on the same project.

Work that may need service, maintenance, correction, repair or replacement, but which is otherwise complete, will be treated as completed.

11. POLLUTION

"Bodily injury" or "property damage" arising out of the actual, alleged or threatened discharge, dispersal, seepage, migration, release or escape of "pollutants":

a. That are, or that are contained in any property that is:

(1) Being transported or towed by, handled, or handled for movement into, onto or from, the covered "auto";

(2) Otherwise in the course of transit by or on behalf of the "insured"; or

(3) Being stored, disposed of, treated or processed in or upon the covered "auto";

b. Before the "pollutants" or any property in which the "pollutants" are contained are moved from the place where they are accepted by the "insured" for movement into or onto the covered "auto"; or

c. After the "pollutants" or any property in which the "pollutants" are contained are moved from the covered "auto" to the place where they are finally delivered, disposed of or abandoned by the "insured".

Paragraph a. above does not apply to fuels, lubricants, fluids, exhaust gases or other similar "pollutants" that are needed for or result from the normal electrical, hydraulic or mechanical functioning of the covered "auto" or its parts, if:

(1) The "pollutants" escape, seep, migrate, or are discharged, dispersed or released directly from an "auto" part designed by its manufacturer to hold, store, receive or dispose of such "pollutants"; and

(2) The "bodily injury", "property damage" or "covered pollution cost or expense" does not arise out of the operation of any equipment listed in paragraphs 6.b. and 6.c. of the definition of "mobile equipment".

Paragraphs b. and c. above of this exclusion do not apply to "accidents" that occur away from premises owned by or rented to an "insured" with respect to "pollutants" not in or upon a covered "auto" if:

(1) The "pollutants" or any property in which the "pollutants" are contained are upset, overturned or damaged as a result of the maintenance or use of a covered "auto"; and

(2) The discharge, dispersal, seepage, migration, release or escape of the "pollutants" is caused directly by such upset, overturn or damage.

12. WAR

"Bodily injury" or "property damage" due to war, whether or not declared, or any act or condition incident to war. War includes civil war, insurrection, rebellion or revolution. This exclusion applies only to liability assumed under a contract or agreement.

13. RACING

Covered "autos" while used in any professional or organized racing or demolition contest or stunting activity, or while practicing for such contest or activity. This insurance also does not apply while that covered "auto" is being prepared for such a contest or activity.

C. LIMIT OF INSURANCE

Regardless of the number of covered "autos", "insureds", premiums paid, claims made or vehicles involved in the "accident", the most we will pay for the total of all damages and "covered pollution cost or expense" combined, resulting from any one "accident" is the Limit of Insurance for Liability Coverage shown in the Declarations.

All "bodily injury", "property damage" and "covered pollution cost or expense" resulting from continuous or repeated exposure to substantially the same conditions will be considered as resulting from one "accident".

No one will be entitled to receive duplicate payments for the same elements of "loss" under this Coverage Form and any Medical Payments Coverage endorsement, Uninsured Motorists Coverage endorsement or Underinsured Motorists Coverage endorsement attached to this Coverage Part.

SECTION III - PHYSICAL DAMAGE COVERAGE

A. COVERAGE

1. We will pay for "loss" to a covered "auto" or its equipment under:

a. Comprehensive Coverage. From any cause except:

(1) The covered "auto's" collision with another object; or

(2) The covered "auto's" overturn.

b. Specified Causes of Loss Coverage. Caused by:

(1) Fire, lightning or explosion;

(2) Theft;

(3) Windstorm, hail or earthquake;

(4) Flood;

(5) Mischief or vandalism; or

(6) The sinking, burning, collision or derailment of any conveyance transporting the covered "auto".

c. Collision Coverage. Caused by:

(1) The covered "auto's" collision with another object; or

(2) The covered "auto's" overturn.

2. **Towing.**

We will pay up to the limit shown in the Declarations for towing and labor costs incurred each time a covered "auto" of the private passenger type is disabled. However, the labor must be performed at the place of disablement.

3. **Glass Breakage - Hitting a Bird or Animal - Falling Objects or Missiles.**

If you carry Comprehensive Coverage for the damaged covered "auto", we will pay for the following under Comprehensive Coverage:

a. Glass breakage;

b. "Loss" caused by hitting a bird or animal; and

c. "Loss" caused by falling objects or missiles.

However, you have the option of having glass breakage caused by a covered "auto's" collision or overturn considered a "loss" under Collision Coverage.

4. Coverage Extension. We will pay up to $15 per day to a maximum of $450 for transportation expense incurred by you because of the total theft of a covered "auto" of the private passenger type. We will pay only for those covered "autos" for which you carry either Comprehensive or Specified Causes of Loss Coverage. We will pay for transportation expenses incurred during the period beginning 48 hours after the theft and ending, regardless of the policy's expiration, when the covered "auto" is returned to use or we pay for its "loss".

B. **EXCLUSIONS**

1. We will not pay for "loss" caused by or resulting from any of the following. Such "loss" is excluded regardless of any other cause or event that contributes concurrently or in any sequence to the "loss".

a. **Nuclear Hazard.**

(1) The explosion of any weapon employing atomic fission or fusion; or

(2) Nuclear reaction or radiation, or radioactive contamination, however caused.

b. **War or Military Action.**

(1) War, including undeclared or civil war;

(2) Warlike action by a military force, including action in hindering or defending against an actual or expected attack, by any government, sovereign or other authority using military personnel or other agents; or

(3) Insurrection, rebellion, revolution, usurped power or action taken by governmental authority in hindering or defending against any of these.

2. We will not pay for "loss" to any covered "auto" while used in any professional or organized racing or demolition contest or stunting activity, or while practicing for such contest or activity. We will also not pay for "loss" to any covered "auto" while that covered "auto" is being prepared for such a contest or activity.

3. We will not pay for "loss" caused by or resulting from any of the following unless caused by other "loss" that is covered by this insurance:

a. Wear and tear, freezing, mechanical or electrical breakdown.

b. Blowouts, punctures or other road damage to tires.

4. We will not pay for "loss" to any of the following:

a. Tapes, records, discs or other similar audio, visual or data electronic devices designed for use with audio, visual or data electronic equipment.

b. Equipment designed or used for the detection or location of radar.

c. Any electronic equipment, without regard to whether this equipment is permanently installed, that receives or transmits audio, visual or data signals and that is not designed solely for the reproduction of sound.

d. Any accessories used with the electronic equipment described in paragraph c. above.

Exclusions **4.c.** and **4.d.** do not apply to:

a. Equipment designed solely for the reproduction of sound and accessories used with such equipment, provided such equipment is permanently installed in the covered "auto" at the time of the "loss" or such equipment is removable from a housing unit which is permanently installed in the covered "auto" at the time of the "loss", and such equipment is designed to be solely operated by use of the power from the "auto's" electrical system, in or upon the covered "auto"; or

b. Any other electronic equipment that is:

(1) Necessary for the normal operation of the covered "auto" or the monitoring of the covered "auto's" operating system; or

(2) An integral part of the same unit housing any sound reproducing equipment described in **a.** above and permanently installed in the opening of the dash or console of the covered "auto" normally used by the manufacturer for installation of a radio.

C. LIMIT OF INSURANCE

The most we will pay for "loss" in any one "accident" is the lesser of:

1. The actual cash value of the damaged or stolen property as of the time of the "loss"; or

2. The cost of repairing or replacing the damaged or stolen property with other property of like kind and quality.

D. DEDUCTIBLE

For each covered "auto", our obligation to pay for, repair, return or replace damaged or stolen property will be reduced by the applicable deductible shown in the Declarations. Any Comprehensive Coverage deductible shown in the Declarations does not apply to "loss" caused by fire or lightning.

SECTION IV - BUSINESS AUTO CONDITIONS

The following conditions apply in addition to the Common Policy Conditions:

A. LOSS CONDITIONS

1. APPRAISAL FOR PHYSICAL DAMAGE LOSS

If you and we disagree on the amount of "loss", either may demand an appraisal of the "loss". In this event, each party will select a competent appraiser. The two appraisers will select a competent and impartial umpire. The appraisers will state separately the actual cash value and amount of "loss". If they fail to agree, they will submit their differences to the umpire. A decision agreed to by any two will be binding. Each party will:

a. Pay its chosen appraiser; and

b. Bear the other expenses of the appraisal and umpire equally.

If we submit to an appraisal, we will still retain our right to deny the claim.

2. DUTIES IN THE EVENT OF ACCIDENT, CLAIM, SUIT OR LOSS

a. In the event of "accident", claim, "suit" or "loss", you must give us or our authorized representative prompt notice of the "accident" or "loss". Include:

(1) How, when and where the "accident" or "loss" occurred;

(2) The "insured's" name and address; and

(3) To the extent possible, the names and addresses of any injured persons and witnesses.

b. Additionally, you and any other involved "insured" must:

(1) Assume no obligation, make no payment or incur no expense without our consent, except at the "insured's" own cost.

(2) Immediately send us copies of any request, demand, order, notice, summons or legal paper received concerning the claim or "suit".

(3) Cooperate with us in the investigation, settlement or defense of the claim or "suit".

(4) Authorize us to obtain medical records or other pertinent information.

(5) Submit to examination, at our expense, by physicians of our choice, as often as we reasonably require.

c. If there is "loss" to a covered "auto" or its equipment you must also do the following:

(1) Promptly notify the police if the covered "auto" or any of its equipment is stolen.

(2) Take all reasonable steps to protect the covered "auto" from further damage. Also keep a record of your expenses for consideration in the settlement of the claim.

(3) Permit us to inspect the covered "auto" and records proving the "loss" before its repair or disposition.

(4) Agree to examinations under oath at our request and give us a signed statement of your answers.

3. LEGAL ACTION AGAINST US

No one may bring a legal action against us under this Coverage Form until:

a. There has been full compliance with all the terms of this Coverage Form; and

b. Under Liability Coverage, we agree in writing that the "insured" has an obligation to pay or until the amount of that obligation has finally been determined by judgment after trial. No one has the right under this policy to bring us into an action to determine the "insured's" liability.

4. LOSS PAYMENT - PHYSICAL DAMAGE COVERAGES

At our option we may:

a. Pay for, repair or replace damaged or stolen property;

b. Return the stolen property, at our expense. We will pay for any damage that results to the "auto" from the theft; or

c. Take all or any part of the damaged or stolen property at an agreed or appraised value.

5. TRANSFER OF RIGHTS OF RECOVERY AGAINST OTHERS TO US

If any person or organization to or for whom we make payment under this Coverage Form has rights to recover damages from another, those rights are transferred to us. That person or organization must do everything necessary to secure our rights and must do nothing after "accident" or "loss" to impair them.

B. GENERAL CONDITIONS

1. BANKRUPTCY

Bankruptcy or insolvency of the "insured" or the "insured's" estate will not relieve us of any obligations under this Coverage Form.

2. CONCEALMENT, MISREPRESENTATION OR FRAUD

This Coverage Form is void in any case of fraud by you at any time as it relates to this Coverage Form. It is also void if you or any other "insured", at any time, intentionally conceal or misrepresent a material fact concerning:

a. This Coverage Form;

b. The covered "auto";

c. Your interest in the covered "auto"; or

d. A claim under this Coverage Form.

3. LIBERALIZATION

If we revise this Coverage Form to provide more coverage without additional premium charge, your policy will automatically provide the additional coverage as of the day the revision is effective in your state.

4. NO BENEFIT TO BAILEE - PHYSICAL DAMAGE COVERAGES

We will not recognize any assignment or grant any coverage for the benefit of any person or organization holding, storing or transporting property for a fee regardless of any other provision of this Coverage Form.

5. OTHER INSURANCE

a. For any covered "auto" you own, this Coverage Form provides primary insurance. For any covered "auto" you don't own, the insurance provided by this Coverage Form is excess over any other collectible insurance. However, while a covered "auto" which is a "trailer" is connected to another vehicle, the Liability Coverage this Coverage Form provides for the "trailer" is:

(1) Excess while it is connected to a motor vehicle you do not own.

(2) Primary while it is connected to a covered "auto" you own.

b. For Hired Auto Physical Damage coverage, any covered "auto" you lease, hire, rent or borrow is deemed to be a covered "auto" you own. However, any "auto" that is leased, hired, rented or borrowed with a driver is not a covered "auto".

c. Regardless of the provisions of paragraph **a.** above, this Coverage Form's Liability Coverage is primary for any liability assumed under an "insured contract".

d. When this Coverage Form and any other Coverage Form or policy covers on the same basis, either excess or primary, we will pay only our share. Our share is the proportion that the Limit of Insurance of our Coverage Form bears to the total of the limits of all the Coverage Forms and policies covering on the same basis.

6. **PREMIUM AUDIT**

a. The estimated premium for this Coverage Form is based on the exposures you told us you would have when this policy began. We will compute the final premium due when we determine your actual exposures. The estimated total premium will be credited against the final premium due and the first Named Insured will be billed for the balance, if any. If the estimated total premium exceeds the final premium due, the first Named Insured will get a refund.

b. If this policy is issued for more than one year, the premium for this Coverage Form will be computed annually based on our rates or premiums in effect at the beginning of each year of the policy.

7. **POLICY PERIOD, COVERAGE TERRITORY**

Under this Coverage Form, we cover "accidents" and "losses" occurring:

a. During the policy period shown in the Declarations; and

b. Within the coverage territory.

The coverage territory is:

a. The United States of America;

b. The territories and possessions of the United States of America;

c. Puerto Rico; and

d. Canada.

We also cover "loss" to, or "accidents" involving, a covered "auto" while being transported between any of these places.

8. **TWO OR MORE COVERAGE FORMS OR POLICIES ISSUED BY US**

If this Coverage Form and any other Coverage Form or policy issued to you by us or any company affiliated with us apply to the same "accident", the aggregate maximum Limit of Insurance under all the Coverage Forms or policies shall not exceed the highest applicable Limit of Insurance under any one Coverage Form or policy. This condition does not apply to any Coverage Form or policy issued by us or an affiliated company specifically to apply as excess insurance over this Coverage Form.

SECTION V - DEFINITIONS

A. "Accident" includes continuous or repeated exposure to the same conditions resulting in "bodily injury" or "property damage".

B. "Auto" means a land motor vehicle, trailer or semitrailer designed for travel on public roads but does not include "mobile equipment".

C. "Bodily injury" means bodily injury, sickness or disease sustained by a person including death resulting from any of these.

D. "Covered pollution cost or expense" means any cost or expense arising out of:

1. Any request, demand or order; or

2. Any claim or "suit" by or on behalf of a governmental authority demanding

that the "insured" or others test for, monitor, clean up, remove, contain, treat, detoxify or neutralize, or in any way respond to, or assess the effects of "pollutants".

"Covered pollution cost or expense" does not include any cost or expense arising out of the actual, alleged or threatened discharge, dispersal, seepage, migration, release or escape of "pollutants":

a. That are, or that are contained in any property that is:

(1) Being transported or towed by, handled, or handled for movement into, onto or from the covered "auto";

(2) Otherwise in the course of transit by or on behalf of the "insured";

(3) Being stored, disposed of, treated or processed in or upon the covered "auto"; or

b. Before the "pollutants" or any property in which the "pollutants" are contained are moved from the place where they are accepted by the "insured" for movement into or onto the covered "auto"; or

c. After the "pollutants" or any property in which the "pollutants" are contained are moved from the covered "auto" to the place where they are finally delivered, disposed of or abandoned by the "insured".

Paragraph a. above does not apply to fuels, lubricants, fluids, exhaust gases or other similar "pollutants" that are needed for or result from the normal electrical, hydraulic or mechanical functioning of the covered "auto" or its parts, if:

(1) The "pollutants" escape, seep, migrate, or are discharged, dispersed or released directly from an "auto" part designed by its manufacturer to hold, store, receive or dispose of such "pollutants"; and

(2) The "bodily injury", "property damage" or "covered pollution cost or expense" does not arise out of the operation of any equipment listed in paragraphs 6.b. or 6.c. of the definition of "mobile equipment".

Paragraphs b. and c. above do not apply to "accidents" that occur away from premises owned by or rented to an "insured" with respect to "pollutants" not in or upon a covered "auto" if:

(1) The "pollutants" or any property in which the "pollutants" are contained are upset, overturned or damaged as a result of the maintenance or use of a covered "auto"; and

(2) The discharge, dispersal, seepage, migration, release or escape of the "pollutants" is caused directly by such upset, overturn or damage.

E. "Insured" means any person or organization qualifying as an insured in the Who Is An Insured provision of the applicable coverage. Except with respect to the Limit of Insurance, the coverage afforded applies separately to each insured who is seeking coverage or against whom a claim or "suit" is brought.

F. "Insured contract" means:

1. A lease of premises;

2. A sidetrack agreement;

3. Any easement or license agreement, except in connection with construction or demolition operations on or within 50 feet of a railroad;

4. An obligation, as required by ordinance, to indemnify a municipality, except in connection with work for a municipality;

5. That part of any other contract or agreement pertaining to your business (including an indemnification of a municipality in connection with work performed for a municipality) under which you assume the tort liability of another to pay for "bodily injury" or "property damage" to a third party or organization. Tort liability means a liability that would be imposed by law in the absence of any contract or agreement;

6. That part of any contract or agreement entered into, as part of your business, pertaining to the rental or lease, by you or any of your employees, of any "auto". However, such contract or agreement shall not be considered an "insured contract" to the extent that it obligates you or any of your employees to pay for "property damage" to any "auto" rented or leased by you or any of your employees.

An "insured contract" does not include that part of any contract or agreement:

a. That indemnifies any person or organization for "bodily injury" or "property damage" arising out of construction or demolition operations, within 50 feet of any railroad property and affecting any railroad bridge or trestle, tracks, roadbeds, tunnel, underpass or crossing; or

b. That pertains to the loan, lease or rental of an "auto" to you or any of your employees, if the "auto" is loaned, leased or rented with a driver; or

c. That holds a person or organization engaged in the business of transporting property by "auto" for hire harmless for your use of a covered "auto" over a route or territory that person or organization is authorized to serve by public authority.

G. "Loss" means direct and accidental loss or damage.

H. "Mobile equipment" means any of the following types of land vehicles, including any attached machinery or equipment:

1. Bulldozers, farm machinery, forklifts and other vehicles designed for use principally off public roads;

2. Vehicles maintained for use solely on or next to premises you own or rent;
3. Vehicles that travel on crawler treads;
4. Vehicles, whether self-propelled or not, maintained primarily to provide mobility to permanently mounted:
   a. Power cranes, shovels, loaders, diggers or drills; or
   b. Road construction or resurfacing equipment such as graders, scrapers or rollers.
5. Vehicles not described in paragraphs **1., 2., 3.,** or **4.** above that are not self-propelled and are maintained primarily to provide mobility to permanently attached equipment of the following types:
   a. Air compressors, pumps and generators, including spraying, welding, building cleaning, geophysical exploration, lighting and well servicing equipment; or
   b. Cherry pickers and similar devices used to raise or lower workers.
6. Vehicles not described in paragraphs **1., 2., 3.** or **4.** above maintained primarily for purposes other than the transportation of persons or cargo. However, self-propelled vehicles with the following types of permanently attached equipment are not "mobile equipment" but will be considered "autos":
   a. Equipment designed primarily for:
      (1) Snow removal;
      (2) Road maintenance, but not construction or resurfacing; or
      (3) Street cleaning;

   b. Cherry pickers and similar devices mounted on automobile or truck chassis and used to raise or lower workers; and
   c. Air compressors, pumps and generators, including spraying, welding, building cleaning, geophysical exploration, lighting or well servicing equipment.

I. "Pollutants" means any solid, liquid, gaseous or thermal irritant or contaminant, including smoke, vapor, soot, fumes, acids, alkalis, chemicals and waste. Waste includes materials to be recycled, reconditioned or reclaimed.

J. "Property damage" means damage to or loss of use of tangible property.

K. "Suit" means a civil proceeding in which:
   1. Damages because of "bodily injury" or "property damage"; or
   2. A "covered pollution cost or expense",

   to which this insurance applies, are alleged.

   "Suit" includes:

   a. An arbitration proceeding in which such damages or "covered pollution costs or expenses" are claimed and to which the "insured" must submit or does submit with our consent; or
   b. Any other alternative dispute resolution proceeding in which such damages or "covered pollution costs or expenses" are claimed and to which the insured submits with our consent.

L. "Trailer" includes semitrailer.

ENDORSEMENT MS #

This endorsement, effective   **12:01 A.M.**   forms a part of

policy No.   issued to

By:

THIS ENDORSEMENT CHANGES THE POLICY. PLEASE READ IT CAREFULLY.
AMENDMENT OF CONDITION: NOTICE TO THE COMPANY

This endorsement modifies insurance provided under the following:
COMMERCIAL AUTOMOBILE COVERAGE PART

It is agreed that when the Insured reports the occurrence of any accident to the insurance company that provides its Workers' Compensation coverage and such loss later develops into a claim involving liability that is covered by the policy to which this endorsement is attached, the failure by the Insured to report such an accident to us at the time of the occurrence shall not be deemed to be in violation of the General Condition entitled Duties In The Event Of Occurrence, Offense, Claim, or Suit. However, the Insured shall give notification to this Company as soon as the Insured is made aware of the fact that the particular loss is a liability case rather than a compensation case.

All other terms, conditions, and exclusions of this policy remain unchanged.

_____

Authorized Representative

POLICY NUMBER: COMMERCIAL AUTO

THIS ENDORSEMENT CHANGES THE POLICY. PLEASE READ IT CAREFULLY.

DRIVE OTHER CAR COVERAGE – BROADENED COVERAGE FOR NAMED INDIVIDUALS

This endorsement modifies insurance provided under the following:

BUSINESS AUTO COVERAGE FORM
GARAGE COVERAGE FORM
TRUCKERS COVERAGE FORM
BUSINESS AUTO PHYSICAL DAMAGE COVERAGE FORM

This endorsement changes the policy effective on the inception date of the policy unless another date is indicated below.

Endorsement effective	
Named Insured	Countersigned by

(Authorized Representative)

SCHEDULE

Name of Individual	Liability		Auto Med. Pay		Uninsured Motorists		Phys. Dam.	
							Comp.	Coll. $50 Ded.
	Limit	Premium	Limit	Premium	Limit	Premium		

Note – When uninsured motorists is provided at limits higher than the basic limits required by a financial responsibility law, underinsured motorists is included, unless otherwise noted.

(If no entry appears above, information required to complete this endorsement will be shown in the Declarations as applicable to this endorsement.)

A. This endorsement changes only those coverages where a premium is shown in the Schedule.

B. CHANGES IN LIABILITY COVERAGE
1. Any "auto" you don't own, hire or borrow is a covered "auto" for LIABILITY COVERAGE while being used by any individual named in the

Schedule or by his or her spouse while a resident of the same household except

a. Any "auto" owned by that individual or by any member of his or her household.

b. Any "auto" used by that individual or his or her spouse while working in a business of selling, servicing, repairing or parking "autos."

2. The following is added to WHO IS AN INSURED:

Any individual named in the Schedule and his or her spouse, while a resident of the same household, are "insureds" while using any covered "auto" described in paragraph B.1. of this endorsement.

C. CHANGES IN AUTO MEDICAL PAYMENTS AND UNINSURED MOTORISTS COVERAGES

The following is added to WHO IS AN INSURED:

Any individual named in the Schedule and his or her "family members" are "insured" while "occupying" or while a pedestrian when being struck by any "auto" you don't own except

Any "auto" owned by that individual or by any "family member."

D. CHANGES IN PHYSICAL DAMAGE COVERAGE

Any private passenger type "auto" you don't own, hire or borrow is a covered "auto" while in the the care, custody or control of any individual named in the Schedule or his or her spouse while a resident of the same household except

1. Any "auto" owned by that individual or by any member of his or her household.

2. Any "auto" used by that individual or his or her spouse while working in a business of selling, servicing, repairing or parking "autos."

E. ADDITIONAL DEFINITION

The following is added to the DEFINITIONS Section:

"Family member" means a person related to the individual named in the the Schedule by blood, marriage or adoption who is a resident of the individual's household, including a ward or foster child.

POLICY NUMBER:

THIS ENDORSEMENT CHANGES THE POLICY. PLEASE READ IT CAREFULLY.

RENTAL REIMBURSEMENT COVERAGE

This endorsement modifies insurance provided under the following:

BUSINESS AUTO COVERAGE FORM
GARAGE COVERAGE FORM
TRUCKERS COVERAGE FORM
BUSINESS AUTO PHYSICAL DAMAGE COVERAGE FORM

This endorsement changes the policy effective on the inception date of the policy unless another date is indicated below.

Endorsement effective	
Named Insured	Countersigned by

(Authorized Representative)

SCHEDULE

			Maximum Payment Each Covered "Auto"			
Coverages	Auto no.	Designation or Description of Covered "Autos" to which this insurance applies	Any One Day	No. of Days	Any one Period	Premium
Comprehensive			$		$	$
			$		$	$
Collision			$		$	$
			$		$	$
Specified			$		$	$
Causes of Loss			$		$	$
			Total Premium			

(If no entry appears above, information required to complete this endorsement will be shown in the Declarations as applicable to this endorsement.)

A. This endorsement provides only those coverages where a premium is shown in the Schedule. It applies only to a covered "auto" described or designated in the Schedule.

B. We will pay for rental reimbursement expenses incurred by you for the rental of an "auto" because of "loss" to a covered "auto". Payment applies in addition to the otherwise applicable amount of each coverage you have on a covered "auto." No deductibles apply to this coverage.

C. We will pay only for those expenses incurred during the policy period beginning 24 hours after the "loss" and ending, regardless of the policy's expiration, with the lesser of the following number of days:

1. The number of days reasonably required to repair or replace the covered "auto." If "loss" is caused by theft, this number of days is added to the number of days it takes to locate the covered "auto" and return it to you.

2. The number of days shown in the Schedule.

D. Our payment is limited to the lesser of the following amounts:

1. Necessary and actual expenses incurred.

2. The maximum payment stated in the Schedule applicable to "any one day" or "any one period."

E. This coverage does not apply while there are spare or reserve "autos" available to you for your operations.

F. If "loss" results from the total theft of a covered "auto" of the private passenger type, we will pay under this coverage only that amount of your rental reimbursement expenses which is not already provided for under the PHYSICAL DAMAGE COVERAGE Coverage Extension.

POLICY NUMBER:

COMMERCIAL AUTO
CA 99 16 12 93

THIS ENDORSEMENT CHANGES THE POLICY. PLEASE READ IT CAREFULLY.

HIRED AUTOS SPECIFIED AS COVERED AUTOS YOU OWN

This endorsement modifies insurance provided under the following:

BUSINESS AUTO COVERAGE FORM
GARAGE COVERAGE FORM
MOTOR CARRIER COVERAGE FORM
TRUCKERS COVERAGE FORM
BUSINESS AUTO PHYSICAL DAMAGE COVERAGE FORM

With respect to coverage provided by this endorsement, the provisions of the Coverage Form apply unless modified by the endorsement.

This endorsement changes the policy effective on the inception date of the policy unless another date is indicated below.

Endorsement effective	
Named Insured	Countersigned by

(Authorized Representative)

SCHEDULE

Description of Auto:

(If no entry appears above, information required to complete this endorsement will be shown in the Declarations as applicable to this endorsement.)

A. Any "auto" described in the Schedule will be considered a covered "auto" you own and not a covered "auto" you hire, borrow or lease under the coverage for which it is a covered "auto".

B. CHANGES IN LIABILITY COVERAGE
The following is added to WHO IS AN INSURED:

While any covered "auto" described in the Schedule is rented or leased to you and is being used by or for you, its owner or anyone else from whom you rent or lease it is an "insured" but only for that covered "auto".

COMMERCIAL AUTO
CA 99 03 12 93

THIS ENDORSEMENT CHANGES THE POLICY. PLEASE READ IT CAREFULLY.

AUTO MEDICAL PAYMENTS COVERAGE

This endorsement modifies insurance provided under the following:

BUSINESS AUTO COVERAGE FORM
GARAGE COVERAGE FORM
MOTOR CARRIER COVERAGE FORM
TRUCKERS COVERAGE FORM

With respect to coverage provided by this endorsement, the provisions of the Coverage Form apply unless modified by the endorsement.

A. COVERAGE

We will pay reasonable expenses incurred for necessary medical and funeral services to or for an "insured" who sustains "bodily injury" caused by "accident". We will pay only those expenses incurred, for services rendered within three years from the date of the "accident".

B. WHO IS AN INSURED

1. You while "occupying" or, while a pedestrian, when struck by any "auto".

2. If you are an individual, any "family member" while "occupying" or, while a pedestrian, when struck by any "auto".

3. Anyone else "occupying" a covered "auto" or a temporary substitute for a covered "auto". The covered "auto" must be out of service because of its breakdown, repair, servicing, loss or destruction.

C. EXCLUSIONS

This insurance does not apply to any of the following:

1. "Bodily injury" sustained by an "insured" while "occupying" a vehicle located for use as a premises.

2. "Bodily injury" sustained by you or any "family member" while "occupying" or struck by any vehicle (other than a covered "auto") owned by you or furnished or available for your regular use.

3. "Bodily injury" sustained by any "family member" while "occupying" or struck by any vehicle (other than a covered "auto") owned by or furnished or available for the regular use of any "family member".

4. "Bodily injury" to your employee arising out of and in the course of employment by you. However, we will cover "bodily injury" to your domestic employees if not entitled to workers' compensation benefits.

5. "Bodily injury" to an "insured" while working in a business of selling, servicing, repairing or parking "autos" unless that business is yours.

6. "Bodily injury" caused by declared or undeclared war or insurrection or any of their consequences.

7. "Bodily injury" to anyone using a vehicle without a reasonable belief that the person is entitled to do so.

8. "Bodily Injury" sustained by an "insured" while "occupying" any covered "auto" while used in any professional racing or demolition contest or stunting activity, or while practicing for such contest or activity. This insurance also does not apply to any "bodily injury" sustained by an "insured" while the "auto" is being prepared for such a contest or activity.

D. LIMIT OF INSURANCE

Regardless of the number of covered "autos", "insureds", premiums paid, claims made or vehicles involved in the "accident", the most we will pay for "bodily injury" for each "insured" injured in any one "accident" is the LIMIT OF INSURANCE for AUTO MEDICAL PAYMENTS COVERAGE shown in the Declarations.

No one will be entitled to receive duplicate payments for the same elements of "loss" under this coverage and any Liability Coverage Form, Uninsured Motorists Coverage endorsement or Underinsured Motorists Coverage endorsement attached to this Coverage Part.

E. CHANGES IN CONDITIONS

The CONDITIONS are changed for AUTO MEDICAL PAYMENTS COVERAGE as follows:

1. The TRANSFER OF RIGHTS OF RECOVERY AGAINST OTHERS TO US Condition does not apply.

2. The reference in OTHER INSURANCE in the Business Auto and Garage Coverage Forms and OTHER INSURANCE - PRIMARY AND EXCESS INSURANCE PROVISIONS in the Truckers and Motor Carrier Coverage Forms to "other collectible insurance" applies only to other collectible auto medical payments insurance.

F. ADDITIONAL DEFINITIONS

As used in this endorsement:

1. "Family member" means a person related to you by blood, marriage or adoption who is a resident of your household, including a ward or foster child.

2. "Occupying" means in, upon, getting in, on, out or off.

ENDORSEMENT MS #

This endorsement, effective   **12:01 A.M.**   forms a part of

policy No.   issued to

By:

CANCELLATION BY US

This endorsement modifies insurance provided under the following:
COMMERCIAL AUTOMOBILE COVERAGE PART

SCHEDULE

Number Of Days   60

Paragraph 2 of CANCELLATION (Common Policy Conditions) is replaced by the following:

2. We may cancel or non-renew this Coverage Part by mailing or delivering to the first Named Insured written notice of cancellation or non-renewal at least:

   a.   10 days before the effective date of cancellation if we cancel for non-payment of premium; or

   b.   The number of days shown in the Schedule before the effective date of cancellation or non-renewal if we cancel or non-renew for any other reason.

All other terms, conditions, and exclusions of this policy remain unchanged.

_____
Authorized Representative

Examples of Bond Forms

- Bid Bond Form
- Performance Bond Form
- Payment Bond Form

THE AMERICAN INSTITUTE OF ARCHITECTS

AIA Document A310

Bid Bond

KNOW ALL MEN BY THESE PRESENTS, that we

(Here insert full name and address or legal title of Contractor)

as Principal, hereinafter called the Principal, and

(Here insert full name and address or legal title of Surety)

a corporation duly organized under the laws of the State of
as Surety, hereinafter called the Surety, are held and firmly bound unto

(Here insert full name and address or legal title of Owner)

as Obligee, hereinafter called the Obligee, in the sum of

Dollars ($),
for the payment of which sum well and truly to be made, the said Principal and the said Surety, bind ourselves, our heirs, executors, administrators, successors and assigns, jointly and severally, firmly by these presents.

WHEREAS, the Principal has submitted a bid for

(Here insert full name, address and description of project)

NOW, THEREFORE, if the Obligee shall accept the bid of the Principal and the Principal shall enter into a Contract with the Obligee in accordance with the terms of such bid, and give such bond or bonds as may be specified in the bidding or Contract Documents with good and sufficient surety for the faithful performance of such Contract and for the prompt payment of labor and material furnished in the prosecution thereof, or in the event of the failure of the Principal to enter such Contract and give such bond or bonds, if the Principal shall pay to the Obligee the difference not to exceed the penalty hereof between the amount specified in said bid and such larger amount for which the Obligee may in good faith contract with another party to perform the Work covered by said bid, then this obligation shall be null and void, otherwise to remain in full force and effect.

Signed and sealed this   day of   19

_____
(Witness)

_____   _____
(Principal)   (Seal)

_____
(Title)

_____
(Witness)

_____   _____
(Surety)   (Seal)

_____
(Title)

THE AMERICAN INSTITUTE OF ARCHITECTS

AIA Document A312

Performance Bond

Any singular reference to Contractor, Surety, Owner or other party shall be considered plural where applicable.

CONTRACTOR (Name and Address):   SURETY (Name and Principal Place of Business):

OWNER (Name and Address):

CONSTRUCTION CONTRACT
    Date:
    Amount:
    Description (Name and Location):

BOND
    Date (Not earlier than Construction Contract Date):
    Amount:
    Modifications to this Bond: ☐ None   ☐ See Page 3

CONTRACTOR AS PRINCIPAL   SURETY
Company: (Corporate Seal)   Company: (Corporate Seal)

Signature: _____   Signature: _____
Name and Title: Name and Title:

(Any additional signatures appear on page 3)

(FOR INFORMATION ONLY—Name, Address and Telephone)
AGENT or BROKER: OWNER'S REPRESENTATIVE (Architect, Engineer or
                                    other party):

1   The Contractor and the Surety, jointly and severally, bind themselves, their heirs, executors, administrators, successors and assigns to the Owner for the performance of the Construction Contract, which is incorporated herein by reference.

2   If the Contractor performs the Construction Contract, the Surety and the Contractor shall have no obligation under this Bond, except to participate in conferences as provided in Subparagraph 3.1.

3   If there is no Owner Default, the Surety's obligation under this Bond shall arise after:

3.1   The Owner has notified the Contractor and the Surety at its address described in Paragraph 10 below that the Owner is considering declaring a Contractor Default and has requested and attempted to arrange a conference with the Contractor and the Surety to be held not later than fifteen days after receipt of such notice to discuss methods of performing the Construction Contract. If the Owner, the Contractor and the Surety agree, the Contractor shall be allowed a reasonable time to perform the Construction Contract, but such an agreement shall not waive the Owner's right, if any, subsequently to declare a Contractor Default; and

3.2   The Owner has declared a Contractor Default and formally terminated the Contractor's right to complete the contract. Such Contractor Default shall not be declared earlier than twenty days after the Contractor and the Surety have received notice as provided in Sub-paragraph 3.1; and

3.3   The Owner has agreed to pay the Balance of the Contract Price to the Surety in accordance with the terms of the Construction Contract or to a contractor selected to perform the Construction Contract in accordance with the terms of the contract with the Owner.

4   When the Owner has satisfied the conditions of Paragraph 3, the Surety shall promptly and at the Surety's expense take one of the following actions:

4.1   Arrange for the Contractor, with consent of the Owner, to perform and complete the Construction Contract; or

4.2   Undertake to perform and complete the Construction Contract itself, through its agents or through independent contractors; or

4.3   Obtain bids or negotiated proposals from qualified contractors acceptable to the Owner for a contract for performance and completion of the Construction Contract, arrange for a contract to be prepared for execution by the Owner and the contractor selected with the Owner's concurrence, to be secured with performance and payment bonds executed by a qualified surety equavalent to the bonds issued on the Construction Contract, and pay to the Owner the amount of damages as described in Paragraph 6 in excess of the Balance of the Contract Price incurred by the Owner resulting from the Contractor's default; or

4.4   Waive its right to perform and complete, arrange for completion, or obtain a new contractor and with reasonable promptness under the circumstances;

.1   After investigation, determine the amount for which it may be liable to the Owner, and, as soon as practicable after the amount is determined, tender payment therefor to the Owner; or

.2   Deny liability in whole or in part and notify the Owner citing reasons therefor.

5   If the Surety does not proceed as provided in Paragraph 4 with reasonable promptness, the Surety shall be deemed to be in default on this Bond fifteen days after receipt of an additional written notice from the Owner to the Surety demanding that the Surety perform its obligations under this Bond, and the Owner shall be entitled to enforce any remedy available to the Owner. If the Surety proceeds as provided in Subparagraph 4.4, and the Owner refuses the payment tendered or the Surety has denied liability, in whole or in part, without further notice the Owner shall be entitled to enforce any remedy available to the Owner.

6   After the Owner has terminated the Contractor's right to complete the Construction Contract, and if the Surety elects to act under Subparagraph 4.1, 4.2, or 4.3 above, then the responsibilities of the Surety to the Owner shall not be greater than those of the Contractor under the Construction Contract, and the responsibilities of the Owner to the Surety shall not be greater than those of the Owner under the Construction Contract. To the limit of the amount of this Bond, but subject to commitment by the Owner of the Balance of the Contract Price to mitigation of costs and damages on the Construction Contract, the Surety is obligated without duplication for:

6.1   The responsibilities of the Contractor for correction of defective work and completion of the Construction Contract;

6.2   Additional legal, design professional and delay costs resulting from the Contractor's Default, and resulting from the actions or failure to act of the Surety under Paragraph 4; and

6.3   Liquidated damages, or if no liquidated damages are specified in the Construction Contract, actual damages caused by delayed performance or non-performance of the Contractor.

7   The Surety shall not be liable to the Owner or others for obligations of the Contractor that are unrelated to the Construction Contract, and the Balance of the Contract Price shall not be reduced or set off on account of any such unrelated obligations. No right of action shall accrue on this Bond to any person or entity other than the Owner or its heirs, executors, administrators or successors.

8   The Surety hereby waives notice of any change, including changes of time, to the Construction Contract or to related subcontracts, purchase orders and other obligations.

9   Any proceeding, legal or equitable, under this Bond may be instituted in any court of competent jurisdiction in the location in which the work or part of the work is located and shall be instituted within two years after Contractor Default or within two years after the Contractor ceased working or within two years after the Surety refuses or fails to perform its obligations under this Bond, whichever occurs first. If the provisions of this Paragraph are void or prohibited by law, the minimum period of limitation avail-

able to sureties as a defense in the jurisdiction of the suit shall be applicable.

10 Notice to the Surety, the Owner or the Contractor shall be mailed or delivered to the address shown on the signature page.

11 When this Bond has been furnished to comply with a statutory or other legal requirement in the location where the construction was to be performed, any provision in this Bond conflicting with said statutory or legal requirement shall be deemed deleted herefrom and provisions conforming to such statutory or other legal requirement shall be deemed incorporated herein. The intent is that this Bond shall be construed as a statutory bond and not as a common law bond.

12 DEFINITIONS

12.1 Balance of the Contract Price: The total amount payable by the Owner to the Contractor under the Construction Contract after all proper adjustments have been made, including allowance to the Con-

tractor of any amounts received or to be received by the Owner in settlement of insurance or other claims for damages to which the Contractor is entitled, reduced by all valid and proper payments made to or on behalf of the Contractor under the Construction Contract.

12.2 Construction Contract: The agreement between the Owner and the Contractor identified on the signature page, including all Contract Documents and changes thereto.

12.3 Contractor Default: Failure of the Contractor, which has neither been remedied nor waived, to perform or otherwise to comply with the terms of the Construction Contract.

12.4 Owner Default: Failure of Owner, which has neither been remedied nor waived, to pay the Contractor as required by the Construction Contract or to perform and complete or comply with the other terms thereof.

MODIFICATIONS TO THIS BOND ARE AS FOLLOWS:

(Space is provided below for additional signatures of added parties, other than those appearing on the cover page.)

CONTRACTOR AS PRINCIPAL		SURETY	
Company:	(Corporate Seal)	Company:	(Corporate Seal)

Signature: _____   Signature: _____
Name and Title: Name and Title:
Address: Address:

THE AMERICAN INSTITUTE OF ARCHITECTS

AIA Document A312

Payment Bond

Any singular reference to Contractor, Surety, Owner or other party shall be considered plural where applicable.

CONTRACTOR (Name and Address):   SURETY (Name and Principal Place of Business):

OWNER (Name and Address):

CONSTRUCTION CONTRACT
   Date:
   Amount:
   Description (Name and Location):

BOND
   Date (Not earlier than Construction Contract Date):
   Amount:
   Modifications to this Bond: ☐ None   ☐ See Page 6

CONTRACTOR AS PRINCIPAL   SURETY
Company: (Corporate Seal)   Company: (Corporate Seal)

Signature: _____   Signature: _____
Name and Title: Name and Title:

(Any additional signatures appear on page 6)

(FOR INFORMATION ONLY—Name, Address and Telephone)
AGENT or BROKER: *OWNER'S REPRESENTATIVE (Architect, Engineer or*
                                                other party):

1 The Contractor and the Surety, jointly and severally, bind themselves, their heirs, executors, administrators, successors and assigns to the Owner to pay for labor, materials and equipment furnished for use in the performance of the Construction Contract, which is incorporated herein by reference.

2 With respect to the Owner, this obligation shall be null and void if the Contractor:

2.1 Promptly makes payment, directly or indirectly, for all sums due Claimants, and

2.2 Defends, indemnifies and holds harmless the Owner from all claims, demands, liens or suits by any person or entity who furnished labor, materials or equipment for use in the performance of the Construction Contract, provided the Owner has promptly notified the Contractor and the Surety (at the address described in Paragraph 12) of any claims, demands, liens or suits and tendered defense of such claims, demands, liens of suits to the Contractor and the Surety, and provided there is no Owner Default.

3 With respect to Claimants, this obligation shall be null and void if the Contractor promptly makes payment, directly or indirectly, for all sums due.

4 The Surety shall have no obligation to Claimants under this Bond until:

4.1 Claimants who are employed by or have a direct contract with the Contractor have given notice to the Surety (at the address described in Paragraph 12) and sent a copy, or notice thereof, to the Owner, stating that a claim is being made under this Bond and, with substantial accuracy, the amount of the claim.

4.2 Claimants who do not have a direct contract with the Contractor:

  .1 Have furnished written notice to the Contractor and sent a copy, or notice thereof, to the Owner, within 90 days after having last performed labor or last furnished materials or equipment included in the claim stating, with substantial accuracy, the amount of the claim and the name of the party to whom the materials were furnished or supplied or for whom the labor was done or performed; and

  .2 Have either received a rejection in whole or in part from the Contractor, or not received within 30 days of furnishing the above notice any communication from the Contractor by which the Contractor has indicated the claim will be paid directly or indirectly; and

  .3 Not having been paid within the above 30 days, have sent a written notice to the Surety (at the address described in Paragraph 12) and sent a copy, or notice thereof, to the Owner, stating that a claim is being made under this bond and enclosing a copy of the previous written notice furnished to the Contractor.

5 If a notice required by Paragraph 4 is given by the Owner to the Contractor or to the Surety, that is sufficient compliance.

6 When the Claimant has satisfied the conditions of Paragraph 4, the Surety shall promptly and at the Surety's expense take the following actions:

6.1 Send an answer to the Claimant, with a copy to the Owner, within 45 days after receipt of the claim, stating the amounts that are undisputed and the basis for challenging any amounts that are disputed.

6.2 Pay or arrange for payment of any undisputed amounts.

7 The Surety's total obligation shall not exceed the amount of this Bond, and the amount of this Bond shall be credited for any payments made in good faith by the Surety.

8 Amounts owed by the Owner to the Contractor under the Construction Contract shall be used for the performance of the Construction Contract and to satisfy claims, if any, under any Construction Performance Bond. By the Contractor furnishing and the Owner accepting this Bond, they agree that all funds earned by the Contractor in the performance of the Construction Contract are dedicated to satisfy obligations of the Contractor and the Surety under this Bond, subject to the Owner's priority to use the funds for the completion of the work.

9 The Surety shall not be liable to the Owner, Claimants or other for obligations of the Contractor that are unrelated to the Construction Contract. The Owner shall not be liable for payment of any costs or expenses of any Claimant under this Bond, and shall have under this Bond no obligations to make payments to, give notices on behalf of, or otherwise have obligations to Claimants under this Bond.

10 The Surety hereby waives notice of any change, including changes of time, to the Construction Contract or to related subcontracts, purchase orders and other obligations.

11 No suit or action shall be commenced by a Claimant under this Bond other than in a court of competent jurisdiction in the location in which the work or part of the work is located or after the expiration of one year from the date (1) on which the Claimant gave the notice required by Subparagraph 4.1 or Clause 4.2 (iii), or (2) on which the last labor or service was performed by anyone or the last materials or equipment were furnished by anyone under the Construction Contract, whichever of (1) or (2) first occurs. If the provisions of this Paragraph are void or prohibited by law, the minimum period of limitation available to sureties as a defense in the jurisdiction of the suit shall be applicable.

12 Notice to the Surety, the Owner or the Contractor shall be mailed or delivered to the address shown on the signature page. Actual receipt of notice by Surety, the Owner or the Contractor, however accomplished, shall be sufficient compliance as of the date received at the address shown on the signature page.

13 When this bond has been furnished to comply with a statutory or other legal requirement in the location where the construction was to be performed, any provision in this Bond conflicting with said statutory or legal requirement shall be deemed deleted herefrom and provisions conforming to such statutory or other legal requirement shall be deemed incorporated herein. The intent is that this

D

Financial Statements for Contractors

William J. Palmer
Ernst & Young LLP

Before accepting a bid on proposed construction work or providing a bid, payment, or surety bond, owners and sureties often require that contractors submit financial statements to reduce risks. With sureties it is a foregone conclusion. In many respects, the surety's interest parallels that of the banker and owners, but the emphasis is different. For example, prospective owners want to know whether contractors will be able to finance, staff, and equip the clients' jobs in addition to keeping up their other work. They also want to see if the contractors have in progress one or more losing jobs that will drain off the progress payment money that owners pay to finance their own jobs. On the other hand, the bondsperson will want to know if the contractor has the money, experience, personnel, and the equipment to complete the job proposed to bid or work upon.

There are other reasons for providing adequate and well-prepared professional financial statements and for which suppliers and subcontractors, stockholders, and partners will most likely want financial statements.

Poorly prepared financial statements may, along with other things, lead to the contractor being denied a bond. Proper and professionally prepared financial statements, along with contractor personnel who can answer questions about the statements in a knowledgeable way, coupled with a CPA who is well known and experienced in construction in the contractor's local area, will go a long way toward helping the contractor not only to get the performance bond but also to increase overall bonding capacity.

Sample Financial Statements

In 1981, the American Institute of Certified Public Accountants published the *Audit and Accounting Guide—Construction Contractors,* including "Statement of Position 81-1 Accounting for Performance of Construction-Type and Certain Production-Type Contracts." In addition to providing information for practicing CPAs on approaches to auditing and construction contractors, the publication explains and illustrates the recommended format for presentation of audited financial statements prepared on the percentage-of-completion method of accounting. In most cases, the surety will want financial statements prepared on the percentage-of-completion basis whereby gross profit on a contract is reflected in the financial statements pro rata over the life of the job, usually in relationship to the progress of cost incurred. Schedule 1 and Schedule 2 present the *Audit and Accounting Guide* and SOP's recommended format for financial statements that have been audited which might be used by the management of a building contractor using the percentage-of-completion method of accounting. Schedule 3 presents the audited financial statements which might be used by the management of the same type of contractor if it were using the completed contract method of accounting. These are good examples of the type of financial statement which a contractor would provide to bonding companies, bankers, owners, and others.

It is rare today to see financial statements prepared on a completed contract method of accounting. Because of the publication of the *Audit and Accounting Guide—Construction Contractors* and the related SOP 81-1, it became more difficult for many contractors to justify using the completed contract method of accounting. Under this method, all costs and progress billings are deferred on the balance sheet until the job is complete. Upon completion of the job, all billings and all costs are transferred to the income statement and the related gross profit is recognized. However, on the rare occasion when completed contracts statements are prepared, a contractor may find that the bonding companies, banks, owners, and others will demand more detail than readily can be gotten from books maintained on the completed contract method of accounting. With this in mind, these statements are presented as well as the percentage-of-completion statements.

SCHEDULE 1 - FINANCIAL STATEMENTS PREPARED ACCORDING TO THE PERCENTAGE-OF-COMPLETION METHOD OF ACCOUNTING

Report of Independent Accountants

To the Board of Directors
Percentage Contractors, Inc.

We have audited the accompanying consolidated balance sheet of Percentage Contractors, Inc. as of October 31, 19X8 and 19X7, and the related consolidated statements of income and retained earnings and changes in financial position for the years then ended. These financial statements are the responsibility of the Company's management. Our responsibility is to express an opinion on these financial statements based on our audits.

We conducted our audits in accordance with generally accepted auditing standards. Those standards require that we plan and perform the audit to obtain reasonable assurance about whether the financial statements are free of material misstatement. An audit includes examining, on a test basis, evidence supporting the amounts and disclosures in the financial statements. An audit also includes assessing the accounting principles used and significant estimates made by management, as well as evaluating the overall financial statement presentation. We believe that our audits provide a reasonable basis for our opinion.

In our opinion, the financial statements referred to above present fairly, in all material respects, the financial position of Percentage Contractors, Inc. at October 31, 19X8 and 19X7, and the results of its operations and its changes in financial position for the years then ended, in conformity with generally accepted accounting principles.

(CPA Firm Signature)

City, State
December 18, 19X9

SCHEDULE 1 - FINANCIAL STATEMENTS PREPARED ACCORDING TO THE PERCENTAGE-OF-COMPLETION METHOD OF ACCOUNTING (cont'd)

Percentage Contractors, Inc.

Consolidated Balance Sheets

	October 31	
	19X8	19X7
Assets		
Cash	$ 264,100	$ 221,300
Certificates of deposit	40,300	-
Contracts receivable *(Note 2)*	3,789,200	3,334,100
Costs and estimated earnings in excess of billings on uncompleted contracts *(Note 3)*	80,200	100,600
Inventory, at lower of cost, on a first-in, first-out basis, or market	89,700	99,100
Prepaid charges and other assets	118,400	83,200
Advances to and equity in joint venture *(Note 4)*	205,600	130,700
Note receivable, related company *(Note 5)*	175,000	150,000
Property and equipment, net of accumulated depreciation and amortization *(Note 6)*	976,400	1,019,200
Total assets	$5,738,900	$5,138,200
Liabilities and shareholders' equity		
Notes payable *(Note 8)*	$ 468,100	$ 578,400
Lease obligations payable *(Note 9)*	197,600	251,300
Accounts payable *(Note 7)*	2,543,100	2,588,500
Billings in excess of costs and estimated earnings on uncompleted contracts *(Note 3)*	242,000	221,700
Accrued income taxes payable	52,000	78,600
Other accrued liabilities	36,600	36,000
Due to consolidated joint venture minority interests	154,200	26,200
Deferred income taxes *(Note 13)*	619,200	408,000
	4,312,800	4,188,700

Contingent liability *(Note 10)*

Shareholders' equity:		
Common stock, $1 par value, 500,000 authorized shares, 300,000 issued and outstanding shares	300,000	300,000
Retained earnings	1,126,100	649,500
Total shareholders' equity	1,426,100	949,500
Total liabilities and shareholders' equity	$5,738,900	$5,138,200

The accompanying notes are an integral part of these financial statements.

SCHEDULE 1 - FINANCIAL STATEMENTS PREPARED ACCORDING TO THE PERCENTAGE-OF-COMPLETION METHOD OF ACCOUNTING (cont'd)

Percentage Contractors, Inc.

Consolidated Statements of Income and Retained Earnings

| | Year ended October 31 | |
	19X8	19X7
Contract revenues earned	$ 22,554,100	$ 16,225,400
Cost of revenues earned	20,359,400	14,951,300
Gross profit	2,194,700	1,274,100
Selling, general and administrative expenses	895,600	755,600
Income from operations	1,299,100	518,500
Other income (expense):		
Equity in earnings from unconsolidated joint venture	49,900	5,700
Gain on sale of equipment	10,000	2,000
Interest expense (net of interest income of $8,800 in 19X8 and $6,300 in 19X7)	(69,500)	(70,800)
	(9,600)	(63,100)
Income before taxes	1,289,500	455,400
Provision for income taxes *(Note 13)*	662,900	225,000
Net income (per share, $2.09 (19X8); $.77 (19X7))	626,600	230,400
Retained earnings, beginning of year	649,500	569,100
	1,276,100	799,500
Less dividends paid (per share, $.50 (19X8); $.50 (19X7))	150,000	150,000
Retained earnings, end of year	$ 1,126,100	$ 649,500

The accompanying notes are an integral part of these financial statements.

SCHEDULE 1 - FINANCIAL STATEMENTS PREPARED ACCORDING TO THE PERCENTAGE-OF-COMPLETION METHOD OF ACCOUNTING (cont'd)

Percentage Contractors, Inc.

Consolidated Statements of Changes in Financial Position

| | Year ended October 31 | |
	19X8	19X7
Source of funds:		
From operations:		
Net income	$ 626,600	$ 230,400
Charges (credits) to income not involving cash and cash equivalents:		
Depreciation and amortization	167,800	153,500
Deferred income taxes	211,200	(75,900)
Gain on sale of equipment	(10,000)	(2,000)
	995,600	306,000
Proceeds from equipment sold	25,000	5,000
Net increase in billings related to costs and estimated earnings on uncompleted contracts	40,700	10,500
Decrease in inventory	9,400	-
Decrease in prepaid charges and other assets	-	16,100
Increase in accounts payable	-	113,200
Increase in other accrued liabilities	600	21,200
Increase in amount due to consolidated joint venture minority interests	128,000	26,200
Total source of funds	1,199,300	498,200
Use of funds:		
Acquisition of equipment:		
Shop and construction equipment	100,000	155,000
Automobile and trucks	40,000	20,000
Dividends paid	150,000	150,000
Increase in contract receivables	455,100	9,100
Increase in inventory	-	3,600
Increase in advances to and equity in joint venture	74,900	15,400
Increase in notes receivable, related company	25,000	50,000
Increase in prepaid charges and other assets	35,200	-
Decrease in notes payable	110,300	90,300
Decrease in lease obligations payable	53,700	9,700
Decrease in accounts payable	45,400	-
Decrease in accrued income taxes payable	26,600	2,400
Total use of funds	1,116,200	505,500
Increase (decrease) in cash and certificates of deposit for year	83,100	(7,300)
Cash and certificates of deposit, beginning of year	221,300	228,600
Cash and certificates of deposit, end of year	$ 304,400	$ 221,300

SCHEDULE 1 - FINANCIAL STATEMENTS PREPARED ACCORDING TO
THE PERCENTAGE-OF-COMPLETION METHOD OF ACCOUNTING (cont'd)

Percentage Contractors, Inc.
Notes to Consolidated Financial Statements
October 31, 19X8

1. Significant Accounting Policies

Company's Activities and Operating Cycle

The Company is engaged in a single industry: the construction of industrial and commercial buildings. The work is performed under cost-plus-fee contracts, fixed-price contracts, and fixed-price contracts modified by incentive and penalty provisions. These contracts are undertaken by the Company or its wholly owned subsidiary alone or in partnership with other contractors through joint ventures. The Company also manages, for a fee, construction projects of others.

The length of the Company's contracts varies but is typically about two years. Therefore, assets and liabilities are not classified as current and noncurrent because the contract-related items in the balance sheet have realization and liquidation periods extending beyond one year.

Principles of Consolidation

The consolidated financial statements include the Company's majority-owned entities, a wholly owned corporate subsidiary and a 75%-owned joint venture (a partnership). All significant intercompany transactions are eliminated. The Company has a minority interest in a joint venture (partnership), which is reported on the equity method.

Revenue and Cost Recognition

Revenues from fixed-price and modified fixed-price construction contracts are recognized on the percentage-of-completion method, measured by the percentage of labor hours incurred to date to estimated total labor hours for each contract.* This method is used because management considers expended labor hours to be the best available measure of progress on these contracts. Revenues from cost-plus-fee contracts are recognized on the basis of costs incurred during the period plus the fee earned, measured by the cost-to-cost method.

Contracts to manage, supervise, or coordinate the construction activity of others are recognized only to the extent of the fee revenue. The revenue earned in a period is based on the ratio of hours incurred to the total estimated hours required by the contract.

---

* There are various other alternatives to the percentage of labor hours method for measuring percentage of completion, which in many cases, may be more appropriate in measuring the extent of progress toward completion of the contract (labor dollars, units of output, and the cost-to-cost method and its variations).

Percentage Contractors, Inc.
Notes to Consolidated Financial Statements (continued)
October 31, 19X8

1. Significant Accounting Policies (continued)

Revenue and Cost Recognition (continued)

Contract costs include all direct material and labor costs and those indirect costs related to contract performance, such as indirect labor, supplies, tools, repairs, and depreciation costs. Selling, general, and administrative costs are charged to expense as incurred. Provisions for estimated losses on uncompleted contracts are made in the period in which such losses are determined. Changes in job performance, job conditions, and estimated profitability, including those arising from contract penalty provisions, and final contract settlements may result in revisions to cost and income and are recognized in the period in which the revisions are determined. Profit incentives are included in revenues when their realization is reasonably assured. An amount equal to contract costs attributable to claims is included in revenues when realization is probable and the amount can be reliably estimated.

The asset, "Cost and estimated earnings in excess of billings on uncompleted contracts," represents revenues recognized in excess of amounts billed. The liability, "Billings in excess of costs and estimated earnings on uncompleted contracts," represents billings in excess of revenues recognized.

Property and Equipment

Depreciation and amortization are provided principally on the straight-line method over the estimated useful lives of the assets. Amortization of leased equipment under capital leases is included in depreciation and amortization.

Pension Plan

The Company has a pension plan covering substantially all employees not covered by union-sponsored plans. Pension costs charged to earnings include current-year costs and the amortization of prior-service costs over 30 years. The Company's policy is to fund the costs accrued.

SCHEDULE 1 - FINANCIAL STATEMENTS PREPARED ACCORDING TO
THE PERCENTAGE-OF-COMPLETION METHOD OF ACCOUNTING (cont'd)

Percentage Contractors, Inc.
Notes to Consolidated Financial Statements (continued)
October 31, 19X8

1. Significant Accounting Policies (continued)

Income Taxes

Deferred income taxes are provided for differences in timing in reporting income for financial statement and tax purposes arising form differences in the methods of accounting for construction contracts and depreciation.

Construction contracts are reported for tax purposes on the completed-contract method and for financial statement purposes on the percentage-of-completion method. Accelerated depreciation is used for tax reporting, and straight-line depreciation is used for financial statement reporting.

Investment tax credits are applied as a reduction to the current provision for federal income taxes using the flow-through method.

2. Contract Receivables

	October 31	
	19X8	**19X7**
Contract receivables:		
Billed:		
Complete contracts	$ 621,100	$ 500,600
Contracts in progress	2,146,100	1,931,500
Retained	976,300	866,200
Unbilled	121,600	105,400
	3,865,100	3,403,700
Less allowances for doubtful collections	75,900	69,600
	$3,789,200	$3,334,100

Contract receivables at October 31, 19X8 include a claim, expected to be collected within one year, for $290,600 arising from a dispute with the owner over design and specification changes in a building currently under construction. The changes were made at the request of the owner to improve the thermal characteristics of the building and, in the opinion of counsel, gave rise to a valid claim against the owner.

The retained and unbilled contract receivables at October 31, 19X8 included $38,600 that was not expected to be collected within one year.

SCHEDULE 1 - FINANCIAL STATEMENTS PREPARED ACCORDING TO THE PERCENTAGE-OF-COMPLETION METHOD OF ACCOUNTING (cont'd)

Percentage Contractors, Inc.
Notes to Consolidated Financial Statements (continued)
October 31, 19X8

3. Costs and Estimated Earnings on Uncompleted Contracts

	October 31	
	19X8	19X7
Costs incurred on uncompleted contracts	$ 15,771,500	$ 12,165,400
Estimated earnings	1,685,900	1,246,800
	17,457,400	13,412,200
Less billings to date	17,619,200	13,533,300
	$ (161,800)	$ (121,100)

Included in accompanying balance sheets under the following captions:

Costs and estimated earnings in excess of billings on uncompleted contracts	$ 80,200	$ 100,600
Billings in excess of costs and estimated earnings on uncompleted contracts	(242,000)	(221,700)
	$ (161,800)	$ (121,100)

SCHEDULE 1 - FINANCIAL STATEMENTS PREPARED ACCORDING TO
THE PERCENTAGE-OF-COMPLETION METHOD OF ACCOUNTING (cont'd)

Percentage Contractors, Inc.
Notes to Consolidated Financial Statements (continued)
October 31, 19X8

4. Advances to and Equity in Joint Venture

The Company has a minority interest (one-third) in a general partnership joint venture formed to construct an office building. All of the partners participate in construction, which is under the general management of the Company. Summary information on the joint venture follows:

| | October 31 | |
	19X8	19X7
Current assets	$ 483,100	$ 280,300
Construction and other assets	220,500	190,800
	703,600	471,100
Less liabilities	236,800	154,000
Net assets	$ 466,800	$ 317,100
Revenue	$ 3,442,700	$ 299,400
Net income	$ 149,700	$ 17,100
Company's interest:		
Share of net income	$ 49,900	$ 5,700
Advances to joint venture	$ 50,000	$ 25,000
Equity in net assets	155,600	105,700
Total advances and equity	$ 205,600	$ 130,700

5. Transactions with Related Party

The note receivable, related company, is an installment note bearing annual interest at 9.25%, payable quarterly, with the principal payable in annual installments of $25,000, commencing October 1, 19Y0.

The major stockholder of Percentage Contractors, Inc. owns the majority of the outstanding common stock of this related company, whose principal activity is leasing land and buildings. Percentage Contractors, Inc. rents land and office facilities from the related company on a ten-year lease ending September 30, 19Y6, for an annual rental of $19,000.

Percentage Contractors, Inc.
Notes to Consolidated Financial Statements (continued)
October 31, 19X8

6. Property and Equipment

| | October 31 | |
	19X8	19X7
Assets		
Land	$ 57,500	$ 57,500
Buildings	262,500	262,500
Shop and construction equipment	827,600	727,600
Automobiles and trucks	104,400	89,100
Leased equipment under capital leases	300,000	300,000
	1,552,000	1,436,700
Accumulated depreciation and amortization		
Buildings	140,000	130,000
Shop and construction equipment	265,600	195,500
Automobiles and trucks	70,000	42,000
Leased equipment under capital leases	100,000	50,000
	575,600	417,500
Net property and equipment	$ 976,400	$ 1,019,200

7. Accounts Payable

Accounts payable include amounts due to subcontractors, totaling $634,900 at October 31, 19X8, and $560,400 at October 31, 19X7, which have been retained pending completion and customer acceptance of jobs. Accounts payable at October 31, 19X8 include $6,500 that are not expected to be paid within one year.

SCHEDULE 1 - FINANCIAL STATEMENTS PREPARED ACCORDING TO
THE PERCENTAGE-OF-COMPLETION METHOD OF ACCOUNTING (cont'd)

Percentage Contractors, Inc.
Notes to Consolidated Financial Statements (continued)
October 31, 19X8

8. Notes Payable

| | October 31 | |
	19X8	19X7
Unsecured note payable to bank, due in quarterly installments of $22,575 plus interest at 1% over prime	$ 388,100	$ 478,400
Note payable to bank, collateralized by equipment, due in monthly installments of $1,667 plus interest at 10% through January 19Y3	80,000	100,000
	$ 468,100	$ 578,400

At October 31, 19X8, the payments due within one year totaled $110,300.

9. Lease Obligations Payable

The Company leases certain specialized construction equipment under leases classified as capital leases. The following is a schedule showing the future minimum lease payments under capital leases by years and the present value of the minimum lease payments as of October 31, 19X8:

Year Ending
December 31

19X9	$ 76,500
19Y0	76,500
19Y1	76,500
Total minimum lease payments	229,500
Less: Amount representing interest	31,900
Present value of minimum lease payments	$ 197,600

At October 31, 19X8, the present value of minimum lease payments due within one year is $92,250.

Total rental expense, excluding payments on capital leases, totaled $86,300 in 19X8 and $74,400 in 19X7.

Percentage Contractors, Inc.
Notes to Consolidated Financial Statements (continued)
October 31, 19X8

10. Contingent Liability

A claim for $180,000 has been filed against the Company and its bonding company arising out of the failure of a subcontractor of the Company to pay its suppliers. In the opinion of counsel and management, the outcome of this claim will not have a material effect on the Company's financial position or results of operations.

11. Pension Plan

Pension costs charged to earnings were $61,400 in 19X8 and $57,300 in 19X7. At October 31, 19X8, the estimated actuarial value of vested benefits exceeded the fund assets (at market) and contribution accruals by $197,600.

12. Management Contracts

The Company manages or supervises commercial and industrial building contracts of others for a fee. These fees totaled $121,600 in 19X8 and $1,700 in 19X7 and are included in contract revenues earned.

13. Income Taxes and Deferred Income Taxes

The provision for taxes on income consists of the following:

| | October 31 | |
	19X8	19X7
Currently payable, net of investment credits of		
$9,400 and $13,800	$ 451,700	$ 300,900
Deferred:		
Contract related	204,200	(80,900)
Property and equipment related	7,000	5,000
	$ 662,900	$ 225,000

SCHEDULE 1 - FINANCIAL STATEMENTS PREPARED ACCORDING TO THE PERCENTAGE-OF-COMPLETION METHOD OF ACCOUNTING (cont'd)

Percentage Contractors, Inc.
Notes to Consolidated Financial Statements (continued)
October 31, 19X8

13. Income Taxes and Deferred Income Taxes (continued)

At October 31 of the respective years, the components of the balance of deferred income taxes were:

	October 31	
	19X8	19X7
Contract related	$ 594,000	$ 389,800
Property and equipment related	25,200	18,200
	$ 619,200	$ 408,000

14. Backlog

The following schedule shows a reconciliation of backlog representing signed contracts, excluding fees from management contracts, in existence at October 31, 19X7 and 19X8.

Balance, October 31, 19X7	$ 24,142,600
Contract adjustments	1,067,100
New contracts, 19X8	3,690,600
	28,900,300
Less: Contract revenue earned, 19X8	22,432,500
Balance, October 31, 19X8	$ 6,467,800

In addition, between November 1, 19X9 and December 18, 19X9, the Company entered into additional construction contracts with revenues of $5,332,800.

SCHEDULE 2 - ADDITIONAL INFORMATION TO AUDITED FINANCIAL STATEMENTS (SCHEDULE 1). THIS INFORMATION IS USUALLY REQUIRED BY BONDING COMPANIES AND BANKS.

Report of Independent Accountants

To the Board of Directors
Percentage Contractors, Inc.

Our audit was made for the purpose of forming an opinion on the basic financial statements taken as a whole. The supplementary information included in the following schedules is presented for purposes of additional analysis and is not a required part of the basic consolidated financial statements. Such information has been subjected to the auditing procedures applied in the audit of the basic consolidated financial statements and, in our opinion, is fairly presented in all material respects in relation to the consolidated financial statements taken as a whole.

(CPA Firm Signature)

City, State
December 18, 19X9

SCHEDULE 2 - ADDITIONAL INFORMATION TO AUDITED FINANCIAL
STATEMENTS (SCHEDULE 1). THIS INFORMATION IS USUALLY
REQUIRED BY BONDING COMPANIES AND BANKS. (cont'd)

Percentage Contractors, Inc.

Schedule 2A

Earnings from Contracts

Year ended October 31, 19X8

| | 19X8 | | | 19X7 |
	Revenues Earned	Cost of Revenues Earned	Gross Profit (Loss)	Gross Profit (Loss)
Contracts completed during the year	$ 6,290,800	$ 5,334,000	$ 956,800	$ 415,300
Contracts in progress at year end	16,141,700	14,636,900	1,504,800	921,400
Management contract fees earned	121,600	51,800	69,800	1,700
Unallocated indirect and warranty costs	-	46,700	(46,700)	(38,100)
Minority interest in joint venture	-	128,000	(128,000)	(26,200)
Charges on prior year contracts	-	162,000	(162,000)	-
	$ 22,554,100	$ 20,359,400	$ 2,194,700	$ 1,274,100

SCHEDULE 2B - ADDITIONAL INFORMATION TO AUDITED FINANCIAL STATEMENTS (SCHEDULE 1). THIS INFORMATION IS USUALLY REQUIRED BY BONDING COMPANIES AND BANKS.

Percentage Contractors, Inc.

Schedule 2

Contracts Completed

Year ended October 31, 19X8

Contract		Contract Totals			Before October 1, 19X8			During the Year ended October 31, 19X8		
Number	Type*	Revenues Earned	Cost of Revenues	Gross Profit (Loss)	Revenues Earned	Cost of Revenues	Gross Profit (Loss)	Revenues Earned	Cost of Revenues	Gross Profit (Loss)
1511	B	$ 5,475,300	$ 4,802,500	$ 672,800	$ 3,223,400	$ 2,932,700	$ 290,700	$ 2,251,900	$ 1,869,800	$ 382,100
1605	A	695,900	880,900	(185,900)	596,100	558,100	38,000	98,900	322,800	(223,900)
1624	A	140,700	150,700	(10,000)	29,600	31,800	(2,200)	111,100	118,900	(7,800)
1711	A	2,725,100	2,391,700	333,400	1,654,100	1,510,000	144,100	1,071,000	881,700	189,300
1791	B	4,770,100	4,288,900	481,200	3,028,500	2,929,600	98,900	1,741,600	1,359,300	382,300
1792	A	635,000	457,900	177,100			-	635,000	457,900	177,100
Small contracts		413,400	349,500	63,900	32,100	25,900	6,200	381,300	323,600	57,700
		$14,854,600	$13,322,100	$ 1,532,500	$ 8,563,800	$ 7,988,100	$ 575,700	$ 6,290,800	$ 5,334,000	$ 956,800

*Contract types:
A - Fixed-price.
B - Cost-plus-fee.

Percentage Contractors, Inc.

Schedule 3

Contracts in Progress

October 31, 19X8

Contract		Total Contract		From Inception to October 31, 19X8						At October 31, 19X8		For the Year ended October 31, 19X8		
Number	Type*	Revenues	Estimated Gross Profit (Loss)	Revenues Earned	Total Costs Incurred	Cost of Revenues	Gross Profit (Loss)	Billed To Date	Estimated Cost to Complete	Costs and Estimated Earnings in Excess of Billings	Billings in Excess of Costs and Estimated Earnings	Revenues Earned	Cost of Revenues	Gross Profit (Loss)
1845	A	$ 6,750,200	$ 877,000	$ 5,890,500	$ 5,244,500	$ 5,143,900	$ 746,600	$ 5,976,000	$ 628,700	$15,100	$ -	$ 5,664,200	$ 4,994,500	$ 679,700
1847	B	1,471,800	127,100	1,250,400	1,139,800	1,139,800	110,600	1,195,800	204,900	54,600	-	962,800	899,000	63,800
1912	A	451,800	(130,100)	108,600	238,700	238,700	(130,100)	98,100	343,200	10,500	-	98,600	191,500	(92,900)
1937	B	11,125,000	847,900	7,337,900	7,045,500	6,721,100	616,800	7,808,000	3,231,600	-	145,700	6,981,900	6,469,900	512,000
1945	A	3,650,100	497,000	2,395,200	2,061,300	2,061,300	333,900	2,491,500	1,091,800	-	96,300	2,395,200	2,061,300	333,900
Small contracts		51,300	8,400	49,800	41,700	41,700	8,100	49,800	1,200			39,000	30,700	8,300
		$23,500,200	$2,227,300	$17,032,400	$15,771,500	$15,346,500	$1,685,900	$17,619,200	$5,501,400	$80,200	$242,000	$16,141,700	$14,636,900	$1,504,800

*Contract types:
A - Fixed-price.
B - Cost-plus-fee.

SCHEDULE 3 - FINANCIAL STATEMENTS PREPARED ON THE COMPLETED-CONTRACT METHOD OF ACCOUNTING. WHEN STATEMENTS ARE PREPARED ON THIS BASIS, BONDING COMPANIES, BANKS, AND OTHERS MAY REQUIRE THE ADDITIONAL INFORMATION SET FORTH IN SCHEDULES 2A, 2B AND 2C.

Report of Independent Accountants

To the Board of Directors
Completed Contractors, Inc.

We have audited the accompanying balance sheet of Completed Contractors, Inc. as of October 31, 19X8 and 19X7, and the related statements of income and retained earnings and changes in financial position for the years then ended. These financial statements are the responsibility of the Company's management. Our responsibility is to express an opinion on these financial statements based on our audits.

We conducted our audits in accordance with generally accepted auditing standards. Those standards require that we plan and perform the audit to obtain reasonable assurance about whether the financial statements are free of material misstatement. An audit includes examining, on a test basis, evidence supporting the amounts and disclosures in the financial statements. An audit also includes assessing the accounting principles used and significant estimates made by management, as well as evaluating the overall financial statement presentation. We believe that our audits provide a reasonable basis for our opinion.

In our opinion, the financial statements referred to above present fairly, in all material respects, the financial position of Completed Contractors, Inc. at October 31, 19X8 and 19X7, and the results of its operations and its changes in financial position for the years then ended, in conformity with generally accepted accounting principles.

(CPA Firm Signature)

City, State
December 18, 19X9

SCHEDULE 3 - FINANCIAL STATEMENTS PREPARED ON THE
COMPLETED-CONTRACT METHOD OF ACCOUNTING. WHEN
STATEMENTS ARE PREPARED ON THIS BASIS, BONDING COMPANIES,
BANKS, AND OTHERS MAY REQUIRE THE ADDITIONAL INFORMATION
SET FORTH IN SCHEDULES 2A, 2B AND 2C. (cont'd)

Completed Contractors, Inc.

Balance Sheets

	October 31	
	19X8	19X7
Assets		
Cash	$ 242,700	$ 185,300
Contract receivables (less allowance for doubtful accounts of $10,000 and $8,000) *(Note 2)*	893,900	723,600
Costs in excess of billings on uncompleted contracts *(Note 3)*	418,700	437,100
Inventories, at lower of cost or realizable value on first-in, first-out basis *(Note 4)*	463,600	491,300
Prepaid expenses	89,900	53,900
Total current assets	2,108,800	1,891,200
Cash value of life insurance	35,800	32,900
Property and equipment, at cost:		
Building	110,000	110,000
Equipment	178,000	163,000
Trucks and autos	220,000	200,000
	508,000	473,000
Less: Accumulated depreciation	218,000	203,200
	290,000	269,800
Land	21,500	21,500
	311,500	291,300
	$2,456,100	$2,215,400

SCHEDULE 3 - FINANCIAL STATEMENTS PREPARED ON THE COMPLETED-CONTRACT METHOD OF ACCOUNTING. WHEN STATEMENTS ARE PREPARED ON THIS BASIS, BONDING COMPANIES, BANKS, AND OTHERS MAY REQUIRE THE ADDITIONAL INFORMATION SET FORTH IN SCHEDULES 2A, 2B AND 2C. (cont'd)

| | October 31 | |
	19X8	19X7
Liabilities and stockholders' equity		
Current liabilities:		
Current maturities, long-term debt *(Note 5)*	$ 37,000	$ 30,600
Accounts payable	904,900	821,200
Accrued salaries and wages	138,300	155,100
Accrued income taxes	53,000	36,200
Accrued and other liabilities	116,400	55,550
Billings in excess of costs on uncompleted contracts *(Note 3)*	34,500	43,700
Total current liabilities	1,284,100	1,142,350
Long-term debt, less current maturities *(Note 5)*	245,000	241,000
	1,529,100	1,383,350
Stockholders' equity:		
Common stock, $10 par value:		
Authorized shares - 50,000		
Issued and outstanding shares - 23,500	235,000	235,000
Additional paid-in capital	65,000	65,000
Retained earnings	627,000	532,050
	927,000	832,050
	$2,456,100	$2,215,400

The accompanying notes are an integral part of these financial statements.

SCHEDULE 3 - FINANCIAL STATEMENTS PREPARED ON THE COMPLETED-CONTRACT METHOD OF ACCOUNTING. WHEN STATEMENTS ARE PREPARED ON THIS BASIS, BONDING COMPANIES, BANKS, AND OTHERS MAY REQUIRE THE ADDITIONAL INFORMATION SET FORTH IN SCHEDULES 2A, 2B AND 2C. (cont'd)

Completed Contractors, Inc.

Statements of Income and Retained Earnings

	Year ended October 31	
	19X8	**19X7**
Contract revenues	$ 9,487,000	$ 8,123,400
Costs and expenses:		
Cost of contracts completed	8,458,500	7,392,300
General and administrative	684,300	588,900
Interest expense	26,500	23,000
	9,169,300	8,004,200
Income before income taxes	317,700	119,200
Income taxes	(164,000)	(54,200)
Net income ($6.54 and $2.77 per share)	153,700	65,000
Retained earnings	532,050	525,800
Balance, beginning of year	685,750	590,800
Dividends paid ($2.50 per share)	(58,750)	(58,750)
Balance, end of year	$ 627,000	$ 532,050

The accompanying notes are an integral part of these financial statements.

SCHEDULE 3 - FINANCIAL STATEMENTS PREPARED ON THE COMPLETED-CONTRACT METHOD OF ACCOUNTING. WHEN STATEMENTS ARE PREPARED ON THIS BASIS, BONDING COMPANIES, BANKS, AND OTHERS MAY REQUIRE THE ADDITIONAL INFORMATION SET FORTH IN SCHEDULES 2A, 2B AND 2C. (cont'd)

Completed Contractors, Inc.

Statements of Changes in Financial Position

	Year ended October 31	
	19X8	19X7
Source of working capital:		
Net income	$ 153,700	$ 65,000
Charge to income not requiring outlay of working capital--depreciation	54,800	50,300
Working capital from operations	208,500	115,300
Proceeds of notes payable	44,000	68,000
	252,500	183,300
Use of working capital:		
Purchase of property and equipment	75,000	53,500
Reduction of long-term debt	40,000	28,000
Payment of dividends	58,750	58,750
Increase in cash value of life insurance	2,900	2,685
	176,650	142,935
Increase in working capital	$ 75,850	$ 40,365
Changes in components of working capital:		
Increase (decrease) in current assets:		
Cash	57,400	(26,435)
Contract receivables	170,300	36,500
Costs in excess of billings on uncompleted contracts	(18,400)	49,100
Inventories	(27,700)	3,400
Prepaid expenses	36,000	(16,500)
	217,600	46,065
Decrease (increase) in current liabilities:		
Current maturities, long-term debt:		
Notes payable, bank	(6,000)	(12,000)
Mortgage payable	(400)	(500)
Accounts payable	(83,700)	(24,600)
Accrued salaries and wages	16,800	(24,300)
Accrued income taxes	(16,800)	6,300
Accrued and other liabilities	(60,850)	33,100
Billings in excess of costs on uncompleted contracts	9,200	16,300
	(141,750)	(5,700)
Increase in working capital	$ 75,850	$ 40,365

The accompanying notes are an integral part of these financial statements.

SCHEDULE 3 - FINANCIAL STATEMENTS PREPARED ON THE
COMPLETED-CONTRACT METHOD OF ACCOUNTING. WHEN
STATEMENTS ARE PREPARED ON THIS BASIS, BONDING COMPANIES,
BANKS, AND OTHERS MAY REQUIRE THE ADDITIONAL INFORMATION
SET FORTH IN SCHEDULES 2A, 2B AND 2C. (cont'd)

Completed Contractors, Inc.

Notes to Financial Statements

October 31, 19X8 and 19X7

1. Significant Accounting Policies

Company's Activities

The Company is a heating and air-conditioning contractor for residential and commercial properties. Work on new structures is performed primarily under fixed-price contracts. Work on existing structures is performed under fixed-price or time-and-material contracts.

Revenue and Cost Recognition

Revenues from fixed-price construction contracts are recognized on the completed-contract method. This method is used because the typical contract is completed in two months or less and financial position and results of operations do not vary significantly from those which would result from use of the percentage-of-completion method. A contract is considered complete when all costs except insignificant items have been incurred and the installation is operating according to specifications or has been accepted by the customer.

Revenues from time-and-material contracts are recognized currently as the work is performed.

Contract costs include all direct material and labor costs and those indirect costs related to contract performance, such as indirect labor, supplies, tools, repairs, and depreciation costs. General and administrative costs are charged to expense as incurred. Provisions for estimated losses on uncompleted contracts are made in the period in which such losses are determined. Claims are included in revenues when received.

Costs in excess of amounts billed are classified as current assets under costs in excess of billings on uncompleted contracts. Billings in excess of costs are classified under current liabilities as billings in excess of costs on uncompleted contracts. Contract retentions are included in accounts receivable.

SCHEDULE 3 - FINANCIAL STATEMENTS PREPARED ON THE COMPLETED-CONTRACT METHOD OF ACCOUNTING. WHEN STATEMENTS ARE PREPARED ON THIS BASIS, BONDING COMPANIES, BANKS, AND OTHERS MAY REQUIRE THE ADDITIONAL INFORMATION SET FORTH IN SCHEDULES 2A, 2B AND 2C. (cont'd)

Completed Contractors, Inc.

Notes to Financial Statements (continued)

1. Significant Accounting Policies (continued)

Inventories

Inventories are stated at cost on the first-in, first-out basis using unit cost for furnace and air-conditioning components and average cost for parts and supplies. The carrying value of furnace and air-conditioning component units is reduced to realizable value when such values are less than cost.

Property and Equipment

Depreciation is provided over the estimated lives of the assets principally on the declining-balance method, except on the building where the straight-line method is used.

Pension Plan

The Company has a pension plan covering all employees not covered by union-sponsored plans. Pension costs charted to income include current-year costs and the amortization of prior-service costs over 30 years. The Company's policy is to fund the costs accrued.

Investment Tax Credit

Investment tax credits are applied as a reduction to the current provision for federal income taxes using the flow-through method.

SCHEDULE 3 - FINANCIAL STATEMENTS PREPARED ON THE
COMPLETED-CONTRACT METHOD OF ACCOUNTING. WHEN
STATEMENTS ARE PREPARED ON THIS BASIS, BONDING COMPANIES,
BANKS, AND OTHERS MAY REQUIRE THE ADDITIONAL INFORMATION
SET FORTH IN SCHEDULES 2A, 2B AND 2C. (cont'd)

Completed Contractors, Inc.

Notes to Financial Statements (continued)

2. Contract Receivables

| | October 31 | |
	19X8	19X7
Completed contracts, including retentions	$ 438,300	$ 408,600
Contracts in progress:		
Current accounts	386,900	276,400
Retentions	78,700	46,600
	903,900	731,600
Less allowance for doubtful accounts	10,000	8,000
	$ 893,900	$ 723,600

Retentions include $10,300 in 19X8, which are expected to be collected after 12 months.

3. Costs and Billings on Uncompleted Contracts

| | October 31 | |
	19X8	19X7
Costs incurred on uncompleted contracts	$ 2,140,400	$ 1,966,900
Billings on uncompleted contracts	1,756,200	1,573,500
	$ 384,200	$ 393,400

Included in accompanying balance sheets under the following captions:		
Costs in excess of billings on uncompleted contracts	$ 418,700	$ 437,100
Billings in excess of costs on uncompleted contracts	(34,500)	(43,700)
	$ 384,200	$ 393,400

SCHEDULE 3 - FINANCIAL STATEMENTS PREPARED ON THE
COMPLETED-CONTRACT METHOD OF ACCOUNTING. WHEN
STATEMENTS ARE PREPARED ON THIS BASIS, BONDING COMPANIES,
BANKS, AND OTHERS MAY REQUIRE THE ADDITIONAL INFORMATION
SET FORTH IN SCHEDULES 2A, 2B AND 2C. (cont'd)

Completed Contractors, Inc.

Notes to Financial Statements (continued)

4. Inventories

	October 31	
	19X8	**19X7**
Furnace and air-conditioning components	$ **303,200**	$ 308,700
Parts and supplies	**160,400**	182,600
	$ **463,600**	$ 491,300

Furnace and air-conditioning components include used items of $78,400 in 19X8 and
$71,900 in 19X7 that are carried at the lower of cost or realizable value.

5. Long-Term Debt

	October 31	
	19X8	**19X7**
Notes payable, bank:		
Notes due in quarterly installments of $2,500, plus		
interest at 8%	$ **140,000**	$ 150,000
Notes due in monthly installments of $1,500, plus		
interest at prime plus 1.5%	**87,000**	58,000
Mortgage payable:		
Due in quarterly payments of $3,500, including		
interest at 9%	**55,000**	63,600
	282,000	271,600
Less current maturities	**37,000**	30,600
	$ **245,000**	$ 241,000

**SCHEDULE 3 - FINANCIAL STATEMENTS PREPARED ON THE
COMPLETED-CONTRACT METHOD OF ACCOUNTING. WHEN
STATEMENTS ARE PREPARED ON THIS BASIS, BONDING COMPANIES,
BANKS, AND OTHERS MAY REQUIRE THE ADDITIONAL INFORMATION
SET FORTH IN SCHEDULES 2A, 2B AND 2C. (cont'd)**

Completed Contractors, Inc.

Notes to Financial Statements (continued)

6. Pension Plans

The total pension expenses for the years 19X8 and 19X7 were $31,200 and $27,300, respectively, including contributions to union-sponsored plans.

At October 31, 19X8, the estimated actuarial value of vested benefits of the Company's plan exceeded the fund assets (at market) and contribution accruals by $48,000.

7. Backlog

The estimated gross revenue on work to be performed on signed contracts was $4,691,000 at October 31, 19X8, and $3,617,400 at October 31, 19X7. In addition to the backlog of work to be performed, there was gross revenue to be reported in future periods under the completed-contract method used by the Company of $2,460,000 at October 31, 19X8, and $2,170,000 at October 31, 19X7.

Personal Financial Statements

When dealing with medium- and smaller-sized, closely held (limited number of shareholders and not publicly traded) construction companies, sureties and bonding companies will usually require financial statements of the individual shareholders as well as for the company itself. This is required for several reasons.

First, many small and closely held companies are undercapitalized and a bonding company wants as many assets as it can go after should the company fail or fail to finish a construction job. They will usually require the individual shareholders to pledge their personal assets as part of the granting of the bid, payment, or surety bond.

Second, in small, closely held companies there is a blurring of personal versus business assets, liabilities, expenses, and income. Decisions can be made quickly to pay out dividends, grants, bonuses, etc., thereby lessening the assets that the bonding company can get at should the contractor fail.

There are major and unique differences between personal financial statements and corporate statements. Appendix D sets forth the recommended format for a construction company presenting its financial statements on the percentage-of-completion method of accounting and the completed contract method of accounting. A quick comparison of the financial statements in this appendix with the company statements in Appendix D will reveal some of the major differences.

One major difference is that the assets and liabilities of personal financial statements are reflected at current value. Corporate financial statements reflect all assets and liabilities and the resulting net worth using historical cost. For example, marketable securities on the company's financial statements will be shown at the cost that was expended at the time the securities were purchased. Either parenthetically or in a footnote, the market value will be revealed. In personal financial statements, the marketable securities will be reflected at their current market price rather than the original cost of the purchase.

Another distinction appears in what is normally referred to as *shareholders' equity*. In personal financial statements, there are two categories shown. First of all, the estimated income taxes on the differences between the estimated current values of assets and the estimated current amount of liabilities and their tax bases are computed. This means that the tax is being estimated on unrealized gains in market value of the assets. The other component of what would normally be the net worth section is simply a line item called *net worth*.

The other major difference is the way the statement of changes in net worth is presented (as opposed to the corporate *statement of income and retained earnings*). There are two components to the change in net worth. The first section presents changes in net worth due to realized increases and decreases in net worth. Examples of items making up realized increase in net worth would be salaries, bonuses, dividends, interest income, and other items received in cash during the year. Examples of realized decreases in net worth are income taxes, interest expense, losses on sales of marketable securities, and, most important, personal living expenses.

The second part of the change in net worth reflects unrealized activity. Examples of unrealized increases in net worth are increases in the value of marketable securities, stock options, investments in businesses, and jewelry. Examples of unrealized decreases in net worth are value decline in market value of marketable securities and decline in value of investment in small businesses.

As with the sound advice that contractors should prepare professional-looking financial statements on the construction business for the surety, they should do the same with personal financial statements for themselves and their families. Most small contractors do not have the capability to prepare financial statements for themselves and will engage their outside accountants to prepare them. Two types of reports usually accompany financial statements prepared by outside accountants. These are the *compilation report* and the *standard review report*.

When outside accountants prepare a compilation report, they are making it clear that while these statements have been prepared "in accordance with statements on Standards for Accounting and Review Services issued by the American Institute of Certified Public Accountants," they go on to "not express an opinion or any other form of assurance on them."

The review report also states that they have reviewed the financial statements, "in accordance with statements on Standards for Accounting and Review Services issued by the American Institute of Certified

Public Accountants." However, a middle paragraph explains the additional procedures performed for a review report as opposed to a compilation report and to point out that it is not an audit in accordance with generally accepted auditing standards. However, the last paragraph of the review report provides a little more assurance on the fairness of the financial statements being presented in that it states "based on our review, we are not aware of any material modifications that should be made to the accompanying financial statements in order for them to be in conformity with generally accepted accounting principles." This gives a great deal more comfort to the reader than the statement, "We have not audited or reviewed the accompanying financial statements and, accordingly, do not express any opinion or any other form of assurance on them," which is the last sentence in a compilation report.

Schedule 1 presents the standard compilation report letter which would appear with financial statements prepared in that manner. Schedule 2 presents the standard review report when statements have been prepared by an accountant who has followed the American Institute of Certified Public Accountants' review standards. Schedule 3 presents the financial statement format for individual income statements.

SCHEDULE 1 - AICPA RECOMMENDED REPORT LETTER FOR PERSONAL FINANCIAL STATEMENTS PREPARED ON A "COMPILATION BASIS."

Standard Compilation Report Letter

We have complied the accompanying statement of financial condition of James and Jane Person as of [*date*], and the related statement of changes in net worth for the [*period*] then ended, in accordance with Statements on Standards for Accounting and Review Services issued by the American Institute of Certified Public Accountants.

A compilation report is limited to presenting in the form of financial statements information that is the representation of the individuals whose financial statements are presented. We have not audited or reviewed the accompanying financial statements and, accordingly, do not express an opinion or any other form of assurance on them.

(CPA Firm Signature)

City, State
March 31, 19X4

SCHEDULE 2 - AICPA RECOMMENDED REPORT LETTER FOR PERSONAL FINANCIAL STATEMENTS PREPARED ON A "REVIEW BASIS."

Standard Review Report Letter

We have reviewed the accompanying statement of financial condition of James and Jane Person as of [*date*], and the related statement of changes in net worth for the [*period*] then ended, in accordance with Statements on Standards for Accounting and Review Services issued by the American Institute of Certified Public Accountants. All information included in these financial statements is the representation of James and Jane Person.

A review of personal financial statements consists principally of inquiries of the individuals whose financial statements are presented and analytical procedures applied to financial data. It is substantially less in scope than an audit in accordance with generally accepted auditing standards, the objective of which is the expression of an opinion regarding the financial statements taken as a whole. Accordingly, we do not express such an opinion.

Based on our review, we are not aware of any material modifications that should be made to the accompanying financial statements in order for them to be in conformity with generally accepted accounting principles.

(CPA Firm Signature)

City, State
March 31, 19X4

SCHEDULE 3 - AICPA RECOMMENDED FINANCIAL STATEMENT FORMAT FOR INDIVIDUALS

James and Jane Person
Statements of Financial Condition

| | December 31 | |
	19X3	19X2
Assets		
Cash	**$3,700**	$15,600
Bonus receivable	**20,000**	10,000
Investments:		
Marketable securities (*Note 2*)	**160,500**	140,700
Stock Options (*Note 3*)	**28,000**	24,000
Kenbruce Associates (*Note 4*)	**48,000**	42,000
Davekar Company, Inc.(*Note 5*)	**550,000**	475,000
Vested interest in deferred profit sharing plan	**111,400**	98,900
Remainder interest in testamentary trust (*Note 6*)	**171,900**	128,800
Cash value of life insurance $43,600 and $42,900), less loans payable to insurance companies ($38,100 and $37,000) (*Note 7*)	**5,500**	5,200
Residence (*Note 8*)	**190,000**	180,000
Personal effects (excluding jewelry) (*Note 9*)	**55,000**	50,000
Jewelry (*Note 9*)	**40,000**	36,500
Total assets	**$1,384,000**	$1,206,700
Liabilities		
Income taxes - current year balance	**$8,800**	$ 400
Demand 10.5% note payable to bank	**25,000**	26,000
Mortgage payable (*Note 10*)	**98,200**	99,000
	132,000	125,400
Contingent liabilities (*Note 11*)		
Estimated income taxes on the difference between the estimated current values of assets and the estimated current amounts of liabilities and their tax bases (*Note 12*)	**239,000**	160,000
Net worth	**1,013,000**	921,300
	$1,384,000	$1,206,700

The accompanying notes are an integral part of these financial statements.

SCHEDULE 3 - AICPA RECOMMENDED FINANCIAL STATEMENT FORMAT FOR INDIVIDUALS (cont'd)

James and Jane Person
Statements of Changes in Net Worth

	19X3	19X2
Realized increases in net worth:		
Salary and bonus	$95,000	$85,000
dividends and interest income	2,300	1,800
Distribution from limited partnership	5,000	4,000
Gains on sales of marketable securities	1,000	500
	103,300	91,300
Realized decreases in net worth		
Income taxes	26,000	22,000
Interest expense	13,000	14,000
Real estate taxes	4,000	3,000
Personal expenditures	36,700	32,500
	79,700	71,500
Net realized increase in net worth	23,600	19,800
Unrealized increases in net worth:		
Marketable securities (net of realized gains on securities sold)	3,000	500
Stock options	4,000	500
Davekar Company, Inc.	75,000	25,000
Kenbruce Associates	6,000	-
Deferred profit sharing plan	12,500	9,500
Remainder interest in testamentary trust	43,100	25,000
Jewelry	3,500	-
	147,100	60,500
Unrealized decrease in net worth:		
Estimated income taxes on the differences between the estimated current values of assets and the estimated current amounts of liabilities and their tax bases	79,000	22,000
Net unrealized increase in net worth	68,100	38,500
Net increase in net worth	91,700	58,300
Net worth at the beginning of year	921,300	863,000
Net worth at end of year	$1,013,000	921,300

The accompanying notes are an integral part of these financial statements

SCHEDULE 3 - AICPA RECOMMENDED FINANCIAL STATEMENT FORMAT FOR INDIVIDUALS. (cont'd)

James and Jane Person
Notes to Financial Statements

Note 1.

The accompanying financial statements include the assets and liabilities of James and Jane Person. Assets are stated at their estimated current values, and liabilities at their estimated current amounts.

Note 2

The estimated current values of marketable securities are either (a) their quoted closing prices or (b) for securities not traded on the financial statement date, amounts that fall within the range of quoted bid and asked prices.

Marketable securities consist of the following:

Stocks	December 31, 19X3		December 31, 19X2	
	Number of Shares or Bonds	Estimated Current Values	Number of Shares or Bonds	Estimated Current Values
Jaiven Jewels, Inc.	1,500	$98,813	-	-
McRae Motors, Ltd.	800	11,000	600	4,750
Parker Sisters, Inc.	400	13,875	200	5,200
Rosenfield Rug Company	-	-	1,200	96,000
Rubin Paint Company	300	9,750	100	2,875
Weiss Potato Chips, Inc.	200	20,337	300	25,075
		153,775		133,900
Bonds				
Jackson Van Lines, Ltd.				
(12% due 7/1/X9)	5	5,225	5	5,100
United Garvey, Inc.				
(7% due 11/15/X9)	2	1,500	2	1,700
		6,725		6,800
		$160,500		$140,700

SCHEDULE 3 - AICPA RECOMMENDED FINANCIAL STATEMENT FORMAT FOR INDIVIDUALS (cont'd)

James and Jane Person
Notes to Financial Statements (continued)

Note 3
Jane Person owns options to acquire 4,000 shares of stock of Winner Corporation at an option price of $5 per share. The option expires on June 30, 19X5. The estimated current value is its published selling price.

Note 4
The investment in Kenbruce Associates is an 8% interest in a real estate limited partnership. The estimated current value is determined by the projected annual cash receipts and payments capitalized at a 12% rate.

Note 5
James Person owns 50% of the common stock of Davekar Company, Inc., a retail mail order business. The estimated current value of the investment is determined by the provisions of a shareholders' agreement, which restricts the sale of the stock and, under current conditions, requires the company to repurchase the stock based on a price equal to the book value of the net assets plus an agreed amount for goodwill. At December 31, 19X3, the agreed amount for goodwill was $112,500, and at December 31, 19X2, it was $100,000.

A condensed balance sheet of Davekar Company, Inc., prepared in conformity with generally accepted accounting principals, is summarized below:

| | December 31 | |
	19X3	19X2
Current assets	$3,147,000	$2,975,000
Plant, property, and equipment, net	165,000	145,000
Other assets	120,000	110,000
Total assets	3,432,000	3,230,000
Current liabilities	2,157,000	2,030,000
Long-term liabilities	400,000	450,000
Total liabilities	2,557,000	2,480,000
Equity	$857,000	$750,000

The sales and net income for 19X3 were $10,500,000 and $125,000 and for 19X2 were $9,700,000 and $80,000.

SCHEDULE 3 - AICPA RECOMMENDED FINANCIAL STATEMENT FORMAT FOR INDIVIDUALS (cont'd)

James and Jane Person
Notes to Financial Statements (continued)

Note 6

Jane Person is the beneficiary of a remainder interest in a testamentary trust under the will of the late Joseph Jones. The amount included in the accompanying statements is her remainder interest in the estimated current value of the trust assets, discounted at 10%.

Note 7

At December 31, 19X3 and 19X2, James Person owned a $300,000 whole life insurance policy.

Note 8

The estimated current value of the residence is its purchase price plus the cost of improvements. The residence was purchased in December 19X1, and improvements were made in 19X2 and 19X3.

Note 9

The estimated current values of personal effects and jewelry are the appraised values of those assets, determined by an independent appraiser for insurance purposes.

Note 10

The mortgage (collateralized by the residence) is payable in monthly installments of $815 a month, including interest at 10% a year through 20Y8.

Note 11

James Person has guaranteed the payment of loans of Davekar Company, Inc. under a $500,000 line of credit. The loan balance was $300,000 at December 31, 19X3 and $400,000 at December 31, 19X2.

SCHEDULE 3 - AICPA RECOMMENDED FINANCIAL STATEMENT FORMAT FOR INDIVIDUALS (cont'd)

James and Jane Person
Notes to Financial Statements (continued)

Note 12

The estimated current amounts of liabilities at December 31, 19X3 and December 31, 19X2 equaled their tax bases. Estimated income taxes have been provided on the excess of the estimated current values of assets over their tax bases as if the estimated current values of the assets had been realized on the statement date, using applicable tax laws and regulations. The provision will probably differ from the amounts of income taxes that eventually might be paid because those amounts are determined by the timing and the method of disposal or realization and the tax laws and regulations in effect at the time of disposal or realization.

The estimated current values of assets exceeded their tax bases by $850,000 at December 31, 19X3, and by $770,300 at December 31, 19X2. The excess of estimated current values of major assets over their tax bases are:

	December 31	
	19X3	19X2
Investment in Davekar Company, Inc.	$430,500	$355,500
Vested interest in deferred profit sharing plan	111,400	98,900
Investment in marketable securities	104,100	100,000
Remainder interest in testamentary trust	97,000	53,900

Index

Strict liability, for design builders, 180
Subchapter S corporation, and tax issues, 36
Subcontractors:
  bonding, 32–33, 40, 47–49
  and CCIPs, 187–188
  and CERCLA, 172–173
  certificates of insurance, 201–202
  and environmental remediation divisions, 163
  hold-harmless clauses, 202
  limiting risk of, 48–49, 158
  and lump-sum insurance settlements, 200–201
  and operational exposure, 180–181
  safety management of, 252–254, 257
Subrogation:
  builder's risk waiver, 113
  recoveries, 158
  general liability waiver, 109
  workers' compensation waiver, 130–131
Substance abuse program for safety management, 258
Succession plans:
  with children, 263, 276–277
  components of, 261–262
  family, 262–263
  and life insurance, 264–265
  for partners, 263
  unfunded agreements, 263–264
  (*See also* Business insurance planning techniques)
Supplementary payments, and Coverage A general liability insurance, 98
Supply bonds, 26
Supply of funds (*see* Premiums)
Surety:
  for design-builders, 181
  performance bonds for design-builders, 184–186
Surety bonds:
  bid bonds, 24–25
  combination bonds, 26
  forms, 39
  indemnity of, 29
  and insurance, 23–24
  labor and materials payment bonds, 25–26
  maintenance bonds, 218–219
  performance bonds, 25
  premiums, 23
  release and retainage bond, 219

Surety bonds (*Cont.*):
  for self-insurance, 142
  supply bonds, 26
  underwriting conditions, 40–41
  underwriting contractor, 26–41
Surety companies, 218–219
  financial reporting to, 44–46
  liability of, 37–38
  and reinsurance, 41–42
Surplus line agents, 217
Survivor policies, 272

T list (*see* Treasury Department list)
Tax returns, for underwriting surety bonds, 33
Taxes:
  and bonding process, 44–45
  estate, 276–277
  and salary continuation plans, 266–267
  and underwriting, 35–36
Term insurance, 272
Terminology, 279–288
Testing, of equipment, and builder's risk insurance, 123
Timing, risk factor, 14–15
Tort liability, and general liability insurance, 108
Training, employee:
  and safety management, 249–250
Transfer for value rule, 275
Treasury Department list (T list), 33, 42
Treasury limit, 42
Treaty, reinsurance, 41
Trust, irrevocable, 277–278

Umbrella liability, 111
Underbilling, assessing for underwriting, 34
Underwriting surety bonds:
  assessing organization, 29–30
  assessing specific jobs, 37–41
  construction history, 30–31
  and continuity planning, 31
  and current operations, 31–33
  and financial condition of contractor, 33–37
  and financial status of owner, 28–29
  process, 26–28
  references of contractor, 37
  (*See also* Bonding)
Unfunded agreements, 263–264